Process Intensification for Sustainable Energy Conversion

Process Intensification for Sustainable Energy Conversion

Edited by

FAUSTO GALLUCCI

and

MARTIN VAN SINT ANNALAND

Department of Chemical Engineering and Chemistry, Eindhoven University of Technology, The Netherlands

WILEY

This edition first published 2015
© 2015 John Wiley & Sons, Ltd

Registered office
John Wiley & Sons Ltd, The Atrium, Southern Gate, Chichester, West Sussex, PO19 8SQ, United Kingdom

For details of our global editorial offices, for customer services and for information about how to apply for permission to reuse the copyright material in this book please see our website at www.wiley.com.

Library of Congress Cataloging-in-Publication Data

Process intensification for sustainable energy conversion / edited by Fausto Gallucci, Chemical Process Intensification, Department of Chemical Engineering and Chemistry, Eindhoven University of Technology, Eindhoven, The Netherlands and Martin van Sint Annaland.
      pages cm
 Includes bibliographical references and index.
 ISBN 978-1-118-44935-6 (cloth)
 1. Chemical processes. 2.   Renewable energy sources. 3.   Green chemistry.   I. Gallucci, Fausto. II. Sint Annaland, Martin van.
 TP155.7.P756 2015
 660′.28–dc23

                                                                                              2015009730

A catalogue record for this book is available from the British Library.

ISBN: 9781118449356

Front Cover image: Membrane assisted fluidized bed chemical looping reforming. The authors would like to thank Joris Garenfeld for the 3D representation of the reactor.

Typeset in 10/12pt TimesLTStd by Laserwords Private Limited, Chennai, India
Printed and bound in Singapore by Markono Print Media Pte Ltd

1   2015

Contents

Preface		**xi**
List of Contributors		**xiii**

1. Introduction   **1**
Fausto Gallucci and Martin van Sint Annaland

References   6

2. Cryogenic CO$_2$ Capture   **7**
M. van Sint Annaland, M. J. Tuinier and F. Gallucci

2.1	Introduction – CCS and Cryogenic Systems		7
	2.1.1	Carbon Capture and Storage	8
	2.1.2	Cryogenic separation	10
2.2	Cryogenic Packed Bed Process Concept		11
	2.2.1	Capture Step	11
	2.2.2	CO$_2$ Recovery Step	12
	2.2.3	H$_2$O Recovery and Cooling Step	13
2.3	Detailed Numerical Model		13
	2.3.1	Model Description	13
	2.3.2	Simulation Results	15
	2.3.3	Simplified Model: Sharp Front Approach	16
	2.3.4	Model Description	16
	2.3.5	Process Analysis	22
	2.3.6	Initial Bed Temperature	24
	2.3.7	CO$_2$ Inlet Concentration	24
	2.3.8	Inlet Temperature	25
	2.3.9	Bed Properties	25
2.4	Small-Scale Demonstration (Proof of Principle)		25
	2.4.1	Results of the Proof of Principle	26
2.5	Experimental Demonstration of the Novel Process Concept in a Pilot-Scale Set-Up		31
	2.5.1	Experimental Procedure	32
	2.5.2	Experimental Results	33

	2.5.3	Simulations for the Proof of Concept	36
	2.5.4	Radial Temperature Profiles	36
	2.5.5	Influence of the Wall	38
2.6		Techno-Economic Evaluation	39
	2.6.1	Process Evaluation	40
	2.6.2	Parametric Study	41
	2.6.3	Comparison with Absorption and Membrane Technology	45
2.7		Conclusions	49
2.8		Note for the Reader	49
		List of symbols	50
		Greek letters	50
		Subscripts	51
		References	51

3. Novel Pre-Combustion Power Production: Membrane Reactors   53
F. Gallucci and M. van Sint Annaland

3.1		Introduction	53
3.2		The Membrane Reactor Concept	55
3.3		Types of Reactors	57
	3.3.1	Packed Bed Membrane Reactors	58
	3.3.2	Fluidized Bed Membrane Reactors	65
	3.3.3	Membrane Micro-Reactors	72
3.4		Conclusions	74
3.5		Note for the reader	75
		References	75

4. Oxy Fuel Combustion Power Production Using High Temperature O_2 Membranes   81
Vesna Middelkoop and Bart Michielsen

4.1		Introduction	81
4.2		MIEC Perovskites as Oxygen Separation Membrane Materials for the Oxy-fuel Combustion Power Production	83
4.3		MIEC Membrane Fabrication	85
4.4		High-temperature ceramic oxygen separation membrane system on laboratory scale	87
	4.4.1	Oxygen permeation measurements and sealing dense MIEC ceramic membranes	87
	4.4.2	$Ba_xSr_{1-x}Co_{1-x}Fe_yO_{3-\delta}$ and $La_xSr_{1-x}Co_{1-y}Fe_yO_{3-\delta}$ Membranes	89
	4.4.3	Chemical Stability of Perovskite Membranes Under Flue-Gas Conditions	96
	4.4.4	CO_2-Tolerant MIEC Membranes	99
4.5		Integration of High-Temperature O_2 Transport Membranes into Oxy-Fuel Process: Real World and Economic Feasibility	103
	4.5.1	Four-End and Three-End Integration Modes	103

4.5.2	Pilot-Scale Membrane Systems	104
4.5.3	Further Scale-Up of O_2 Production Systems	106
References		109

5. Chemical Looping Combustion for Power Production   117
V. Spallina H. P. Hamers, F. Gallucci and M. van Sint Annaland

5.1	Introduction	117
5.2	Oxygen carriers	120
	5.2.1 Nickel-based OCs	122
	5.2.2 Iron-based OCs	122
	5.2.3 Copper-based OCs	122
	5.2.4 Manganese-based OCs	123
	5.2.5 Other Oxygen Carriers	123
	5.2.6 Sulfur Tolerance	123
5.3	Reactor Concepts	124
	5.3.1 Interconnected Fluidized Bed Reactors	124
	5.3.2 Packed Bed Reactors	132
	5.3.3 Rotating Reactor	143
5.4	The Integration of CLC Reactor in Power Plant	144
	5.4.1 Natural Gas Power Plant with CLC	144
	5.4.2 Coal-Based Power Plant with CLC	148
	5.4.3 Comparison between CLC in packed beds and circulated fluidized beds	162
5.5	Conclusions	164
Nomenclature		167
Subscripts		168
References		168

6. Sorption-Enhanced Fuel Conversion   175
G. Manzolini, D. Jansen and A. D. Wright

6.1	Introduction	175
6.2	Development in Sorption-Enhanced Processes	176
	6.2.1 Enhanced Steam Methane Reformer	177
	6.2.2 SEWGS	177
6.3	Sorbent Development	180
	6.3.1 Sorbent for Sorption-Enhanced Reforming	180
	6.3.2 Sorbent for Enhanced Water-Gas Shift	182
6.4	Process Descriptions	188
	6.4.1 Fluidised Beds	189
	6.4.2 Fixed Beds	190
	6.4.3 Design Optimisation of Fixed Bed Processes	195
6.5	Sorption-Enhanced Reaction Processes in Power Plant for CO_2 Capture	196
	6.5.1 SER	196
	6.5.2 SEWGS case	199

6.6 Conclusions 203
Nomenclature 204
References 204

7. Pd-Based Membranes in Hydrogen Production for Fuel cells 209
Rune Bredesen, Thijs A. Peters, Tim Boeltken and Roland Dittmeyer

7.1 Introduction 209
7.2 Characteristics of Fuel Cells and Applications 211
7.3 Centralized and Distributed Hydrogen Production for Energy
Applications 213
7.4 Pd-Based Membranes 216
7.5 Hydrogen Production Using Pd-Based Membranes 216
7.5.1 Hydrogen from Natural Gas and Coal 217
7.5.2 Hydrogen from Ethanol 219
7.5.3 Hydrogen from Methanol 220
7.5.4 Hydrogen from Other Hydrocarbon Sources 221
7.5.5 Hydrogen from Ammonia 221
7.6 Process Intensification by Microstructured Membrane Reactors 221
7.7 Integration of Pd-Based Membranes and Fuel Cells 229
7.8 Final Remarks 231
Acknowledgements 231
References 232

8. From Biomass to SNG 243
Luca Di Felice and Francesca Micheli

8.1 Introduction 243
8.2 Current Status of Bio-SNG Production and Facilities in Europe 244
8.3 Bio-SNG Process Configuration 245
8.3.1 The Gasification Step 247
8.3.2 Gas Cleaning 248
8.3.3 The Synthesis Step 250
8.4 Catalytic Systems 251
8.5 The Case Study 253
8.5.1 The Feeding Composition 254
8.5.2 Heat Exchangers 256
8.5.3 Scrubber Tar Removal 257
8.5.4 Ammonia Absorber 258
8.5.5 HCl and H_2S Removal 259
8.5.6 Compression Section 259
8.5.7 Separation Section: H_2O and CO_2 Removal 259
8.5.8 Methanation Section Case 1: Adiabatic Fixed Bed with
Intermediate Cooling 260
8.5.9 Methanation Section Case 2: Isothermal Fluidized Bed 262
8.6 Chemical Efficiency 263
8.7 Conclusions 263
References 264

9. Blue Energy: Salinity Gradient for Energy Conversion   **267**
Paolo Chiesa, Marco Astolfi and Antonio Giuffrida

9.1   Introduction   267
9.2   Fundamentals of Salinity Gradient Exploitation   268
9.3   Pressure Retarded Osmosis Technology   270
        9.3.1   Operating Principles   271
        9.3.2   Plant Layout and Components   272
        9.3.3   Design Criteria and Optimization   276
        9.3.4   Technology Review   277
        9.3.5   Pilot Testing   278
9.4   The Reverse Electrodialysis Technology   279
        9.4.1   Operating Principles and Plant Layout   279
        9.4.2   RED Technology Review   282
9.5   Other Salinity Gradient Technologies   284
        9.5.1   Reverse Vapor Compression   284
        9.5.2   Hydrocratic Generator   288
9.6   Osmotic Power Plants Potential   290
        9.6.1   Site Criteria for Osmotic Power Plants   292
9.7   Conclusions   294
References   296

10. Solar Process Heat and Process Intensification   **299**
Bettina Muster and Christoph Brunner

10.1   Solar Process Heat – A Short Technology Review   299
        10.1.1   Examples of solar process heat system concepts   301
        10.1.2   Solar process heat collector development   302
10.2   Potential of Solar Process Heat in Industry   305
10.3   Bottlenecks for Integration of Solar Process Heat
        in Industry   305
        10.3.1   Introduction   305
        10.3.2   Bottlenecks of the Industrial Process to Integrate Solar
                 Heat Supply   306
        10.3.3   Bottlenecks of the Solar Process Heat System   308
        10.3.4   Engineering Intensified Process Systems for Renewable
                 Energy Integration   308
10.4   PI – A Promising Approach to Increase the Solar Process Heat
        Potential?   309
        10.4.1   Intensifying the Industrial Process and Possible
                 Effects on Solar Process Heat   311
10.5   Conclusion   328
References   328

11. Bioenergy – Intensified Biomass Utilization   **331**
Katia Gallucci and Pier Ugo Foscolo

11.1   Introduction   331

11.2   Biomass Gasification: State-of-the-Art Overview   332
        11.2.1 Cold Gas Cleaning and Conditioning: Current Systems   335
11.3   Hot Gas Cleaning   343
        11.3.1 Contaminant Problems Addressed   343
        11.3.2 Dust Filtration   349
        11.3.3 Catalytic Conditioning   352
        11.3.4 The *UNIQUE* Concept for Gasification and Hot Gas
                Cleaning and Conditioning   363
11.4   Conclusions   376
References   377

Index   **387**

Preface

Process Intensification (PI) is a hot topic in both industrial and academic worlds and is widely considered a key enabling approach to improve the competitiveness of the chemical industry. In essence, Process Intensification aims at the design of innovative (reactor) concepts for significantly smaller, safer, more efficient and cheaper processes.

The need for more efficient processes and more flexible engineering designs with, at the same time, increased safety and decreased environmental footprint is pushing the chemical industry towards novel research in this field. This is also reflected in a large number of funding initiatives focussing on PI worldwide. An example in Europe is the large emphasis of PI in the new research framework Horizon 2020 http://ec.europa.eu/programmes/horizon2020/ and in particular in the subprogram SPIRE http://www.spire2030.eu/. Following the increasing interest in PI, various interesting books have been published discussing new reactor concepts for the chemical and process industries.

Energy is the driver for many developments, and energy demands are foreseen to further increase with increasing population and fast developments in Asian countries. This creates many opportunities for process intensification in the energy sector and, in particular, the reduction of anthropogenic CO_2 emissions associated with fossil fuel conversion can help in the long transition period towards renewable energy conversion scenarios. The aim of this book is to compensate for the lack of academic/scientific books concerning the application of process intensification strategies to sustainable energy conversion.

In this book, we have collected, in 11 chapters, information on novel possible intensified methods and reactors for sustainable energy conversion, including – but not limited to – novel concepts for chemical looping combustion, PI concepts for CO_2 capture, oxy-fuel and oxygen permeable membranes, blue energy and biomass conversion.

The book has been written for academicians, PhD students, researchers and engineers curious about novel trends of PI applied for a more sustainable energy conversion.

Fausto Gallucci,
Martin van Sint Annaland

List of Contributors

Marco Astolfi, Energy Department, Politecnico di Milano, Italy

Tim Boeltken, Karlsruhe Institute of Technology (KIT), Institute for Micro Process Engineering (IMVT), Germany

Rune Bredesen, SINTEF Materials and Chemistry, Norway

Christoph Brunner, AEE - Institut für Nachhaltige Technologien, Austria

Paolo Chiesa, Energy Department, Politecnico di Milano, Italy

Luca Di Felice, Chemical Process Intensification, Department of Chemical Engineering and Chemistry, Eindhoven University of Technology, The Netherlands

Roland Dittmeyer, Karlsruhe Institute of Technology (KIT), Institute for Micro Process Engineering (IMVT), Germany

Pier Ugo Foscolo, Department of Industrial Engineering, University of L'Aquila, Italy

Fausto Gallucci, Chemical Process Intensification, Department of Chemical Engineering and Chemistry, Eindhoven University of Technology, The Netherlands

Katia Gallucci, Department of Industrial Engineering, University of L'Aquila, Italy

Antonio Giuffrida, Energy Department, Politecnico di Milano, Italy

Paul H. Hamers, Chemical Process Intensification, Department of Chemical Engineering and Chemistry, Eindhoven University of Technology, The Netherlands

Daniel Jansen, Energy & Resources, Copernicus Institute of Sustainable Development, Faculty of Geosciences, Utrecht University, The Netherlands

Giampaolo Manzolini, Energy Department, Politecnico di Milano, Italy

Francesca Micheli, Chemical Engineering Department, University of L'Aquila, Italy

Bart Michielsen, Materials Technology, Flemish Institute for Technological Research, Belgium

Vesna Middelkoop, Materials Technology, Flemish Institute for Technological Research, Belgium

Bettina Muster, AEE - Institut für Nachhaltige Technologien, Austria

Thijs A. Peters, SINTEF Materials and Chemistry, Norway

Martin van Sint Annaland, Chemical Process Intensification, Department of Chemical Engineering and Chemistry, Eindhoven University of Technology, The Netherlands

Vincenzo Spallina, Chemical Process Intensification, Department of Chemical Engineering and Chemistry, Eindhoven University of Technology, The Netherlands

Martin Tuinier, Chemical Process Intensification, Department of Chemical Engineering and Chemistry, Eindhoven University of Technology, The Netherlands

Andrew David Wright, Energy Technology, Air Products PLC, UK

1

Introduction

Fausto Gallucci and Martin van Sint Annaland
Eindhoven University of Technology, Chemical Process Intensification,
Department of Chemical Engineering and Chemistry,
Eindhoven, The Netherlands

It is expected that in the current century, the theme "energy" will become increasingly more important and will pose some serious challenges to our society and our way of living, but it may also create opportunities.

On the one hand, the combination of a rapidly growing world population and increasing energy consumption per capita requires large investments to secure sufficient energy supply at affordable prices. On the other hand, fossil fuel reserves are shrinking, while the transition toward a world economy based on energy supply via sustainable or renewable resources is still in its infancy. According to the World Energy Outlook 2013 of the International Energy Agency (IEA), the world energy demand will increase by more than 30% by 2035 (compared with 2011) and the demand for oil alone will still be more than 57% in 2035. Oil and gas reserves are increasingly concentrated in a few countries that control them through monopoly companies. The dependence of Europe on imported oil and gas is growing: we import 50% of our energy, and it will be 55% by 2035 (Bp Outlook 2035), if we do not act.

The relevance of this issue is even higher when one relates the increase in anthropogenic CO_2 emissions by the use of fossil fuels to the evident changes in the Earth's climate. The International Panel on Climate Change (IPCC) has collected results of substantial research efforts to obtain a comprehensive scientific framework describing the evolution of the climate over very long time periods, the observed deviations from this behavior in recent times, the interpretation of both natural and anthropogenic causes and their effect on the increase of the greenhouse effect, the consequences of global warming in the past, present and future and possible solutions to combat further climate changes. In its 2013

Process Intensification for Sustainable Energy Conversion, First Edition.
Edited by Fausto Gallucci and Martin van Sint Annaland.
© 2015 John Wiley & Sons, Ltd. Published 2015 by John Wiley & Sons, Ltd.

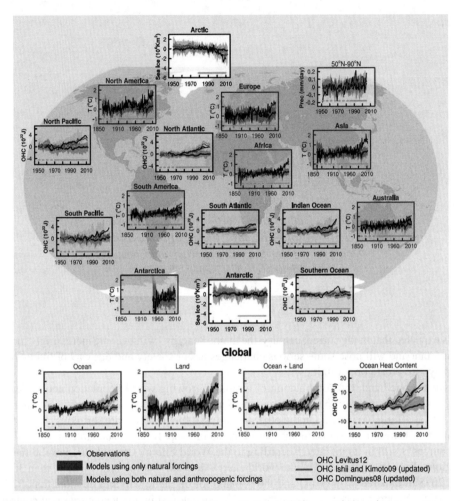

Figure 1.1 *Detection and attribution signals in some elements of the climate system, at regional scales (top panels) and global scales (bottom four panels). Brown panels are land surface–temperature–time series, green panels are precipitation–time series, blue panels are ocean heat content–time series and white panels are sea ice–time series. Observations are shown on each panel in black or black and shades of grey. Blue shading is the model time series for natural forcing simulations and pink shading is the combined natural and anthropogenic forcings. The dark blue and dark red lines are the ensemble means from the model simulations. All panels show the 5–95% intervals of the natural forcing simulations and the natural and anthropogenic forcing simulations. (Source: Extracted from the IPCC report 2013) (For a color version of this figure, please see color plate section.)*

Assessment Report, IPCC conclude that (Climate Change 2013: The Physical Science Basis), see Figure 1.1:

> *"From up in the stratosphere, down through the troposphere to the surface of the Earth and into the depths of the oceans there are detectable signals of change such that the assessed likelihood of a detectable, and often quantifiable, human contribution ranges from likely to extremely likely for many climate variables."*

According to IPCC, the effect of human activities on changes in the climate is very likely to have been dominating natural variations (due to, e.g., variations in solar irradiance) especially in the past 50 years. Since the beginning of the industrial revolution, the concentrations of the relevant greenhouse gases (especially carbon dioxide, methane, nitrous oxide, and halocarbons) have increased substantially and now by far exceed natural ranges encountered in the past 650,000 years [1].

On the short term, significant reductions of carbon dioxide emissions may be attained from energy savings, for example, via efficiency improvements both in power production and consumer products and as a consequence of increased public awareness. However, strong economic growth anticipated in especially the developing countries is expected to impede a net decrease in anthropogenic emissions. On the longer term, the use of fossil fuels for energy supply will need to be phased out not only to stabilize greenhouse gas concentrations but also to avoid shortages in raw materials for the production of, for example, bulk chemicals.

The transition towards a world economy based on energy supply via sustainable sources such as wind-, hydro- and solar energy, or nuclear power (of which fission still suffers from a bad public image caused by concerns over nuclear waste and proliferation, whereas fusion has so far failed to live up to its potential) is therefore expected to be a lengthy process that cannot be expected to be solely responsible for the stabilization of atmospheric greenhouse gas concentrations in this century. Rather, a combination of many of the mitigation alternatives will need to be adopted to significantly curb CO_2 emissions.

In this respect, novel concepts based on process intensification can help to reduce CO_2 emissions and can lead the transition towards a more sustainable energy scenario. Indeed, according to Ramshaw [2], process intensification is a strategy for making dramatic reductions in the size of a chemical plant so as to reach a given production objective. As such, applying process intensification to the energy sector can result in a dramatic decrease in the production of wastes including greenhouse gas emissions.

According to Stankiewicz and Moulijn [3], the whole field of process intensification can be classified into two main categories:

1. Process-intensifying equipment:
   These include novel reactors and intensive mixing, heat-transfer and mass-transfer devices, and so on.
2. Process-intensifying methods:
   These include new or hybrid separations, integration of reaction and separation, heat exchange, or phase transition (in multifunctional reactors), techniques using alternative energy sources (light, ultrasound, etc.) and new process-control methods (like intentional unsteady-state operation).

Clearly, as also indicated by Stankiewicz and Moulijn, there is a big overlap between the two areas. For instance, membrane reactors are an example of process-intensifying equipment (novel reactor) making use of process-intensifying methods (integration of reaction and separation).

Since the "invention" of the term *process intensification*, many articles and books appeared on the same topic. An interested reader is referred to the book of Reay *et al.* [4] for an overview of the various process intensification methods. In the present book, a selection of different, novel process intensification methods and reactors are presented and discussed with the focus on sustainable energy conversion.

In particular, in Chapter 2 the development of a new cryogenic separation technology based on dynamic operation of packed bed columns is described. When it is possible to exploit the cold available at, for example, LNG regasification stations, this new technology could be used as an efficient post-combustion CO_2 capture technology. In the chapter, the technology is described to freeze-out CO_2 from flue gases at atmospheric pressures. The dynamic operation and the effects of the operating conditions have been analyzed in detail using modelling and an experimental proof of principle at laboratory scale and small pilot scale is provided. Finally, a techno-economic analysis shows the great potential of the technology over other post-combustion capture processes such as amine scrubbing and membrane separation, when cold duty is available at low prices or when high CO_2 capture efficiencies are required. This makes the cryogenic technology also particularly interesting as an auxiliary unit downstream of other post-combustion technologies.

Chapter 3 describes the application of membrane reactors in pre-combustion CO_2 capture technologies. Different membrane reactor configurations are described, among which the fluidized bed membrane reactor configuration seems to have the most potential. In this concept, hydrogen perm-selective membranes are submerged in a fluidized suspension. Thus, mass and heat transfer coefficients are much improved compared to packed bed membrane reactor configurations (decreasing problems in heat management and concentration polarization), while maintaining a relatively large amount of catalyst combined with a relatively low pressure drop in comparison with micro-membrane reactors. The chapter also describes a hybrid concept integrating both membrane reactors and chemical looping combustion for autothermal operation with integrated CO_2 capture. With this new concept, high hydrogen efficiency can be obtained at lower temperatures compared with other concepts, while the amount of membrane area required is kept to a minimum.

Chapter 4 focuses on the possibility to apply high-temperature oxygen-selective membranes in oxy-fuel power production systems. These perovskite-like or mixed ionic electronic conducting materials present an infinite perm-selectivity for oxygen compared with other gases and can thus be used to separate oxygen from air at high temperatures. The chapter describes the main features of oxygen selective membranes, their production methods and their integration in membrane (reactor) modules. The chapter also reports on the progress of research projects on oxygen selective membranes.

A different kind of oxy-combustion can be achieved by exploiting the air separation through a solid material that is alternating oxidized (with air) and reduced (with a fuel). This solid material is called oxygen carrier and is the "catalyst" of a new concept called chemical looping combustion. Chapter 5 describes chemical looping combustion (CLC) concepts for power production. The oxidation and reduction stages can be achieved with different reactor technologies. In particular, it is possible to circulate the solid material between two fluidized bed reactors, where one reactor operates as oxidation reactor (air reactor) and a second reactor operates as reduction stage (fuel reactor). Another possibility is to keep the solid in fixed position and to alternately switch the gas feed streams; in this case, the concept is based on dynamically operated packed beds. Finally, the solid can be kept in a fixed position and the reactor can be rotated. The chapter reports all the possible configurations for CLC and compares the efficiencies of different concepts when exploited for power production.

Another interesting concept that can also be used for power production with integrated CO_2 capture is sorption-enhanced fuel conversion and is described in Chapter 6. In this concept, a solid sorbent is used together with the catalyst such that the CO_2 produced during the reforming (or water-gas shift) of a fuel can be directly captured and separated, while the other products, often hydrogen, can be used for downstream power production. Also with sorption-enhanced processes, the efficiency of the CO_2 capture can be significantly increased, because the hydrogen is generally produced at high pressure, whereas in many other concepts, such as the membrane reactor concepts presented in Chapter 3, the hydrogen is produced at lower pressures. On the other hand, the CO_2 pressure is higher in membrane processes compared with sorption-enhanced processes. The selection between these two concepts is thus related to many different parameters, and the efficiency should be evaluated separately for each individual case.

Chapter 7 reports on the hydrogen production for fuel cell applications. Also for this application, membrane reactors are described in detail. The need for smart reactor designs in order to reduce or circumvent concentration polarization (or bulk-to-membrane mass transfer resistances) and improve the heat management is pointed out, which places stringent requirements on membrane stability, catalyst activity, sealing technology, support materials and module design. The chapter describes the ongoing research on micro-structured membrane reactors which will be a step forward towards low cost and high productivity units for hydrogen production.

All these chapters describe the smart use of reactor design and process integration to increase the efficiency and reduce the emissions when using fossil fuels as energy source. Of course, intensified systems can also be used for bio-based energy sources. Chapter 8 reports the possibility to convert biomass into substitute natural gas. The chapter describes how both packed beds and fluidized bed reactors can effectively be used to improve the methanation reaction so that the products of biomass gasification can be converted into a more sustainable methane stream.

Chapter 9 describes how to efficiently make use of a salinity gradient for power production. This is a completely CO_2-free power production system, so that the concept is often referred to as blue energy. The concept is based on the fact that chemical potential is associated with a difference in salt concentration, so that electric power can be produced by exploiting the salinity gradient between freshwater of rivers and seawater. When exploited at large scale, this concept can supply a large part of the electricity required worldwide. However, the concept is not that easy as conventional hydropower electricity production. The chapter describes the fundamentals of the salinity gradient technology and the attainable energy efficiencies associated with this energy conversion technology.

Chapter 10 describes how process intensification can be applied to efficiently make use of the most abundant energy source: solar energy. The question is addressed whether an intensified process layout can increase the potential of solar process heat and its efficiency. The chapter describes how to make efficient use of solar heat by designing the heat profile in a certain process, where the importance of process modeling in this respect is stressed.

Finally, Chapter 11 reports on intensified processes for biomass utilization. In particular, the authors describe the broad field of power and combined heat and power (CHP) generation from biomass: more specifically, advances in biomass gasification technology aimed at increases in overall conversion and efficiency and hence in a decreased cost of electricity. Poly-generation strategies (for combined heat, power and chemical production applications) are also considered, with particular reference to recent technological innovations in hot gas cleaning and conditioning; these have been developed to achieve the required improvements in syngas quality and have been validated under industrially relevant conditions.

We surely know that many other intensified processes can be designed for efficient power production, or in general for energy conversion. We hope that the content of this book will stimulate the design, implementation and testing of novel integrated reactor concepts for a more sustainable energy future.

References

1. IPCC report 2007: Climate Change 2007: Mitigation of Climate Change
2. Ramshaw, C. (1995) The Incentive for Process Intensification, Proceedings, 1st Intl. Conf. Proc. Intensif. for Chem. Ind., 18, BHR Group, London, p. 1
3. Stankiewicz, A.I. and Moulijn, J.A. (2000) *Transforming Chemical Engineering*, Chemical Engineering Progress, AiCHE.
4. Reay, D., Ramshaw, C. and Harvey, A. (2013) *Process Intensification*, Second edn, Butterworth-Heinemann, Oxford. ISBN: 9780080983042

2

Cryogenic CO$_2$ Capture

M. van Sint Annaland, M. J. Tuinier and F. Gallucci
Eindhoven University of Technology, Chemical Process Intensification,
Department of Chemical Engineering and Chemistry, P.O. Box 513, 5612 AZ,
Eindhoven, The Netherlands

2.1   Introduction – CCS and Cryogenic Systems

There is a growing worldwide awareness of the fact that the earth's surface temperatures are changing globally. Although the climate of our planet has been altering continuously during its history, the current changes are taking place at an unprecedented pace and are expected to have dramatic consequences on human kind [1]. Since the Industrial Revolution in the late 19th century, fossil fuels started to play an important role in our energy supply for transportation, heating and electricity. The combustion of fossil fuels results in large amounts of CO$_2$, which are emitted into the atmosphere. The increase in CO$_2$ concentrations coincides with the increase in global temperatures. There is a consensus among most scientists that the rise in CO$_2$ concentrations is responsible for the observed increase in temperatures. In 1988, Intergovernmental Panel on Climate Change (IPCC) was established in order to evaluate the risks of climate change caused by human activities.

Based on the evaluations by the IPCC, it is expected that global temperatures will keep on rising in the next century. In order to prevent or at least minimize further temperature increases, it is necessary to reduce anthropogenic CO$_2$ emissions not only for health and safety reasons but also for economic reasons. A study on the economic effects of climate change by Stern [2] states that the costs of mitigating climate change can be limited to around 1% of global GDP per year. Doing nothing ('business as usual') and facing the consequences of climate change will be equivalent to losing 5% to possibly 20% of global GDP each year. Therefore, immediate actions to reduce CO$_2$ emissions are essential.

Process Intensification for Sustainable Energy Conversion, First Edition.
Edited by Fausto Gallucci and Martin van Sint Annaland.
© 2015 John Wiley & Sons, Ltd. Published 2015 by John Wiley & Sons, Ltd.

CO_2 emission reduction can be achieved in several ways, in the first place by improving efficiencies. Developments in the automobile industry, for example, are leading to more and more economic engines, consuming less fuel. The efficiency of power plants is also increasing, and chemical industry is able to save energy by, for example, heat integration. These developments will contribute to CO_2 emission reductions. However, it is not expected that these efficiency improvements will be sufficient to bring down our CO_2 emissions to acceptable levels, mainly because of increased energy demands by the developing countries such as China and India. Therefore, more measures are required. A key measure is to switch our energy supply to renewable energy sources such as biomass, solar and wind energy not only to reduce CO_2 emissions but also to bring down our dependency on scarce fossil fuels. However, at this point, power supply by renewable energy sources is still under development and is not yet competitive with conventional power generation based on fossil fuels. A third possible route to emission reduction is nuclear power, but safety issues and nuclear waste disposal are causing moral and political concerns. Due to the aforementioned reasons, it is expected that fossil fuels will continue to play a significant role in our energy supply for the next decades.

It is therefore considered to be necessary to introduce carbon capture and storage (CCS) to mitigate anthropogenic CO_2 emissions as a midterm solution until a full transition to energy supply by renewables can be realized.

2.1.1   Carbon Capture and Storage

The goal of CCS is to remove CO_2 from flue gases and to store it for the long term. This process is schematically represented in Figure 2.1. Fossil fuels are normally combusted

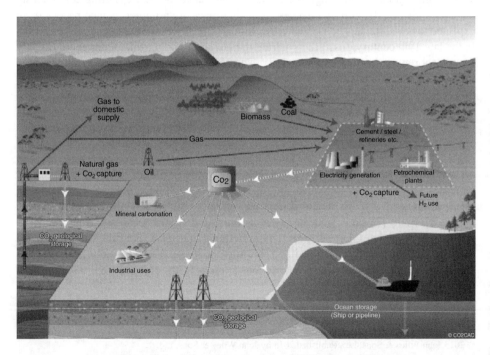

Figure 2.1   *Schematic overview of CCS process [3]*

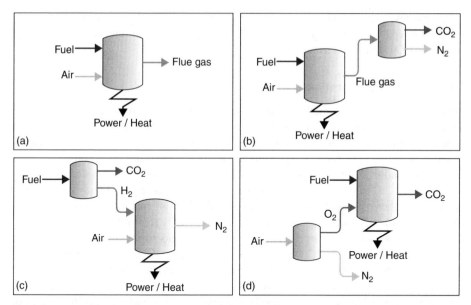

Figure 2.2 *Pathways to CO$_2$ capture: (a) conventional combustion process without capture, (b) post-combustion, (c) pre-combustion, (d) oxy-fuel*

using air. The flue gas is therefore composed of a large amount of N$_2$ and 5–20 vol% CO$_2$. Furthermore, it contains H$_2$O and impurities such as sulphur and nitrous oxides, depending on the feedstock and process. Compressing and storing the entire flue gas, including N$_2$, will be too costly. Therefore, it is necessary to obtain CO$_2$ in purified form first, before it can be stored in geological formations. About 75% of the costs involved in CCS are associated with the capture step [4], and, therefore, many research projects focus on the development or optimization of capture technologies.

CO$_2$ capture technologies are often classified into oxy-fuel, pre- and post-combustion processes, which are schematically represented in Figure 2.2. In oxy-fuel processes, fossil fuels are combusted using pure oxygen, circumventing dilution of CO$_2$ with N$_2$. The disadvantage is that an energy-intensive air separation unit is required to obtain pure O$_2$, although this could be avoided by using chemical-looping combustion (see the other chapters of this book and a.o. Ishida and Jin [5], Noorman *et al.* [6]).

In pre-combustion processes, fossil fuels are first converted into H$_2$ and CO$_2$ via (autothermal) reforming or partial oxidation and water-gas shift, CO$_2$ is subsequently captured and H$_2$ is fed to the combustion chamber or fuel cell. The advantage is that the separation of CO$_2$ and H$_2$ can be carried out at high pressure, resulting in a high driving force for the separation. The disadvantage of pre-combustion is that these processes can only be applied in new plants and not to the many existing operational facilities.

Post-combustion processes are based on capturing CO$_2$ from flue gases from conventional air-fired combustion processes. A disadvantage is that the CO$_2$ is diluted and, at low pressures, reducing the driving force for separation. However, this technology can be retrofitted to already operating power plants and industries. For this reason, post-combustion is considered the most realistic technology on the short term, even though the efficiency of alternatives could be higher [7].

Several post-combustion technologies are under development, such as amine scrubbing, membrane separation, adsorption. Among these technologies, an interesting option is the cryogenic CO_2 separation as reported in the following section.

2.1.2   Cryogenic separation

Cryogenic separation is another option for separating CO_2 from gas mixtures. The advantages are that no chemical absorbents or adsorbents or large pressure differences are needed and that high-purity products can be obtained. However, cryogenic CO_2 capture is not included in most (economic) comparison studies, as it has been considered as an unrealistic candidate for post-combustion CO_2 capture in the first place due to expected high cooling costs and because it has been considered as a gas–liquid separation [4, 8]. At atmospheric pressures, CO_2 will go directly from its gas phase to its solid phase (desublimation). In order to be able to carry out the CO_2 removal from flue gases as a gas–liquid separation, it is necessary to compress the gas to pressures above the triple point of CO_2, which is at 5.2 bar and $-56.6\,°C$ for pure CO_2, as shown in the phase diagram of pure CO_2 in Figure 2.3. Compressing flue gases to high pressures for CO_2 capture is too energy (and therefore cost) intensive.

On the other hand, expensive refrigeration can possibly be avoided by exploiting the cold duty available at liquefied natural gas (LNG) regasification sites. Currently, LNG is being regasified by using seawater or water baths that are heated by burning a fuel gas [9]. The global LNG market is strongly growing [10]; therefore, integration of LNG regasification with a cryogenic CO_2 capture process could be beneficial.

Clodic and Younes [11, 12] have developed a cryogenic CO_2 capture process, in which CO_2 is desublimated as a solid onto surfaces of heat exchangers, which are cooled by

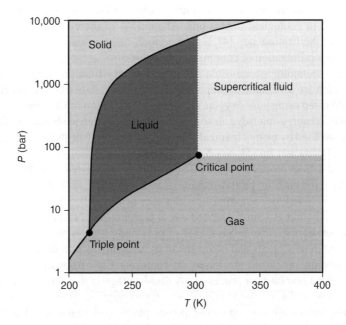

Figure 2.3   *Phase diagram of pure CO_2*

evaporating a refrigerant blend. With calculations and experimental tests, they showed that their process could compete with other post-combustion CO_2 capture processes. The main disadvantage of their system is that the water content in the feed stream to the cooling units should be minimal in order to prevent plugging by ice or an unacceptably high rise in pressure drop during operation. Therefore, several costly steps are required to remove all water traces from the flue gas. In addition, the increasing layer of solid CO_2 on the heat exchanger surfaces during the capture cycle will adversely affect the heat transfer, thereby reducing the process efficiency. Moreover, the costly heat exchangers have to be switched to regeneration cycles operated at a different temperature, which should be carried out with great care to avoid excessive mechanical stresses. To overcome these disadvantages, a new process concept is developed to separate CO_2 from flue gases at atmospheric pressures.

2.2 Cryogenic Packed Bed Process Concept

In this section, a new process concept based on dynamically operated packed beds is described in detail, where the flue gas is represented as a mixture of N_2, CO_2 and H_2O to simplify the description. Continuous separation of these components can be obtained when three packed beds are operated in parallel in three different steps: a capture, recovery and cooling step. These three steps are discussed consecutively in the following sections focusing on the evolution of axial temperature and mass deposition profiles (see Figure 2.4).

2.2.1 Capture Step

When a gas mixture consisting of N_2, CO_2 and H_2O is being fed at a relatively high temperature $T_{c,in}$ to an initially cryogenically refrigerated packed bed (at T_0), an effective separation between these components can be accomplished, due to differences in dew and sublimation points. The gas mixture will cool down and the packing material will heat up, until H_2O starts to condense on the packing surface. A certain amount of H_2O per volume of the packing material (indicated as m_{H_2O} in Figure 2.4a) will condense, until a local equilibrium is reached (at a temperature T_{H_2O}). Actually a very small part of the H_2O at the front will be frozen to ice, but simulations have revealed that this is a very small part of the H_2O and has negligible influence on the resulting axial temperature and mass deposition profiles. The cold energy stored in the packing will be consumed, and a front of condensing H_2O will move through the bed towards the outlet of the bed. At the same time, previously condensed H_2O will evaporate due to the incoming relatively hot gas mixture. Therefore, two fronts of evaporating and condensing water will move through the bed, with a faster moving condensing front. After all water is condensed, the gas mixture will be cooled further until CO_2 starts to change its phase.

At atmospheric pressure, CO_2 will desublimate directly from gas to solid, and, therefore, solid CO_2 is deposited on the packing surface. Similar to H_2O, two CO_2 fronts will move through the bed: an evaporation front and a desublimation front. Again an equilibrium is reached, and a certain amount of CO_2 (m_{CO_2}) is deposited on the packing surface at a temperature of T_{CO_2}. Note that an effective separation between CO_2 and H_2O can also be accomplished in this way. N_2 will not undergo any phase change (as long as T_0 is not chosen

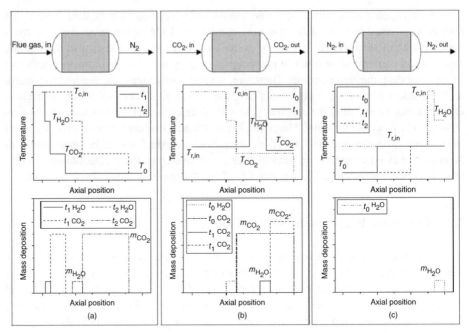

Figure 2.4 *Schematic axial temperature and corresponding mass deposition profiles for the capture (a), recovery (b) and cooling (c) steps, respectively*

too low) and will therefore move through the bed unaffected. When the CO_2 desublimation front reaches the end of the bed, CO_2 may break through and the bed should be switched to a recovery step just before that.

2.2.2   CO$_2$ Recovery Step

The first zone of the bed is heated to $T_{c,in}$ during the capture step. This heat is used in the recovery step to evaporate the condensed H_2O and frozen CO_2. A gas flow consisting of pure CO_2 is fed to the bed. When feeding a pure CO_2 gas flow at a temperature $T_{r,in}$ to the packed bed, the gas will be heated up to $T_{c,in}$ and all fronts will move through the bed, as illustrated in Figure 2.4b. However, during the initial period of the recovery step, the ingoing CO_2 will deposit on the packing. Due to the increase in CO_2 partial pressure compared to that in the capture step, more CO_2 is able to desublimate on the packing surface (from m_{CO_2} to $m_{CO_2}^*$), and the bed temperature will slightly increase to $T_{CO_2}^*$.

Pure CO_2 is obtained at the outlet of the bed after this new equilibrium is reached. Part of the outgoing CO_2 should be compressed for transportation and sequestration, while the other part can be recycled to the inlet of the bed at a temperature $T_{r,in}$, which is slightly higher than $T_{CO_2}^*$ due to the heat production associated with compression in the recycle blower and some unavoidable heat leaks. When all CO_2 has been recovered, the bed is switched to a step in which H_2O is removed and the bed is cooled simultaneously.

Alternatively, the deposited CO_2 could be recovered as liquid, avoiding expensive compression costs required for transportation and storage. This could be accomplished by closing the valves connected to the bed and by introducing heat into the bed. CO_2 evaporation occurs and pressure builds up until the system reaches the triple point of pure CO_2, and liquid CO_2 will be formed. The drawbacks of this alternative process are that pressure vessels are required and that heat should be introduced into the bed, for example, by means of internal tubes. Both measures will result in a significant increase in capital costs. Furthermore, not all liquid CO_2 might be recovered from the packing due to the static liquid hold up in the bed. This process is not further explored in this chapter.

2.2.3 H_2O Recovery and Cooling Step

In the last step, the bed is cooled down by using a gas flow refrigerated before to temperature T_0. The cleaned flue gas can be used for this purpose. Cooling can be performed using a cryogenic refrigerator or by evaporating LNG. H_2O is evaporated and removed from the bed during the first period of the cooling step. The N_2/H_2O mixture can be released to the atmosphere, and when all H_2O is recovered, the outgoing flow can be recycled to the inlet of the bed, via a cooler. Temperature and mass deposition profiles are shown in Figure 2.4c. It should be noted that it is not required to cool the entire bed to T_0. The last zone can be kept at $T_{r,in}$, as during the capture step, this last part will be cooled down by the cleaned flue gas.

2.3   Detailed Numerical Model

2.3.1   Model Description

The prevailing heat and mass transfer processes in the periodically operated packed beds have been investigated with a pseudo-homogeneous, one-dimensional plug flow model with superimposed axial dispersion.

The modelling is based on the following main assumptions:

- Heat losses to the environment are small (i.e., adiabatic operation) and additionally a uniform velocity profile exists in the absence of radial temperature and concentration gradients, allowing the consideration of the axial temperature and concentration profiles only.
- Possible heat transfer limitations between the solid packing and the bulk of the gas phase are accounted for via effective axial heat dispersion (pseudo-homogeneous model).
- The rate of mass deposition and sublimation of CO_2 is assumed to be proportional to the local deviation from the phase equilibrium, estimating the equilibration time constant (g) at 1×10^{-6} s/m, which is assumed independent of temperature. The rate of sublimation of previously deposited CO_2 is assumed to approach a first-order dependency on the mass deposition when this mass deposition approaches zero [13].

The mass and energy conservation equations have been listed in Table 2.1. The constitutive equations for the transport parameters and the mass deposition rate have been summarized in Tables 2.2 and 2.3, respectively. The gas phase (mixture) properties have been computed according to Reid *et al.* [14], using the pure component data supplied by Daubert and Danner [15]. Uniform initial temperature profiles are taken without any mass deposited

Table 2.1   *Model equations for the 1-D pseudo-homogeneous model*

Component mass balance for the gas phase:

$$\varepsilon_g \rho_g \frac{\partial \omega_{i,g}}{\partial t} = -\varepsilon_g \rho_g \frac{\partial \omega_{i,g}}{\partial z} + \frac{\partial}{\partial z_i}\left(\rho_g D_{eff} \frac{\partial \omega_{i,g}}{\partial z}\right) - \dot{m}_i'' a_s + \omega_{i,g} \sum_{i=1}^{n_c} \dot{m}_i'' a_s$$

Component mass balance for the solid phase:

$$\frac{\partial m_i}{\partial t} = \dot{m}_i'' a_s$$

Total continuity equation for the gas phase:

$$\frac{\partial \varepsilon_g \rho_g}{\partial t} = -\frac{\partial \rho_g v_g}{\partial z} - \sum_{i=1}^{n_c} \dot{m}_i'' a_s$$

Energy balance (gas and solid phase)

$$(\varepsilon_g \rho_g C_{p,g} + \rho_s(1-\varepsilon_g)C_{p,s})\frac{\partial T}{\partial t} = -C_{p,g}\rho_g v_g \frac{\partial T}{\partial z} + \frac{\partial}{\partial z}\left(\lambda_{eff}\frac{\partial T}{\partial z}\right) + \sum_{i=1}^{n_c} \dot{m}_i'' a_s \Delta H_i$$

Pressure drop over the packing:

$$\frac{\partial P}{\partial z} = -4\frac{f}{d_h}\frac{1}{2}\rho_g v_g^2 \quad \text{with} \quad f = \frac{14.9}{Re}\sqrt{1+0.0445 Re\frac{d_h}{L}}$$

Table 2.2   *Heat and mass transfer coefficients for a monolith packing*

Effective axial heat dispersion:

$$\lambda_{eff} = (1-\varepsilon_g)\lambda_s + \left(\frac{\rho_g v_g C_{p,g}}{\varepsilon_g}\right)^2 \frac{1}{\alpha_{g,s} a_s}$$

Gas to solid heat transfer coefficient:

$$\alpha_{g,s} = \frac{\lambda_g}{d_h}2.978\left(1-0.095 Re\,Pr\frac{d_h}{L_c}\right)^{0.45} \quad \text{with} \quad Re = \frac{\rho_g v_g d_h}{\eta_g \varepsilon_g}$$

Axial mass dispersion:

$$D_{ax} = D_{eff,i} + \frac{v_g^2 d_{h,c}^2}{192 D_{eff,i}} \quad \text{with} \quad D_{eff,i} = \frac{1}{\sum_{j=1}^{n_c} \frac{y_{j,g}}{D_{i,j}}}$$

on the solid packing, where the gas phase in the bed is initially N_2. Furthermore, the usual Danckwerts-type boundary conditions are applied at the inlet and outlet of the beds.

A system of strongly non-linear, coupled partial differential equations is solved using a very efficient finite volume discretization technique, using a second-order SDIRK (Singly Diagonally Implicit Runge–Kutta) scheme for the accumulation terms, an explicit fifth-order WENO (Weighted Essentially Non-Oscillatory) scheme for the convection terms (with implicit first-order upwind treatment using the deferred correction method), second-order standard implicit central discretization for the dispersion terms and the

Table 2.3   *Mass deposition rate*

Mass deposition rate:

$$\dot{m}_i'' = \begin{cases} g\left(y_{i,s}P - P_i^\sigma\right) & \text{if} \quad y_{i,s}P \geq P_i^\sigma \\ g(y_{i,s}P - P_i^\sigma)\dfrac{m_i}{m_i + 0.1} & \text{if} \quad y_{i,s}P < P_i^\sigma \end{cases}$$

Gas–solid equilibrium:

$$P_{CO_2}^\sigma(T) = \exp\left(10.257 - \frac{3082.7}{T} + 4.08 \ln T - 2.2658 \cdot 10^{-2}T\right)$$

$$\Delta H_{CO_2}^{sub} = 5.685 \cdot 10^5 \, \text{J/kg}$$

standard Newton–Raphson technique for the linearly implicit treatment of the source terms. Moreover, time-step adaptation and local grid refinement procedures have been implemented, making effective use of the WENO smoothness indicators and interpolation polynomials [16]. The steep temperature and mass deposition gradients in combination with the strongly non-linear sublimation kinetics require a very efficient and stable numerical implementation using higher order implicit schemes.

2.3.2   Simulation Results

Simulations have been carried out for all three process steps. The bed properties and process conditions used are listed in Tables 2.4 and 2.5, respectively. A stainless steel monolithic structure is chosen as packing material, because axial dispersion and pressure drop are minimal for this type of packing material while the volumetric heat capacity is relatively high. The axial temperature and mass deposition profiles during the capture step are shown in Figure 2.5a and d, respectively.

Table 2.4   *Bed properties used in the numerical study*

Length bed (m)	8
Diameter bed (m)	3
Packing type	Steel monolith
Solid density (kg/m^3)	7750
Channel diameter (m)	6.76×10^{-4}
Wall thickness (m)	1.34×10^{-4}
Porosity (–)	0.7
Surface area (m^2/m^3)	4124
Heat capacity (J/kg/K)	$\sum\limits_{i=0}^{5} c_i T^i$
c_0	-203.75
c_1	6.4335
c_2	-2.4320×10^{-2}
c_3	4.6266×10^{-5}
c_4	-4.2721×10^{-8}
c_5	1.5296×10^{-11}

Table 2.5   *Conditions used in the numerical study*

	Capture	Recovery	Cooling
T_{in} (°C)	250	−70	−140
Φ_{in} (kg/s)	20	100	70
$y_{CO_2,in}$ (−)	0.1	1.0	0.0
$y_{H_2O,in}$ (−)	0.01	0.0	0.0

At the chosen initial bed temperature of $-140\,°C$, more than 99% of CO_2 is recovered. After 600 seconds, the CO_2 desublimation front reaches the end of the bed and the capture cycle should be stopped. The conditions at 600 seconds are used as initial conditions for the simulation of the recovery step. Figure 2.5b and e show that during the recovery step extra CO_2 will be deposited on the packing surface and that all deposited CO_2 is removed after again 600 seconds.

Now the data at the end of the recovery step are used as initial conditions for the cooling step. A refrigerated N_2 flow is being fed to the bed and profiles will develop as illustrated in Figure 2.5c and f. Note that not the entire bed is cooled down in this step, the last zone is cooled down during the capture step. The results show that CO_2 and H_2O capture can be integrated in one single bed.

However, the temperature profile during the recovery step shows that the heat stored in the first zone during the capture step is only sufficient to remove H_2O from the bed. The hot zone is moved through the bed, but due to axial heat dispersion this hot zone will be spread out over the bed, which is clearly visible in Figure 2.5b.

When feeding the gas mixture at realistic flue gas temperatures (which are generally lower than $250\,°C$) during the capture step, insufficient heat is stored in the packing to evaporate previously condensed water again. A possibility would be to introduce extra heat into the bed in the initial period of the recovery step. However, more practical is to carry out the H_2O capture step in a separate smaller bed, which can be cooled down to temperatures much higher than the initial bed temperature of the CO_2 capture bed.

2.3.3   Simplified Model: Sharp Front Approach

The process concept can be described by the advanced numerical model as detailed in the previous section. However, by assuming that the fronts that are formed during the different process steps are perfectly well defined (sharp), a simplified and relatively easy-to-solve and fast model can be developed, referred to as the 'sharp front approach'. In the first place, this approach is a very useful tool to quickly investigate the influence of process parameters on process behaviour. Furthermore, it can be used in conceptual design studies. This section describes this sharp front approach and compares the outcomes with the more advanced numerical model. Finally, the influences of several process parameters are studied.

2.3.4   Model Description

2.3.4.1   Capture Step

The model is derived for capturing a component i from a binary gas mixture consisting of components i and j (but could be easily extended to multicomponent mixtures). When

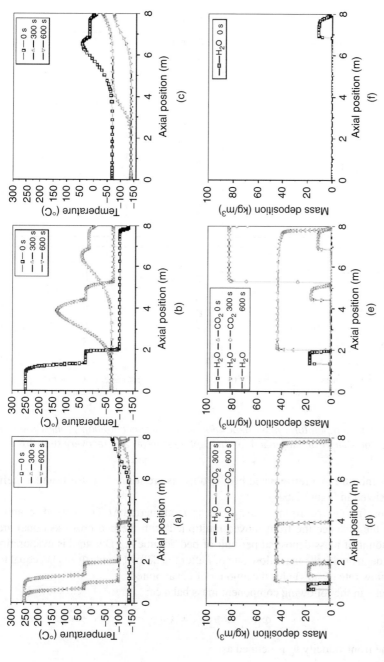

Figure 2.5 *Simulated axial temperature (a–c) and mass deposition (d–f) profiles for the capture, recovery and cooling step. Bed properties and operating conditions can be found in tables 2.4 and 2.5, respectively*

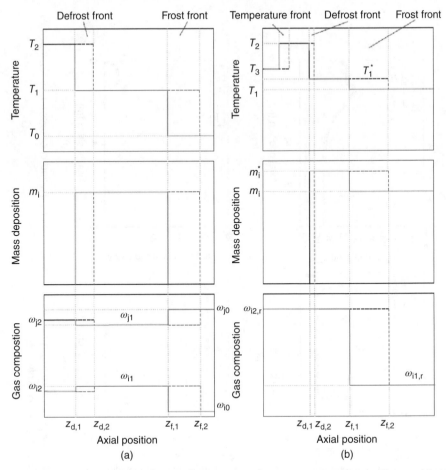

Figure 2.6 *Axial temperature, mass deposition and gas concentration profiles used in the derivation of the sharp front approach for the capture step (a) and recovery step (b)*

feeding this mixture to a refrigerated bed, two fronts are formed: a 'frost' and a 'defrost' front as depicted in Figure 2.6a.

The defrost front moves from $z_{d,1}$ to $z_{d,2}$ during a time period Δt. The mass of i evaporated is equal to the distance the front moved multiplied with the bed cross-sectional area A and the amount of mass deposited per unit of bed volume m_i. Due to this evaporation of previously deposited i, the mass flow rate of i after the defrost front ($\Phi_1\omega_{i1}$) is equal to the inlet mass flow rate ($\Phi_2\omega_{i2}$) plus the amount of i evaporated.

This results in the following component mass balance:

$$\Phi_1\omega_{i1} = \Phi_2\omega_{i2} + Am_iv_d$$

in which the front velocity v_d is defined as

$$v_d = \frac{(z_{d,2} - z_{d,1})}{\Delta t}$$

Due to evaporation of previously frosted i, the mass fraction after the first front (ω_{i1}) will be higher than the inlet mass fraction (ω_{i2}). An overall mass balance can also be formulated. The total mass flow rate after the defrost front (Φ_1) is equal to the inlet flow rate (Φ_2) plus the amount of component i evaporated:

$$\Phi_1 = \Phi_2 + Am_i v_d$$

At the frost front, i is deposited on the packing surface; therefore, the outlet mass flow of component i is equal to the inlet flow minus the amount of i deposited in time Δt, which can be written as follows:

$$\Phi_0 \omega_{i0} = \Phi_1 \omega_{i1} + Am_i v_f$$

The velocity of this frost front v_f is described as

$$v_f = \frac{(z_{f,2} - z_{f,1})}{\Delta t}$$

Again, an overall mass balance can be formulated. The total mass flow after the frost front (Φ_0) is equal to Φ_1 minus the amount of component i deposited on the packing surface:

$$\Phi_0 = \Phi_1 + Am_i v_f$$

Energy balances can be formulated for both fronts. At the defrost front, heat is required to heat up the packing material from the saturation temperature (T_1) to the inlet temperature (T_2). Furthermore, heat is consumed due to the (endothermic) sublimation of component i. The required energy is provided by the feed gas, which is cooled down from T_2 to T_1. The energy balance results in

$$Av_d(\rho_s C_{p,s}(T_2 - T_1) + m_i \Delta H_i) = \Phi_2(T_2 - T_1)(\omega_{i2} C_{p,i} + \omega_{j2} C_{p,j})$$

At the frost front, the exothermic desublimation is included in the energy balance, resulting in

$$Av_f(\rho_s C_{p,s}(T_1 - T_0) + m_i \Delta H_i) = \Phi_0(T_1 - T_0)(\omega_{i0} C_{p,i} + \omega_{j0} C_{p,j})$$

So, three balances have been derived for each front (component mass, overall mass and energy balances) giving a total of six balances. There are eight unknowns (T_1, Φ_0, Φ_1, ω_{i0}, ω_{i1}, v_d, v_f and m_i). The two mass fractions of component i after the defrost front (ω_{i1}) and at the outlet (ω_{i0}) are related to the temperature by the phase equilibrium:

$$\omega_i = \frac{P_i^\sigma(T)}{P_{tot}} \frac{M_i}{M}$$

where M is the average molar weight.

Finally, eight equations and eight unknowns are obtained. The gas and solid heat capacities are dependent on temperature and are normally described by polynomial correlations, and vapor pressures as a function of temperature are normally described by exponential relations. Thus, a system of non-linear equations is obtained, which can be solved by standard root seeking methods such as the Newton–Raphson technique.

2.3.4.2   Recovery Step

During the recovery step, the bed is fed with pure component i. Three fronts will develop, as illustrated in Figure 2.6b. Initially, additional i will deposit on the packing surface, due to the higher pressure of i compared to the capture step. Therefore, a front is formed and moves with a velocity $v_{f,r}$ towards the outlet. The new amount of mass deposited per unit of bed volume is now indicated as m_i^*.

The component mass, the overall mass balance and the energy balance for this frost front are listed as follows:

$$\Phi_{1,r}\omega_{i1,r} = \Phi_{1,r}^*\omega_{i1,r}^* - A(m_i^* - m_i)v_{f,r}$$

$$\Phi_{1,r} = \Phi_{1,r}^* - A(m_i^* - m_i)v_{f,r}$$

$$Av_{f,r}\left[\rho_s C_{p,s}\left(T_1^* - T_1\right) - (m_i^* - m_i)\Delta H_i\right] = \Phi_{1,r}(T_1^* - T_1)(\omega_{i1,r}C_{p,i} + \omega_{j1,r}C_{p,j})$$

Also, a defrost front will move through the bed during the recovery step, for which the next mass and energy balances can be formulated (note that $\omega_{i1,r}^*$ and $\omega_{i2,r}$ are both unity when the bed is recovered with pure i):

$$\Phi_{1,r}^*\omega_{i1,r}^* = \Phi_{2,r}\omega_{i2,r} + Am_i v_{d,r}$$

$$\Phi_{1,r}^* = \Phi_{2,r} + Am_i v_{d,r}$$

$$Av_{d,r}\left[\rho_s C_{p,s}\left(T_2 - T_1^*\right) + m_i^*\Delta H_i\right] = \Phi_{2,r}(T_2 - T_1^*)(\omega_{i2,r}C_{p,i} + \omega_{j2,r}C_{p,j})$$

Similar to the capture step, a system of non-linear equations is obtained, which can be solved using a root seeking technique. During the recovery step, a third front is formed, the front closest to the inlet (moving with velocity v_r), due to the difference in the inlet temperature (T_3) and the temperature of the bed in the initial zone after the capture step (T_2). No phase change of component i takes place at this front; therefore, the front velocity can be described with

$$v_r = \frac{\Phi_{3,r}(\omega_{i3,r}C_{p,i} + \omega_{j3,r}C_{p,j})}{A\rho_s C_{p,s}}$$

When the heat capacities of the gas and solid phases are independent of temperature, the front velocity is not a function of temperature.

2.3.4.3   Cooling Step

During the cooling step, no phase change is involved. The bed is cooled down with an inert gas fed at temperature T_0. A temperature front moves through the bed, with a velocity of

$$v_c = \frac{\Phi_{0,r}(\omega_{i0,r}C_{p,i} + \omega_{j0,r}C_{p,j})}{A\rho_s C_{p,s}}$$

All equations for the three steps for the sharp front approach have been summarized in Table 2.6.

Table 2.6 *Equations for the sharp front approach*

Capture step

Defrost front

$$\Phi_1\omega_{i1} = \Phi_2\omega_{i2} + Am_i v_d$$

$$\Phi_1 = \Phi_2 + Am_i v_d$$

$$Av_d(\rho_s C_{p,s}(T_2 - T_1) + m_i\Delta H_i) = \Phi_2(T_2 - T_1)(\omega_{i2}C_{p,i} + \omega_{j2}C_{p,j})$$

Frost front

$$\Phi_0\omega_{i0} = \Phi_1\omega_{i1} + Am_i v_f$$

$$\Phi_0 = \Phi_1 + Am_i v_f$$

$$Av_f(\rho_s C_{p,s}(T_1 - T_0) + m_i\Delta H_i) = \Phi_0(T_1 - T_0)(\omega_{i0}C_{p,i} + \omega_{j0}C_{p,j})$$

Recovery step

Temperature front

$$v_r = \frac{\Phi_{3,r}(\omega_{i3,r}C_{p,i} + \omega_{j3,r}C_{p,j})}{A\rho_s C_{p,s}}$$

Defrost front

$$\Phi^*_{1,r}\omega^*_{i1,r} = \Phi_{2,r}\omega_{i2,r} + Am_i v_{d,r}$$

$$\Phi^*_{1,r} = \Phi_{2,r} + Am_i v_{d,r}$$

$$Av_{d,r}[\rho_s C_{p,s}(T_2 - T^*_1) + m^*_i\Delta H_i] = \Phi_{2,r}(T_2 - T^*_1)(\omega_{i2,r}C_{p,i} + \omega_{j2,r}C_{p,j})$$

Frost front

$$\Phi_{1,r}\omega_{i1,r} = \Phi^*_{1,r}\omega^*_{i1,r} - A(m^*_i - m_i)v_{f,r}$$

$$\Phi_{1,r} = \Phi^*_{1,r} - A(m^*_i - m_i)v_{f,r}$$

$$Av_{f,r}[\rho_s C_{p,s}(T^*_1 - T_1) - (m^*_i - m_i)\Delta H_i] = \Phi_{1,r}(T^*_1 - T_1)(\omega_{i1,r}C_{p,i} + \omega_{j1,r}C_{p,j})$$

Cooling step

Temperature front

$$v_c = \frac{\Phi_{0,r}(\omega_{i0,r}C_{p,i} + \omega_{j0,r}C_{p,j})}{A\rho_s C_{p,s}}$$

2.3.4.4 Simulation Results

The outcomes of the sharp front approach are presented in this section and are compared to the simulation results of the advanced numerical model. Figure 2.7 shows axial temperature and mass profiles, which are formed after 400 seconds, when feeding a binary N$_2$/CO$_2$ mixture at an inlet temperature of 150 °C. Other conditions are equal to those listed in Tables 2.4 and 2.5. It can be observed that the front positions, equilibrium temperature and the amount of mass deposited per unit of bed volume match very well between the two approaches. Although heat and mass dispersion is included in the advanced model, the fronts are reasonably sharp during the capture step. Especially, the frost front is well

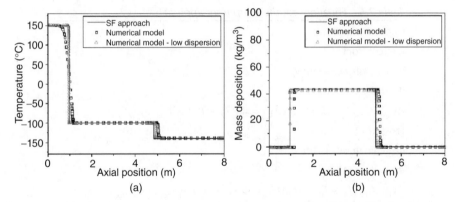

Figure 2.7 *Simulated axial temperature (a) and mass deposition (b) profiles for the capture step. The line indicated with 'Low Dispersion' shows the profile calculated with the advanced model, in which the axial mass and heat dispersion coefficients have been decreased by a factor 100*

defined, which can be attributed to a 'self-sharpening' effect of exothermic desublimation. In order to demonstrate that the outcomes of the advanced model approach the simplified model even closer, an additional simulation has been carried out using the detailed model in which the actual heat and mass dispersion coefficients have been decreased by a factor of 100. The excellent agreement between the two models can be discerned from Figure 2.7.

Also for other inlet compositions and initial bed temperatures, the two models agree very well (results are not included here). The recovery step has also been simulated using the two models. As already explained earlier, additional CO_2 will deposit on the packing in the initial phase of the recovery step. This is described by both models and again matches well, as observed in the temperature and mass deposition profiles after 10 seconds in Figure 2.8a and b, respectively.

It is observed again that when assuming low axial heat and mass dispersion, the solution of the advanced model is approaching the sharp front approach. Finally, the temperature and mass deposition profiles have been computed for the recovery step after 400 seconds, as illustrated in Figure 2.8c and d. It can be observed that dispersion is playing a more prominent role during the recovery step, which is related to the higher flow rates in comparison to the capture step. For that reason, the sharp front approach is especially suited to describe the capture step.

2.3.5   Process Analysis

This section aims at giving an overview of the influences of several process parameters on the process performance, using the sharp front approach. The influences of the initial bed temperature, inlet composition, inlet temperature and packing material are analyzed on the basis of two aspects: the amount of CO_2 deposited per unit of bed volume and the required specific cooling duty, which is defined as follows:

$$Q = \frac{V_{bed}(1 - \varepsilon_g)\rho_s C_{p,s}(T_s - T_0)}{(\Phi_{CO_2,in} - \Phi_{CO_2,out})t_{step}} \quad (J/kg_{CO_2})$$

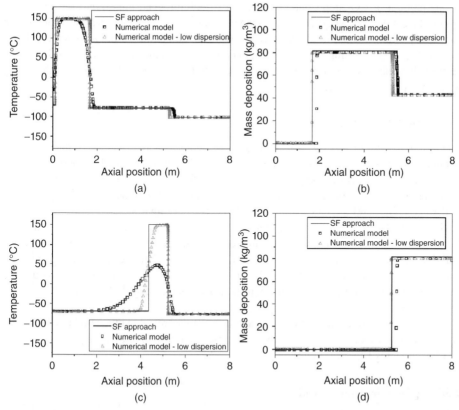

Figure 2.8 *Simulated axial temperature and mass deposition profiles for the recovery step, after 10 s (a), (b) and 400 s (c), (d). The lines indicated with 'Low Dispersion' show the profiles calculated with the advanced model, In which the axial mass and heat dispersion coefficients have been decreased by a factor 100*

The numerator of the equation represents the amount of energy required to cool down the bed after the recovery step from T_s to temperature T_0, which is the initial bed temperature before starting the capture step. The denominator gives the amount of CO$_2$ captured during the capture step. $\Phi_{CO_2,in}$ and $\Phi_{CO_2,out}$ are the inlet and outlet mass flow rates of CO$_2$ during the capture step, respectively.

The temperature T_s is assumed to be $-70\,^{\circ}$C for all cases. Initially, the bed temperature at the zone where CO$_2$ is deposited will increase to $-78\,^{\circ}$C during the recovery step, as pure CO$_2$ is being fed to the system and is being deposited additionally on the packing surface. However, the outlet flow during the recovery step is recycled and will increase slightly in temperature and is assumed to be at $-70\,^{\circ}$C, which will therefore be the initial bed temperature before the cooling step is started. In practice, not the entire bed will be at this temperature at the end of the recovery step. The zone located close to the outlet of the bed will have a slightly higher temperature. On the other hand, as explained earlier, it is actually not necessary to cool down the entire bed to T_0 during the cooling step, and, therefore, the numerator gives a realistic value for the required cooling duty.

The amount of CO_2 deposited per unit of bed volume is an important indicator of the required capital costs, while the calculated specific cooling duty gives a good indication of the energy requirements for different cases. Blowers and compressors are responsible for part of the power consumption. The techno-economic evaluation described in the following sections includes these requirements and discusses the economic feasibility of the newly developed process concept.

2.3.6   Initial Bed Temperature

The required specific cooling duty and CO_2 mass deposition as a function of the initial bed temperature are illustrated in Figure 2.9. Below $-130\,°C$ the initial bed temperature hardly affects the specific cooling duty; at the same time, the additional energy required to cool the bed to lower temperatures will result in a correspondingly higher CO_2 storing capacity of the bed and, therefore, causing a constant specific cooling duty.

However, at initial bed temperatures above $-120\,°C$, the specific cooling duty will increase strongly. This is related to the decreasing amount of CO_2 being captured, which will decrease exponentially above $-120\,°C$. For example, when feeding a N_2/CO_2 mixture containing 10 vol% CO_2 to a bed cooled at $-120\,°C$, 90% of the fed CO_2 is recovered. However, when feeding the same mixture to a bed cooled at $-110\,°C$, only 12% CO_2 is recovered.

Therefore, a relatively low amount of extra cooling will result in much higher CO_2 recovery rates. This effect is directly related to the exponential temperature dependency of the equilibrium CO_2 vapor pressure.

2.3.7 CO_2 Inlet Concentration

When the CO_2 fraction in the flue gas feed decreases, the specific cooling duty will increase, as shown in Figure 2.9a. This can be explained by the fact that less CO_2 is stored in the bed at lower inlet CO_2 concentrations (see Figure 2.9b), while the energy required to cool down the bed remains the same.

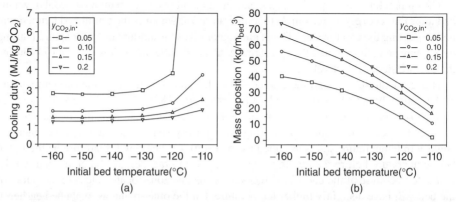

Figure 2.9 *Specific cooling duty (a) and mass deposition (b) as a function of the initial bed temperature for different inlet CO_2 fractions*

2.3.8 Inlet Temperature

A higher gas inlet temperature will cause the defrost front to move faster, because more heat is available for desublimation. At the same time, this causes the CO_2 concentration to increase after the defrost front. Also, more CO_2 will deposit per unit of packing volume and the frost front moves faster, and, consequently, the step time reduces. Therefore, the specific cooling duty will increase when increasing the inlet temperature.

However, the specific cooling duty reaches an asymptotic value when increasing the temperature. This is related to the fact that the heat required to heat up the packing (and cool down the gas) relative to the heat involved in the evaporation of previously deposited CO_2 is playing a dominant role at higher inlet temperatures. When, for example, feeding a mixture containing 10% CO_2 to a bed refrigerated to $-140\,°C$ at an inlet temperature of $150\,°C$, the amount of heat required for sublimation is only 8.2% of the amount of energy involved in cooling down the gas. It should be noted that the contribution of desublimation at the frost front does play a significant role in the energy balance.

2.3.9 Bed Properties

The packing properties (i.e., material or porosity) do not influence the specific cooling duty. When the heat capacity of a packing material is, for example, doubled (and the bed dimensions remain the same), the amount of cooling will also be doubled. However, the amount of CO_2 deposited per unit of bed volume will change proportionally, and, therefore, the specific cooling costs will not be influenced. However, bed dimensions are influenced when changing the packing material and keeping step times constant, and, therefore, capital costs are influenced by the packing choice. On the other hand, the packing choice will influence the pressure drop and axial dispersion.

2.4 Small-Scale Demonstration (Proof of Principle)

To provide a proof of principle and investigate the performance of the new concept, a small experimental set-up was constructed. A flowsheet and a picture of the experimental set-up are shown in Figures 2.10 and 2.11, respectively. A glass tube ($OD \times ID \times L = 40 \times 35 \times 300\,mm$) surrounded by a glass vacuum jacket ($OD \times ID = 60 \times 55\,mm$) was filled with mono-disperse blue spherical glass beads ($d_p = 4.04\,mm$, $\rho_s = 2547\,kg/m^3$). The bed was cooled with a N_2 gas flow, which was refrigerated in a coil positioned in a liquid nitrogen bath. After cooling, the feed was switched to a N_2/CO_2 mixture with or without the addition of H_2O.

These mixtures were prepared by controlling the gas flow rates of N_2 and CO_2 with mass flow controllers (Bronkhorst El-flow). This mixture could then be fed to a saturator filled with demineralized water. The water content of the feed flow could be varied by changing the temperature in the saturator. The flow from the saturator could be further overheated by using an electrically traced line. The temperatures in the bed were measured along the bed length in the radial centre with 11 thermocouples (Thermo-Electric K-type) at every 3 cm in axial direction. The pressure at the inlet of the bed was monitored by using an analogue pressure indicator.

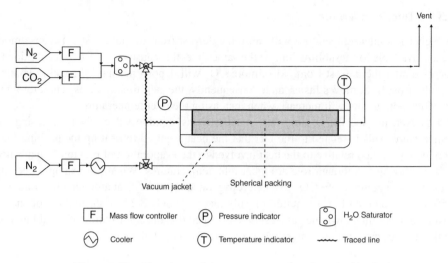

Figure 2.10   *Flowsheet of the vacuum insulated packed bed*

The CO_2 content in the outlet stream was analyzed with an IR-analyzer (Sick-Maihak, s610, 0–3 vol%). The front of sublimated CO_2 was visually inspected with a camera. Axial temperature profiles have been measured for different initial bed temperatures and inlet mole CO_2 and H_2O fractions (see Table 2.7).

2.4.1   Results of the Proof of Principle

The bed was cooled using refrigerated N_2 until the bed reached a stationary temperature profile. Due to heat radiation into the system, this initial profile is slightly increasing (almost linearly) from the inlet. The temperature difference between the inlet and outlet is typically about 20 °C, which is relatively small compared to the temperature difference between the refrigerated bed and the gas being fed during the capture cycle. It should be noted that during experiments no pressure excursions were observed, and the pressure drop was very small.

2.4.1.1 N_2/CO_2 Mixtures

When feeding N_2/CO_2 mixtures to the refrigerated bed, a moving front of deposited CO_2 was observed visually as depicted in Figure 2.12. The axial temperature profiles at several time steps for experiment #1 are shown in Figure 2.13a. After approximately 200 seconds, the frost front reached the end of the bed, and CO_2 breakthrough was detected in the outlet stream, as shown in Figure 2.14. The mixture was fed through the same inlet tube as the refrigerated N_2 during the cooling step. Therefore, the temperature of the packing at the inlet did not attain ambient temperatures immediately, but increased slowly as can be seen from Figure 2.13a.

The CO_2 content in experiment #2 was increased to 30 vol% CO_2. The increased CO_2 inlet concentration resulted in a higher saturation temperature (−91.5 °C versus −94.5 °C), and, therefore, the packing storage capacity slightly increased. However, due to the higher

Figure 2.11   *Picture of the experimental set-up for the proof of principle*

Table 2.7   *Experimental conditions*

Experiment	#1	#2	#3	#4	#5	#6
Initial bed temperature (°C)	−140	−140	−140	−140	−120	−140
Inlet CO_2 mole fraction (−)	0.2	0.3	0.2	0.1	0.2	0.2
Inlet H_2O mole fraction (−)	0	0	0.022	0.020	0.021	0.046
Total mass flow (10^{-4} kg/s)	2.55	2.68	2.35	2.23	2.35	2.39

Figure 2.12   *CO_2 ice formed at the packing surface during a capture step*

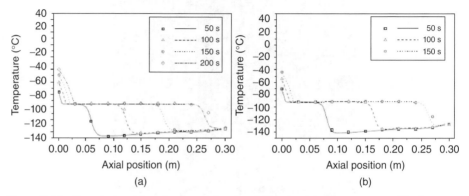

Figure 2.13   *Experimental (markers) and simulated (lines) axial temperature profiles for experiments #1 (a) and #2 (b)*

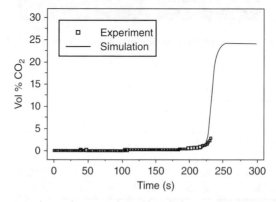

Figure 2.14   *Experimental (markers) and simulated (line) CO_2 concentration at the outlet during experiment #1*

molar CO_2 feed flow rate, the front velocity of the frost front increased. Figure 2.13b shows that CO_2 breakthrough already occurred after about 150 seconds.

2.4.1.2 $N_2/CO_2/H_2O$ Mixtures

Figure 2.15a shows axial temperature profiles measured during experiment #3. Again deposition of CO_2 occurs and a CO_2 front develops, similar to experiment #1. The gas feed mixture also contains H_2O during this experiment, resulting in H_2O condensation at the packing surface. In between the zones where H_2O is condensed and CO_2 is desublimated, a small amount of H_2O ice is also observed. This is caused by the decrease in the H_2O concentration and the temperature at the condensing front. At some point, the mixture reaches conditions below the triple point of H_2O, and desublimation of H_2O is observed.

The amount of desublimated H_2O remains constant during the capture step, and no influence on the temperature profiles is observed. The effect of H_2O condensation on the axial temperature profiles is small in the first 200 seconds of the measurement. However, after CO_2 breakthrough, the gas flow was not stopped in order to show further development of the H_2O front. Figure 2.15b shows that a water front is also moving through the bed and that an equilibrium is formed at a temperature of approximately 28 °C. Lowering the CO_2 inlet concentration from 20% to 10% in experiment #4 results in a lower equilibrium temperature for CO_2 (−98 °C), as illustrated in Figure 2.15c. The front of freezing

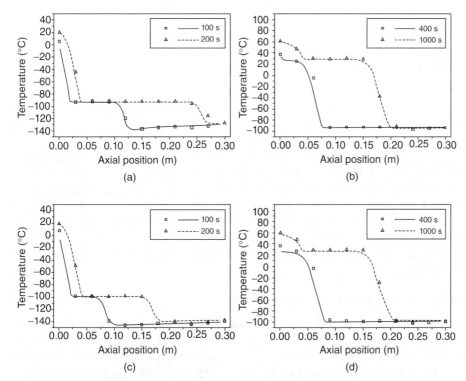

Figure 2.15 *Experimental (Markers) and Simulated (Lines) axial temperature profiles for experiments #3 (a) (b) and #4 (c) (d)*

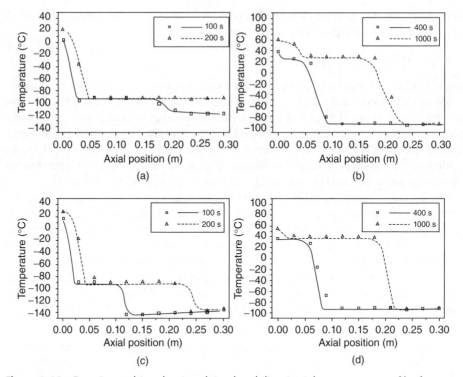

Figure 2.16 *Experimental (markers) and simulated (lines) axial temperature profiles for experiments #5 (a) (b) and #6 (c) (d)*

CO_2 will move slower in this case, and a breakthrough is observed after approximately 350 seconds. The development of the H_2O front is not influenced by the decrease in the CO_2 inlet concentration, as shown in Figure 2.15d.

Figure 2.16a and b show axial temperature profiles for experiment #5 with a higher initial bed temperature. This higher bed temperature results in shorter cycle times, because less CO_2 is stored per unit of bed volume. Moreover, the purity of the cleaned flue gas will be slightly lower (98.8% N_2 compared to 99.9% in the other experiments). Finally, the effect of the increased H_2O inlet concentration can be observed in Figure 2.16c and d. The evolution of the CO_2 fronts is not much influenced, but a clear effect is visible for the H_2O front. The equilibrium temperature reaches a higher value of approximately 40 °C, compared to 28 °C in other experiments.

2.4.1.3   Simulations for the Proof of Principle

The experimental results have been compared with profiles calculated by the advanced numerical model, which was described earlier. The packing material used for the experiments consisted of spherical particles. Therefore, the correlations used above, which are valid for a structured monolith packing, are not applicable here, and are replaced by the correlations for spherical particles [13]. As no information on the sublimation rates is available in the literature, the equilibrium time constant (g) was determined by comparing the simulation results with the experimental findings. Results are best described when using a constant of about 1×10^{-6} s/m. For a more detailed discussion on desublimation rates, the interested reader is referred to the PhD thesis of Martin Tuinier (Eindhoven 2011).

The initial temperature profile used in the simulations is taken from the experiments. As mentioned earlier, the initial bed temperature is not totally uniform, but is increasing slightly from the inlet towards the outlet. This is caused by the heat leak into the system. To account for this heat leak in more detail, the tube was first cooled down, then cooling was stopped and the temperature rise was measured as a function of time in the radial centre and close to the tube wall. It was found that the temperature difference between these two locations was minimal and the temperature rise could be well described by an additional radiative energy influx. Therefore, the following contribution was added to the energy balance:

$$\Phi_{rad} = \frac{4}{d_{tube}} \frac{\sigma(T_h^4 - T_c^4)}{\frac{1}{\varepsilon_c} + \frac{A_c}{A_h}\left(\frac{1}{\varepsilon_h} - 1\right)} \quad (J/m^3 \, s)$$

where σ is the Stefan–Boltzmann constant, ϱ is the integral emissivity and the subscripts h and c stand for the hot and cold sides. In the used experimental set-up, both the hot side (jacket) and cold sides (tube) are made of glass, and, therefore, an equal integral emissivity of 0.9 is assumed.

Furthermore, it was found that the inlet temperature of the gas was only increasing slowly in time after switching to the capture step, due to cold stored at the inlet of the tube, insulation material, piping, etc. Therefore, the inlet temperature (and the corresponding composition) used in simulations was based on the measured inlet temperature. Finally, it should be noted that the heat capacity of the glass wall was included in simulations (in the accumulation term).

The resulting calculated temperature profiles have been plotted in the same figures in which the experimental profiles were shown. It can be concluded that the developed model is very well capable of describing all experimental findings.

2.5 Experimental Demonstration of the Novel Process Concept in a Pilot-Scale Set-Up

A schematic flow diagram of the experimental rig is shown in Figure 2.17. Three identical stainless steel vacuum insulated vessels (OD $\times$ ID $\times L = 106 \times 100 \times 500$ mm) were filled with spherical glass beads ($d_p = 3$ mm). All three beds were equipped with 9 K-type thermocouples at several axial and radial positions in order to monitor the development of temperature profiles within the beds.

Pressure sensors were installed both at the inlet and outlet of all three beds to measure the pressure drop over the beds. Cooling of the beds was performed by feeding a cold N$_2$ gas flow, which was refrigerated before using a cryogenic cooler (Stirling SPC-1). The cooler was connected to the beds by using polystyrene-insulated pipes. Mixtures of N$_2$ and CO$_2$ were fed to the refrigerated bed, whose composition was controlled by using two mass flow controllers (Brooks).

Before feeding the gas mixture to the beds, the gas was heated up by using a tracing line wound around the feed tube. The recovery step was performed by recycling CO$_2$ by using

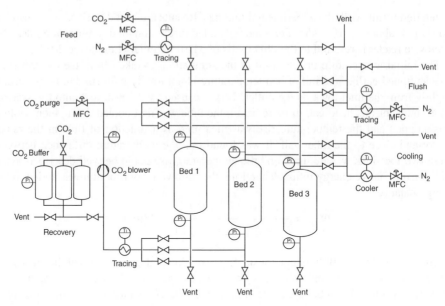

Figure 2.17   *Simplified flow scheme of the pilot set-up, in which TI, PI and FI stand for temperature, pressure and flow indication, respectively*

a 3 kW side channel blower (VACOM SC 425). At initial stages of the recovery step, some additional amount of CO_2 deposited on the packing as explained earlier. Therefore, a CO_2 buffer was required, which consists of three stainless steel vessels (total volume of 36 l). Low-temperature CO_2 was not allowed to enter the blower; therefore, a tracing line was installed around the blower inlet tube to heat up the incoming gas to ambient temperature. Furthermore, all beds could be flushed with heated N_2, if so required.

All valves are pneumatic ball valves, and special cryogenic valves (Meca-inox) were used at those positions where low temperatures were prevailing. Signal processing and process control was carried out by NI Labview. A picture of the experimental set-up is shown in Figure 2.18.

2.5.1   Experimental Procedure

Before an actual measurement was started, the N_2/CO_2 mixtures were prepared and heated up, the cooler was started up and the CO_2 buffer vessels were filled. After these preparations, the first bed (bed #1) was cooled down until the average temperature of two thermocouples ($r = 0$ cm and $r = 4$ cm) at the outlet of the bed reached a certain set-point temperature. At that point, the second bed (bed #2) was fed with the cold N_2 flow and bed #1 was fed with the N_2/CO_2 gas mixture.

The temperature at the outlet of a bed will increase when CO_2 starts to break through. Therefore, again a certain set-point temperature was used to monitor whether a bed finished the capture step. When CO_2 breakthrough occurred in bed #1, and bed #2 was cooled down, bed #1 was switched to the recovery step, bed #2 to the capture step and bed #3 to the cooling step.

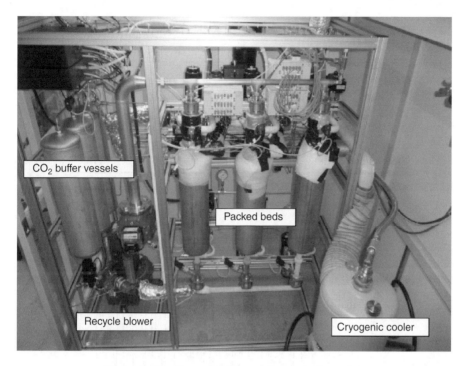

Figure 2.18 *Picture of the experimental pilot set-up*

The recovery step was started by switching on the recycle blower for CO_2. Due to the additional CO_2 that desublimates initially on the packing surface in a relatively short time, the pressure decreases. This is counteracted by opening the valve connected to the CO_2 buffer vessels. Fresh CO_2 is fed into the recovery piping, until the additional desublimation of CO_2 stopped and the pressure reached a certain set point. At that point CO_2 was recycled, and the previously deposited CO_2 in the bed was evaporated and removed from the bed.

The outlet flow rate is larger than the inlet flow rate; therefore, the pressure starts to increase. This is again counteracted, in this case by opening a mass flow controller, controlled using a PID control loop. Again, the temperature at the outlet of the bed is used to determine the point to switch.

When all CO_2 is removed from the packing, the temperature will immediately start to increase and at that point the bed was switched again to the cooling step. In case one of the beds finished its step before other beds were finished, the bed was put on a standby mode and switched again to the next process step as soon as other beds were finished.

2.5.2 Experimental Results

Experiments have been carried out using the conditions listed in Table 2.8. Long-term runs of more than 10 h proved that the system is able to run stable for a long period. The average temperature of the thermocouples at $r = 0$ cm and $r = 2$ cm at the outlets of the three beds as a function of time is plotted for 6 h in Figure 2.19a. Similar repeating temperature

Table 2.8 *Conditions used for the proof of concept at pilot scale*

	Capture	Recovery	Cooling
N_2 flow (NL/min)	22	–	285
CO_2 flow (NL/min)	5.5	220	–
Total flow (NL/min)	27.5	220	285
CO_2 gas mole fraction (–)	0.2	1	0
Inlet gas temperature (°C)	100	20	−155
Set-point temperature (°C)	−100	−70	−130
Step time (min)	15	15	15

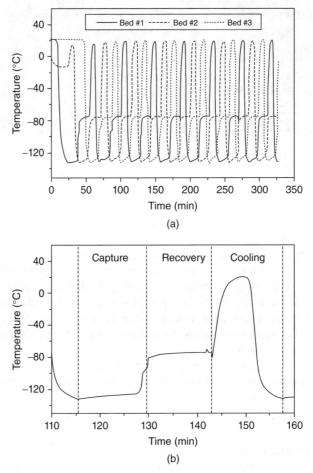

Figure 2.19 *Evolution of temperature at the outlets of the three beds for a long run (a) and zoomed in for one process cycle for one bed (b)*

patterns for the three beds can be observed. To have a closer look, the temperature profile is zoomed in for one bed, showing the cooling, capture and recovery steps in more detail (see Figure 2.19b).

During the capture step, the temperature at the outlet of the bed is almost stable at the initial bed temperature (approximately $-130\,^{\circ}$C), but increases strongly as soon as CO$_2$ starts to breakthrough. When the bed is at a temperature of $-100\,^{\circ}$C, the capture step is stopped. When the recovery step is started, it can be observed that the temperature increases quickly to a temperature of approximately $-76\,^{\circ}$C corresponding to a saturation temperature of pure CO$_2$ at the operating pressure (1.2 bar).

Again it can be observed that the temperature increases at the end of the recovery step. At the point that the average outlet temperature reaches $-70\,^{\circ}$C, the bed is switched to the cooling step. During the cooling step, the outlet temperature first rises to a temperature of about $20\,^{\circ}$C. This is the temperature at which the CO$_2$ is fed to the bed during the recovery step. Note that for commercial operation it is better to feed the recovery CO$_2$ gas flow at lower temperatures ($-70\,^{\circ}$C) to save cooling duty, but due to practical limitations of the blower used, this could not be realized in the experimental situation. After this relatively hot zone of $20\,^{\circ}$C has moved through the bed, the temperature at the outlet of the bed decreases, until the set point is reached ($-130\,^{\circ}$C).

Figure 2.20a shows the pressure at the outlet of one bed, again zoomed in for one process cycle with the three steps. During the capture step, the bed is at atmospheric pressure. As soon as the bed is switched to the recovery step, the pressure suddenly drops to a low value. As explained earlier, this is related to the gaseous CO$_2$ present in the piping of the recovery step, which is depositing on the packing surface. Immediately the valve connected to the CO$_2$ buffer vessels is opened to allow the pressure to increase again, additional CO$_2$ is deposited on the packing. At that point the valve is closed, the pressure increases and is controlled at 1.2 bar, by allowing CO$_2$ to leave the system via a mass flow controller. The flow through this mass flow controller is plotted in Figure 2.20b. Note that the pressure is initially well maintained at 1.2 bar, but at some point the pressure increases further to approximately 1.28 bar. This is related to the fact that the mass flow controller is fully opened (15 NL/min) as observed in Figure 2.20b. During the cooling step, a pressure of

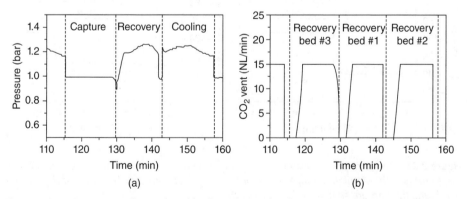

Figure 2.20 *Pressure at the outlet of bed #1 during the three process steps (a) and the CO$_2$ gas flow towards the vent during the recovery steps of the three beds (b)*

1.2 bar is measured, which is slightly changing during the step and is likely related to the changing gas phase temperature in the bed.

During all steps, the pressures at the inlet and outlet of the beds have been monitored. It is observed that the pressures at the inlet and outlet are virtually equal during all the three steps, meaning that the pressure drop over the packed bed is negligible compared to the pressure drop over piping/valves, etc.

2.5.3   Simulations for the Proof of Concept

In this section, the experimental results are studied in more detail and have been compared to the simulation results of the aforementioned the numerical model. Subsequently, the model will be extended with a separate heat balance for the wall to provide a better understanding of the observed behaviour.

2.5.4   Radial Temperature Profiles

The experimental temperature development during the capture step is shown for two different radial positions ($r=0$ cm and $r=4$ cm) for two axial locations: $z=16$ cm (Figure 2.21a) and $z=28$ cm (Figure 2.21b). It can be observed that the temperature close to the wall ($r=4$ cm) is initially higher than that in the radial centre.

The effect on the temperature development is that the front of freezing CO_2 is moving faster through the bed close to the wall. These experimental results are compared with simulations with the one-dimensional model. The resulting breakthrough times match well with the experimental outcomes and are exactly positioned between the two measured temperatures.

The evolution of the temperature has also been plotted for the recovery step for $z=16$ cm (Figure 2.22a) and $z=28$ cm (Figure 2.22b). A large difference in the temperature profiles between the two radial positions can be observed. The simulated temperature profile is again in between the measured profiles. Remarkably, the CO_2 is removed faster from the

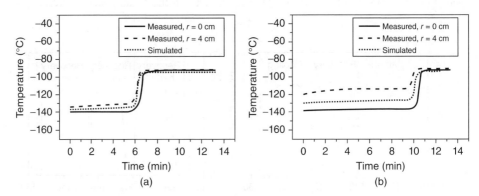

Figure 2.21   *Temperatures measured in the radial centre ($r=0$ cm) and close to the wall ($r=4$ cm) and the simulated temperature for two axial locations: $z=16$ cm (a) and $z=28$ cm (b) during the capture step*

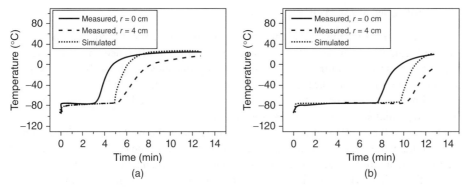

Figure 2.22 *Temperatures measured in the radial centre (r = 0 cm) and close to the wall (r = 4 cm) and the simulated temperature for two axial locations: z = 16 cm (a) and z = 28 cm (b) during the recovery step*

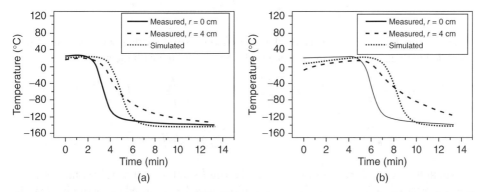

Figure 2.23 *Temperatures measured in the radial centre (r = 0 cm) and close to the wall (r = 4 cm) and the simulated temperature for two axial locations: z = 16 cm (a) and z = 28 cm (b) during the cooling step*

radial centre than from the zone closer to the walls. Based on the capture step, one would expect lower mass deposition of CO_2 close to the walls (because of the lower initial bed temperature) and therefore faster removal. Apparently, the wall is playing a significant role, which is also visible for the cooling step, discussed in the following sections.

Finally, the temperature development is shown for two axial locations: $z = 16$ cm (Figure 2.23a) and $z = 28$ cm (Figure 2.23b) for the cooling step. The temperature measured close to the wall shows a much more dispersed temperature profile. Therefore, the zones close to the wall require much longer cooling times in order to reach a low temperature. The simulation outcomes show that the temperature development is again in between the profiles measured at the two radial positions, but the profile is as steep as the profile measured in the radial centre. This is attributed to the effect of the vessel wall, which is studied in more detail in the following section.

2.5.5   Influence of the Wall

The previous section showed that the one-dimensional numerical model can predict the average breakthrough times but is not able to explain the differences between the temperature measurements in the radial centre and the zone close to the wall of the bed. The difference in temperature profiles between these locations is particularly pronounced for the cooling step. The temperature development close to the wall is very disperse, which cannot be predicted by the one-dimensional model.

It is very well possible that the axial heat conduction in the wall has an influence on the temperature in the region close to the wall. A detailed description of the temperature in the bed as a function of the radial and axial positions can be obtained by extending the model to two dimensions. However, it is also possible to get more insights into the behaviour by adjusting the one-dimensional model. Instead of accounting for the heat capacity of the wall in the accumulation term, an extra energy balance for the wall is included in the model, in which heat is exchanged with the packed bed:

$$\rho_w C_{p,w} \frac{\partial T_w}{\partial t} = \frac{\partial}{\partial z} \left(\lambda_w \frac{\partial T}{\partial z} \right) + \alpha_w \alpha_{gw} (T_w - T_g)$$

The energy balance for the bed (both the gas and the solid phases) will therefore also be extended with an extra contribution due to the heat exchange with the wall:

$$(\varepsilon_g \rho_g C_{p,g} + \rho_s (1 - \varepsilon_g) C_{p,s}) \frac{\partial T_g}{\partial t} = -C_{p,g} \rho_g v_g \frac{\partial T_g}{\partial z} + \frac{\partial}{\partial z} \left(\lambda_{eff} \frac{\partial T_g}{\partial z} \right)$$

$$- \sum_{i=1}^{n_c} \dot{m}_i'' a_s \Delta H_i + a_b \alpha_w (T_w - T_g)$$

in which the expressions used to calculate the bed-to-wall heat transfer coefficient (α_{gw}) are summarized by Tiemersma [17]. Simulations have been carried out assuming that the initial wall temperature is equal to the temperature measured close to the wall and that the packing has an initial temperature as measured in the radial centre.

The temperature development for both the wall and the bed has been plotted as a function of time for $z = 16$ cm in Figure 2.24. It can be clearly observed that the temperature within the wall is very dispersed, which is explained by a much higher value of the conductivity of the solid steel wall compared to the effective conductivity within the bed. Based on this result, the dispersed measured temperature close to the wall can be better understood. Accounting for the wall separately also leads to a better prediction of the temperature within the packing in the radial centre of the bed.

The wall will also influence the temperature development during the capture and recovery steps. It is very well possible that CO_2 will deposit at the tube wall (which is not accounted for in the extended model). This could well explain the slower removal of CO_2 during the recovery step close to the wall. Locally there will be more CO_2 deposited, due to the heat capacity of the wall, and, therefore, it will take longer to remove solid CO_2 from those zones. An additional effect may be that the pressure drop will locally increase due to the higher amount of deposited CO_2, resulting in lower flow rates and therefore even slower solid CO_2 removal rates.

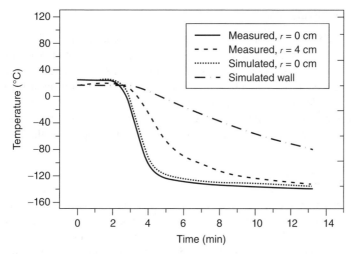

Figure 2.24 *Effect of accounting for the wall separately in simulations for the temperature development at z = 16 cm during the cooling step*

2.6 Techno-Economic Evaluation

Several economic studies on CO_2 capture methods have been published in the literature. Resulting costs vary strongly, as they are highly influenced by the system boundaries such as the CO_2 source and, therefore, inlet concentration, whether or not transport and storage is included, the level of maturity and cost measures and assumptions. For example, the optimized costs in a study by Abu-Zahra *et al.* [18, 19] for CO_2 capture by 30% MEA absorption from a 600 MW bituminous coal–fired power plant have been estimated to be 33 €/ton CO_2 avoided. On the other hand, Merkel *et al.* [20] evaluated a process based on CO_2 capture by using membranes and calculated CO_2 capture costs of \$39/ton CO_2. In a report by McKinsey [21], the development of costs for CCS (including storage costs) is analyzed over the next 20 years. They expect that early demonstration projects will operate at 60–90 €/ton, but that costs could come down to 30–45 €/ton in 2030, a price level that is expected to make CCS economically self-sustaining. More research is required to bring down the costs. Although many studies focus on reducing operational costs, for example, by finding novel and more efficient solvents for amine scrubbing, it is at least as important to reduce capital costs in order to reduce CO_2 avoidance costs [22].

The aim of the work described in this chapter is to evaluate the cryogenic packed bed (CPB) concept, both on technical aspects and on economic performance. Furthermore, the economics of the CPB concept are compared to other post-combustion technologies, namely, amine scrubbing and membrane technology, investigating the importance of various process assumptions.

The chapter is organized as follows: first a base case is defined and the costs per ton of CO_2 emissions avoided are calculated. Subsequently, a sensitivity analysis of some key process parameters is discussed. Finally, the results are compared to the results of economic studies on CO_2 capture via amine scrubbing [23] and membrane technology [20].

2.6.1   Process Evaluation

In order to be able to compare the CPB concept with other technologies, a basic design is made for a capture plant treating flue gases typically generated by a 600-MW coal-fired power plant, which is often used as a base case in literature studies. The bed dimensions and required process conditions are obtained by carrying out simulations by using the detailed model, which is described in the previous sections. The capital and operation costs are then estimated and the costs per ton of CO_2 avoided are calculated. This section ends with a parameter study, in which the influence of several key parameters on the capture costs is evaluated.

2.6.1.1   Base Case

To simplify the comparison, only CO_2 capture is taken into account without impurities and H_2O removal. The assumed flue gas conditions and composition are shown in Table 2.9. The bed dimensions and properties of the base case are detailed in Table 2.10.

The initial bed temperature was set at $-150\,°C$, which resulted in more than 99.9% CO_2 recovery. A breakthrough time (duration of each step) of 600 seconds was chosen.

The required flow rates, pressures and inlet temperatures are listed in Table 2.11. The resulting pressure drops over the beds (also shown in Table 2.11) are rather small due to the nature of the selected packing: a structured monolith.

Table 2.9   *Flue gas conditions and composition*

Temperature (°C)	150	
Pressure (bar)	1.013	
	Vol%	Flow (kg/s)
N_2	86.5	510
CO_2	13.5	125
Total		635

Table 2.10   *Bed dimensions and properties*

Diameter (m)	8.5
Length (m)	4.25
Number of beds (–)	21 (7 per step)
Packing	Steel monolith structure
Density solids (kg/m³)	7750
Porosity (–)	0.697

Table 2.11   *Process parameters for the base case*

	Capture	Recovery	Cooling
T_{in} (°C)	162	−66	−150
P_{in} (mbar)	1100	1200	1200
Flow/bed (kg/s)	91	564	357
ΔP packing (mbar)	16.9	82.5	56.7
Total ΔP (mbar)	100	200	200

However, gas distribution over the beds, piping and valves will cause an additional pressure drop; therefore, a total pressure drop of 100 mbar is assumed for the capture step and 200 mbar for the cooling and recovery steps (because of the higher volumetric flow rates). During the recovery step, the outgoing CO_2 flow has a temperature of $-78\,°C$ and will be partly recycled to the inlet. Due to a temperature increase during compression by the recycle blower, the inlet temperature during the recovery step is increased to $-66\,°C$. The flue gas temperature is estimated to be 150 °C, but will increase to 162 °C, because of compression. The resulting axial temperature and mass deposition profiles are shown in Figure 2.25. It can be observed that the heat stored in the bed during the capture step is being used during the recovery step to evaporate the previously deposited CO_2. Furthermore, it can be observed that during the cooling step not the entire bed has to be cooled down, as the last part of the bed will be cooled down during the capture step.

2.6.1.2 Costs Base Case

In order to calculate the CO_2 avoidance costs, the capital investment costs are first calculated using a conceptual cost estimation method with an accuracy of 40%. In this method, the main equipment costs are estimated. The costs for blowers, the heat exchanger and the columns have been calculated using correlations reported by Seider *et al.* [24] and Loh and Lyons [25] and are updated to costs in 2010 using the Chemical Engineering Plant Cost Index (CEPCI).

The packing costs are calculated using a steel price of $1200/ton steel (market price of $600/ton multiplied with a factor of 2 for packing construction). The module costs, including piping, installation, and so on, are then calculated by multiplying the equipment costs with a Hand factor. When all the module costs are summed up, 25% is added for contingencies. The total direct investment is subsequently calculated, and an allocated investment (for storage, utilities and environmental provisions), start-up investments and working capital are added. Finally, the total fixed capital is calculated; the results are shown in Table 2.12. The operational costs include the electricity costs required for the blowers. The CO_2 emitted due to the additional power required by the blowers could also be captured, but in this study it is assumed to be emitted into the atmosphere. For this base case, the cooling is provided by the evaporation of LNG and no additional costs are assumed.

Depreciation, interest, labor and maintenance are calculated using 20% of the total capital charge per year. The used cost parameters can be found in Table 2.13. The operational and final CO_2 avoidance costs are summarized in Table 2.14. The total costs per ton of CO_2 emission avoided amount to $52.8. The capital costs ($28.9/ton CO_2 avoided) and the operational costs ($23.9/ton CO_2 avoided) have a similar share in the avoidance costs.

2.6.2 Parametric Study

2.6.2.1 Initial Bed Temperature

The process has been evaluated for different initial bed temperatures. Both the CO_2 avoidance costs and LNG consumption are shown in Figure 2.26. At a higher initial bed temperature, less CO_2 is deposited per unit of bed volume. Therefore, more bed volume is required to maintain similar breakthrough times, resulting in increasing capital costs. Larger flow rates during the recovery and cooling steps are required as well, in order to finish in 600 seconds. A larger flow rate during the cooling cycle results in a higher

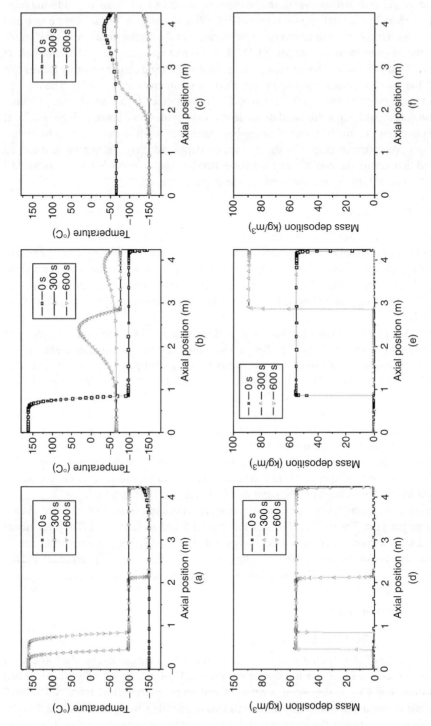

Figure 2.25 *Simulated axial temperature (a–c) and mass deposition (d–f) profiles for the capture, recovery and cooling steps*

Table 2.12 *Capital investment costs for the base case*

Equipment	Equipment costs [M$]	Hand factor	Module costs [M$]
Column for packed bed (21)	0.39	4	32.9
Packing [22]	0.67	4	56.5
Flue gas compressor	1.12	2.5	2.8
CO_2 recycle blower	10.09	2.5	25.2
N_2 cooling blower	10.55	2.5	26.4
CO_2 product compressor	15.09	2.5	37.7
LNG heat exchanger	1.03	4.8	4.9
Contingencies	25%		46.6
Total direct investment (TDI)	233		
Total allocated investment	40% TDI	93	
Start-up investment	5% TDI	12	
Total process investment (TPI)	338		
Working capital	2% TPI	7	
Total fixed capital	345		

Table 2.13 *Cost evaluation parameters*

Operation h/year	7000
Capital charge	0.2
Blower/compressor/pump efficiency	0.72
Electricity price [$/kW h]	0.06
CO_2 emission due to additional power [ton/MW h]	0.8042
CO_2 product pressure (bar)	140

Table 2.14 *Operational and total costs for the base case*

	MW	$/h	$/ton CO_2 avoided
CO_2 captured (ton/h)			
CO_2 emitted due to additional power (ton/h)			
CO_2 avoided (ton/h)			
Flue gas blower	8.2	490	1.4
CO_2 recycle blower	34.4	2063	6.1
N_2 cooling blower	36.4	2181	6.4
CO_2 product compressor	56.9	3412	10.0
Total electricity blowers	135.8	8145	23.9
Capital/maintenance/labor charge		9850	28.9
Total costs			52.8

LNG consumption. An initial bed temperature of −160 °C results in even more efficient use of the beds and therefore slightly lower costs and LNG consumption. However, the temperature difference between LNG (−162 °C) and the refrigerated N_2 becomes too small.

It can be concluded that a lower bed temperature results in more efficient CO_2 capture. It should be noted that this conclusion cannot be drawn when part of the cooling is generated

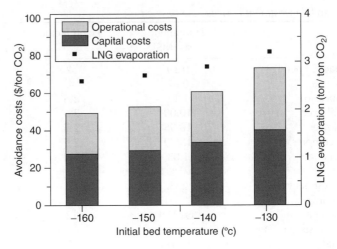

Figure 2.26   Avoidance costs and LNG consumption as a function of the initial bed temperature

by refrigeration. The efficiency of a refrigerator decreases with decreasing temperatures and results in higher cooling costs.

2.6.2.2   *CO_2 Concentration in Flue Gas*

The CO_2 concentration in flue gases depends on the used feedstock and process. A concentration of about 5 vol% is, for example, encountered in natural-gas-fired combined-cycle power plants. The effect of the CO_2 concentration on the performance of the CPB concept is summarized in Figure 2.27. The amount of the flue gas (635 kg/s) is kept constant for

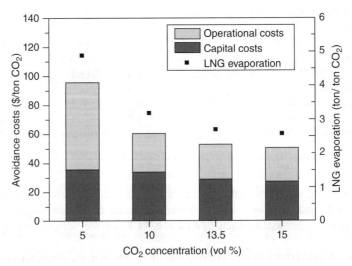

Figure 2.27   Avoidance costs and LNG consumption as a function of the CO_2 inlet concentration

all cases. The front of desublimating CO_2 will move slower through the bed at decreasing inlet CO_2 concentrations.

Therefore, smaller equipment can be used (when maintaining an equal breakthrough time), and consequently, lower flow rates are required for the recovery and cooling steps. However, at the same time, the amount of CO_2 captured will decrease, due to the lower CO_2 content in the gas.

The reduction in equipment size and required flows is cancelled out by this decrease. An inlet concentration of 5 vol% results in avoidance costs of \$95.7/ton, which are substantially higher than that for the base case (\$52.8/ton). The increase in costs is especially high when going to even lower concentrations, which is related to the CO_2 emissions caused by the extra power required. The ratio of the additional required power to the amount of CO_2 captured becomes high at low concentrations. Figure 2.27 also shows that a CO_2 inlet concentration of 15% results in lower avoidance costs. At even higher CO_2 concentrations, recovery of the beds becomes more difficult, as the heat stored in the first zone of the bed during the capture step becomes insufficient. Additional heat has to be supplied to the process to recover CO_2 in those cases.

2.6.2.3 *Pressure Drop*

A pressure drop of 100 mbar for the capture step and 200 mbar for the recovery and cooling steps are assumed. The actual pressure drop depends on packing type, tubing diameters but also possibly to a large extent on the gas distribution over shallow packed beds. For a better distribution, a larger pressure drop is required. A non-uniform distributed feed might result in different freezing/evaporating front velocities at different radial positions and therefore result in a non-optimal use of the bed volume.

Earlier or less sharp breakthrough might be observed, resulting in a lower capture rate of CO_2 or higher LNG consumption. The amount of maldistribution, which is still acceptable, is unknown and requires further studies. To indicate the influence of the pressure drop on the process performance and economics, two cases with 50% higher and lower pressure drops were evaluated. Figure 2.28 shows that the pressure drop over the beds has a significant effect on the operational costs, which is explained by higher compression costs. Also, the amount of required LNG changes slightly, which is related to the heat generated by compression. It should be noted that some CO_2 bypass might be tolerated, since often 90% capture is deemed sufficient.

2.6.3 Comparison with Absorption and Membrane Technology

The economics of the CPB concept is compared to absorption and membrane technology in this section. In order to present a comparison as fair as possible, costs are calculated based on the same cost parameters as that used in the evaluation of the cryogenic concept.

2.6.3.1 *Absorption Technology*

The required input for the evaluation of CO_2 capture costs via absorption technology is obtained from the Integrated Environmental Control Model developed by Rubin [23]. A 600 MW power plant with monoethanolamine (MEA) absorption was simulated, resulting in a flue gas of 666 kg/s containing 14 vol% CO_2 (on a dry basis), which is similar to the flue gas composition that used in the evaluation of the cryogenic concept. The costs for all

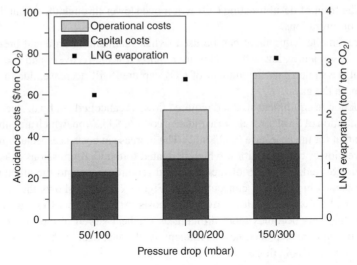

Figure 2.28 *Avoidance costs and LNG consumption as a function of the pressure drop during the capture step (left value) and recovery and cooling steps (right value)*

purification steps upstream the capture process (NO_x, SO_2 and particulates removal) are not taken into account. The equipment costs and the electricity, steam, MEA and corrosion inhibitor consumption are taken from the model. Steam required for stripping is generated with an auxiliary boiler in the simulation, but capital and operational costs are not taken into account. The resulting values are presented in Tables 2.15 and 2.16.

Table 2.15 *Capital investment costs for amine scrubbing and membrane technology*

Amine scrubbing Equipment	Module costs (M$)	Membrane Equipment	Module costs (M$)
Flue gas blower	5.7	Membranes	150
CO_2 absorber	81.3	Compressors/expanders	100
Heat exchangers	6.2	Installation factor (60%)	150
Circulation pumps	12,7		
Sorbent regenerator	46.7		
Reboiler	22.3		
Sorbent reclaimer	15.4		
Drying/compression unit	49.1		
Contingencies	59.9	Contingencies	100
Total direct investment	299.3	Total direct investment	500
Total allocated investment	119.7	Total allocated investment	200
Start-up investment	15.0	Start-up investment	25
Total process investment	434	Total process investment	725
Working capital	8.7	Working capital	15
Total fixed capital	443	Total fixed capital	740

2.6.3.2   Membrane Technology

Merkel *et al.* [20] carried out a basic study on the economics of CO$_2$ capture with membrane technology, treating a flue gas of 602 kg/s containing 12.9 vol% CO$_2$. In their study, two process alternatives are evaluated. In the first option, the driving force for permeation is generated by a vacuum on the permeate side. In the second option, an air sweep is used, which is then fed to the boiler of the power plant.

Although the second alternative is more efficient and looks promising, it will not be taken into account in this study, as it will influence the combustion process and might be more difficult to retrofit to existing facilities. The calculated equipment costs only consist of the membrane costs and compressors/expanders, but are multiplied with an installation factor. The capital and operational costs adjusted with the parameters used in this study are shown in Tables 2.15 and 2.16.

2.6.3.3   Comparison

The CO$_2$ avoidance costs for all three technologies are compared in Figure 2.29. Amine scrubbing and the cryogenic concept have comparable costs, while membranes are significantly more expensive in this evaluation. The results are highly dependent on the assumptions, especially on the availability of utilities. In the amine case, it was assumed that steam is available at no costs, which is unrealistic. When no steam is available at all and the operational and capital costs and additional CO$_2$ emissions related to steam generation in the auxiliary boiler are taken into account, the avoidance costs for scrubbing become high ($133.4/ton).

This is related to the large amount of heat required during stripping of MEA (4.5 MJ/kg CO$_2$). The model developed by Rubin [23] also offers the possibility to select a novel amine, which is used in Fluors Econamine RFG + process. The resulting avoidance costs are substantially lower ($84.2/ton), related to less steam required for regeneration and lower degradation rates.

Table 2.16 *Operational and total costs for amine scrubbing and membrane technology*

	Amine scrubbing	Membrane
CO$_2$ captured (ton/h)	439	369
CO$_2$ emitted due to additional power (ton/h)	58	120
CO$_2$ avoided (ton/h)	381	249
	$/ton CO$_2$ avoided	
Sorbent	7.0	
Inhibitor	1.4	
Reclaimer waste disposal	1.4	
Total chemicals	9.9	
Flue gas blower	2.1	
CO$_2$ product compressor	9.1	
Solvent pump	0.2	
Total power costs	11.3	36.0
Capital charge (20% total fixed capital/year)	33.2	84.9
Total costs	54.5	120.9

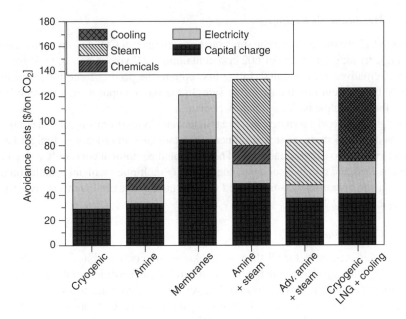

Figure 2.29   *Avoidance costs for different technologies*

The cryogenic concept is attractive when the cold liberated during regasification of LNG could be used for free. If no LNG is available and the entire required cooling capacity should be generated using cryogenic refrigerators, the electricity consumption of the refrigerators would be in the same order of magnitude as the electricity production of the power plant and can therefore be considered unrealistic. Furthermore, evaporation of LNG could be integrated with other processes; therefore, LNG might be only available at certain costs. When comparing the avoidance costs of the advanced amine scrubbing, a maximum price of $8.7/ton LNG can be allowed.

The required cooling power is 248 MW, which corresponds to an LNG consumption of 2.7 kg LNG/kg CO_2 avoided. An average-sized LNG terminal evaporates about 5 million ton/year. Based on an operation of 7000 hours per year, a total amount of 8.6 million ton of LNG would be required for cooling, which corresponds to more than one terminal. When only one terminal is available and the remaining cooling duty has to be generated by refrigeration, the avoidance costs will be $314.4/ton avoided, which is still excessively high.

CO_2 removal using membrane technology is more costlier than cryogenics and scrubbing, due to its high capital costs. When the costs of the membrane modules could be reduced in the future, this option may become competitive, especially when cold or steam utilities are not available or are expensive.

The cryogenic concept shows the advantage that deep CO_2 removal can be obtained, generating a very pure cleaned flue gas as CO_2 product. When cooling to $-150\,^{\circ}$C, the vapor pressure of CO_2 is only 8 Pa, resulting in more than 99.9% CO_2 removal, compared to 90% for the other technologies. To quantify the exact advantage of this 'deep' CO_2 removal, the costs for CO_2 emissions should be known. The removal of impurities is not incorporated in this study. The cryogenic concept has the potential to remove water and, for example, sulphur-containing impurities simultaneously, as vapor pressures are low at the used temperatures. In that case, it could be necessary to install a small separate bed, which allows separate regeneration. Future work will focus on these aspects.

2.7   Conclusions

This chapter discussed a novel process concept for cryogenic CO$_2$ capture based on dynamically operated packed beds proposed by Tuinier in his PhD work. When feeding a flue gas to a refrigerated packed bed, a separation between CO$_2$ and N$_2$ can be obtained. To describe this process, two models were developed: a detailed one-dimensional pseudo-homogeneous model taking axial dispersion effects into account and a simplified, but fast 'sharp front' model. These models were validated by using experiments in a small-scale experimental set-up. This set-up was used to demonstrate the capture step for N$_2$/CO$_2$/H$_2$O gas mixtures, showing simultaneous separation of CO$_2$ and H$_2$O. The entire process, including the recovery and cooling steps, was demonstrated in a continuous fully automated experimental pilot set-up.

The costs of cryogenic CO$_2$ capture using dynamically operated packed beds depend strongly on initial bed temperatures and CO$_2$ concentrations in the feed gas. At lower initial temperatures, the cold stored in the bed can be used more efficiently, resulting in more CO$_2$ deposited per unit of bed volume. At low CO$_2$ inlet concentrations, the relative costs for the amount of CO$_2$ avoided increase strongly. Due to high flow rates required during the process, the pressure drops over the system substantially influence the CO$_2$ avoidance costs. It is expected that required gas distribution plays an important role in the resulting pressure drop. Future research should focus on the effects of gas (mal)distribution (and hence the required pressure drop over the gas distributor) on the process performance.

In comparison with other technologies, it was found that the preferred technology depends heavily on the availability of utilities. The cryogenic concept requires a cold source, such as the evaporation of LNG at a regasification terminal, while amine scrubbing requires low-pressure steam in order to strip the solvent. When both LNG and steam are not available at low costs, membrane technology shows advantages. When steam is available at low costs, especially when using an advanced amine, scrubbing is the preferred technology. The cryogenic concept could be the preferred option, when LNG is available at low costs. Especially when pressure drops can be decreased and the simultaneous removal of impurities can be incorporated in one process, the concept could become a serious candidate for capturing CO$_2$ from flue gases. It is believed that the proposed technology is particularly suited as a second stage in the CO$_2$ capture.

2.8   Note for the Reader

The work presented in this chapter has been published in the PhD dissertation of M. J. Tuinier (Eindhoven 2011) and in his publications on this topic. In the thesis and several journal publications also, the extension of the developed technology to biogas purification is discussed.

List of symbols

a	specific surface, m^2/m^3
c	concentration, mol/m^3
C_p	heat capacity, kJ/kg/K
D_{ax}	axial dispersion coefficient, m^2/s
D_{eff}	effective diffusion in the mass average reference velocity frame, m^2/s
d	diameter, m
d_p	particle diameter, m
$D_{i,j}$	binary diffusivity of component i, m^2/s
H_i	enthalpy of component i, J/kg
L	length, m
m_i''	mass deposition rate per unit surface area for component i, $kg/m^2/s$
M_i	mole mass of component i, kg/mol
n_i	mass flux of component i, $kg/m^2/s$
Nu	Nusselt number ($\alpha_{gs} d_p/\lambda_g$)
p, P	(partial) pressure, Pa
Pe_{ax}	Peclet number for axial heat dispersion ($\rho_g v_g d_p C_{p,g}/\lambda_{ax}$)
Pr	Prandtl number ($C_{p,g}\eta_g/\lambda_g$)
Re	Reynolds number ($\rho_g v_g d_p/\eta_g$)
Sc	Schmidt number ($\eta_g/\rho_g/D$)
Q	heat, W
r	radial coordinate, m
r_p	particle radius, m
R	universal gas constant
t	time, s
T	temperature, K
v	velocity, m/s
x	mole fraction, –
T_{bulk}	temperature of the bulk gas phase, K
T_w	temperature of the wall, K
V	volume, m^3
x	mole fraction
y	dimensionless radial coordinate, –
Y	yield, –
z	axial coordinate, m

Greek letters

α_{bw}	bed-to-wall heat transfer coefficient, $W/m^2/K$
δ	thickness, m
ε	porosity, –
$\phi_{i,tot}$	total mass flow of component i, –
φ_i'	Thiele modulus of component i, –
ϕ	mass flux, $kg/m^2/s$

$\Delta H_{r,j}$ reaction enthalpy of reaction j, J/mol
ΔT_{ad} adiabatic temperature rise, K
η dynamic viscosity, kg/m/s
$v_{i,j}$ stoichiometric coefficient, $-$
$\lambda_{eff,s}$ effective solids heat conductivity, W/m/K
λ_g gas phase heat conductivity, W/m/K
ρ density, kg/m³
ω weight fraction, $-$
σ Stefan–Boltzmann constant, $[J/s/m^2/K^4]$

Subscripts

ax   Axial
c   Cold side
g   gas phase
h   Hot side
inert   inert porous layer in catalyst particle
o   Outer
p   particle
r   reactor
s   solid phase
w   wall

References

1. IPCC (2007) *Climate Change 2007: The Physical Science Basis. Contribution of Working Group I to the Fourth Assessment Report of the Intergovernmental Panel on Climate Change* (eds S. Solomon, D. Qin, M. Manning *et al.*), Cambridge University Press, Cambridge, United Kingdom and New York, NY, USA.
2. Stern, N. (2007) *The Economics of Climate Change. the Stern Review*, Cambridge University Press, Cambridge, United Kingdom and New York, NY, USA.
3. IPCC (2005) *IPCC Special Report of Carbon Dioxide Capture and Storage* (eds B. Metz, O. Davidson, H.C. De Coninck *et al.*), Cambridge University Press, Cambridge, United Kingdom and New York, NY, USA.
4. Ebner, A.D. and Ritter, J.A. (2009) State-of-the-art adsorption and membrane separation processes for carbon dioxide production from carbon dioxide emitting industries. *Separation Science and Technology*, **44** (6), 1273–1421.
5. Ishida, M. and Jin, H. (1994) A new advanced power-generation system using chemical-looping combustion. *Energy*, **19** (4), 415–422.
6. Noorman, S., van Sint Annaland, M. and Kuipers, J.A.M. (2007) Packed bed reactor technology for chemical-looping combustion. *Industrial and Engineering Chemistry Research*, **46** (12), 4212–4220.
7. Kvamsdal, H.M., Jordal, K. and Bolland, O. (2007) A quantitative comparison of gas turbine cycles with CO₂ capture. *Energy*, **32** (1), 10–24.

8. Aaron, D. and Tsouris, C. (2005) Separation of CO_2 from flue gas: A review. *Separation Science and Technology*, **40** (1–3), 321–348.

9. Ertl, B., Durr, C., Coyle, D., Mohammed, I. and Huang, S. (2006) New lng receiving terminal concepts. In World Petroleum Congress Proceedings, vol. 2006.

10. John, A. and Robertson, S. (2008) Lng: World: Strong growth forecast for global lng expenditure. *Petroleum Review*, **62** (732), 34–35.

11. Clodic, D. and Younes, M. (2005) CO_2 capture by anti-sublimation – thermoeconomic process evaluation. In Fourth Annual Conference on Carbon Capture & Sequestration (Alexandria, USA).

12. Clodic, D. and Younes, M. (2002) A new method for CO_2 capture: frosting CO_2 at atmospheric pressure. In: Sixth International Conference on Green house Gas Control Technologies, GHGT6, Kyoto, October2002, pp. 155–160.

13. Tuinier, M.J. (2011) *Novel process concept for cryogenic CO_2 capture*, Technische Universiteit Eindhoven, Eindhoven, The Netherlands.

14. Reid, R.C., Prausnitz, J.M. and Poling, B.E. (1987) *The properties of gases and liquids*, 4th edn, McGraw-Hill, Inc, New York.

15. Daubert, T.E. and Danner, R.P. (1985) *Data compilation tables of properties of pure compounds*, American Institute of Chemical Engineers, New York.

16. Smit, J., van Sint Annaland, M. and Kuipers, J.A.M. (2005) Grid adaptation with weno schemes for non-uniform grids to solve convection-dominated partial differential equations. *Chemical Engineering Science*, **60** (10), 2609–2619.

17. Tiemersma, T.P. (2010) Integrated autothermal reactor concepts for combined oxidative coupling and reforming of methane. PhD thesis, University of Twente, The Netherlands.

18. Abu-Zahra, M.R.M., Niederer, J.P.M., Feron, P.H.M. and Versteeg, G.F. (2007a) CO_2 capture from power plants. part ii. a parametric study of the economical performance based on monoethanolamine. *International Journal of Greenhouse Gas Control*, **1** (2), 135–142.

19. Abu-Zahra, M.R.M., Schneiders, L.H.J., Niederer, J.P.M. *et al.* (2007b) CO_2 capture from power plants. part i. a parametric study of the technical performance based on monoethanolamine. *International Journal of Greenhouse Gas Control*, **1** (1), 37–46.

20. Merkel, T.C., Lin, H., Wei, X. and Baker, R. (2010) Power plant post-combustion carbon dioxide capture: An opportunity for membranes. *Journal of Membrane Science*, **359** (1–2), 126–139.

21. McKinsey (2008). Carbon capture & storage: Assessing the economics. Technical report, McKinsey Climate Change Initiative.

22. Schach, M., Schneider, R., Schramm, H. and Repke, J. (2010) Techno-economic analysis of postcombustion processes for the capture of carbon dioxide from power plant flue gas. *Industrial and Engineering Chemistry Research*, **49** (5), 2363–2370.

23. Rubin, E.S. (2010) Integrated environmental control model. Technical report, Center for Energy and Environmental Studies, Carnegie Mellon University, Pittsburgh.

24. Seider, W.D., Seader, J.D. and Lewin, D.R. (2004) *Product & Process Design Principles*, 2nd edn, John Wiley & Sons, New York.

25. Loh, H.P. and Lyons, J. (2002) Process equipment cost estimation. Technical report, DOE.

3

Novel Pre-Combustion Power Production: Membrane Reactors

F. Gallucci and M. van Sint Annaland
Eindhoven University of Technology, Chemical Process Intensification,
Chemical Engineering and Chemistry, Eindhoven, The Netherlands

3.1 Introduction

There is a growing consensus that anthropogenic greenhouse gas emissions (mainly CO_2) contribute to climate change and should, therefore, be significantly decreased in the near future (Ref. [1–3] report), while the expected energy demand is strongly increasing, and that the increase in energy production can only be realised in the coming decades using mainly fossil fuel sources. Among the various strategies proposed, carbon capture and sequestration (CCS) seems to be a good candidate for CO_2 emissions mitigation (Ref. [4] project report), where CCS has to be regarded as a mid-term solution: CCS is to be used as long as other technologies such as large-scale exploitation of renewable sources and nuclear fusion are still under development. CCS consists of two challenging processes, namely, carbon dioxide capture and its sequestration (mineralization or storage). In this chapter, only carbon capture is discussed, while sequestration (perhaps in offshore geological formations) is a process with its own technological and societal challenges.

The easiest way to introduce CCS to mitigate anthropogenic emissions is to focus on point sources with concentrated CO_2 emissions. In a report by IPCC (Ref. [1] Chapter 2), the concentrated sources for CO_2 emissions (those producing more than 1 Mt CO_2/year) have been inventorized (see Table 3.1). From this table, it is evident that the main effort should be dedicated to capture CO_2 from fossil fuel-driven power plants.

Three main technological paths have been proposed for CCS from fossil fuel-fired power stations, namely, the post-combustion capture, the pre-combustion decarbonization and

Process Intensification for Sustainable Energy Conversion, First Edition.
Edited by Fausto Gallucci and Martin van Sint Annaland.
© 2015 John Wiley & Sons, Ltd. Published 2015 by John Wiley & Sons, Ltd.

Table 3.1   *Distribution of worldwide large stationary CO_2 sources originating from fossil fuel [IPCC Chapter 2]*

Process/industry	Emissions (Mt CO_2/yr)	% on the total
Power production	10,539	78.8
Cement production	932	7.0
Refineries	798	6.0
Iron and steel industry	646	4.8
Petrochemical industry	379	2.8
Oil and gas processing	50	0.4
Other sources	33	0.2

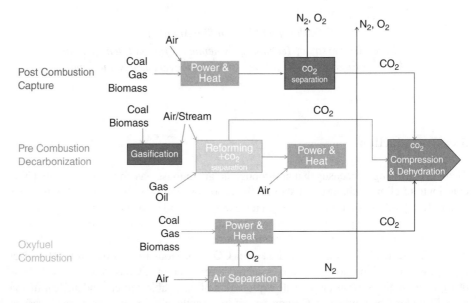

Figure 3.1   *The three main CO_2 capture routes (Source: Adapted from IPCC – chapter 3)*

the oxyfuel combustion route (see Figure 3.1). In each of these processes, an additional separation step is required, which brings an efficiency penalty to the whole power plant. Research and development in CCS is devoted to decrease this energy penalty as much as possible.

In particular, during post-combustion capture, carbon dioxide is removed after the combustion of the fuel with air. This resulting flue gas stream mainly contains CO_2, H_2O and N_2. Although the low concentration of CO_2 (5–20% depending on the fuel used) in the flue gas makes the capture more costly, the post-combustion route is important because it can in principle be used to retrofit almost all existing fossil fuel power plants with a tremendous market perspective. For post-combustion capture, several approaches have been proposed, including concepts based on chemical and physical absorption, adsorption processes using solid sorbents, cryogenic separation and membrane processes. While a great scientific effort is paid to membrane separation, the most mature process up to now is surely the one based on amine scrubbing. This technology relies on the reaction of carbon dioxide with an absorption liquid, typically an amine in an absorption column. Subsequently, the absorption liquid is regenerated in a desorber, where the carbon dioxide is stripped from the solvent using pressure and temperature differences, and recycled to the absorber. The most important challenges in this process are to decrease the large regeneration energy input and to decrease the sensitivity of the absorption liquids to oxygen, water and other impurities present in the flue gas stream.

As indicated in Figure 3.1, the pre-combustion decarbonization consists of the conversion of a fuel into a mixture of hydrogen and carbon dioxide and the subsequent separation of the carbon dioxide from the hydrogen, which is then used for power generation. The first step (conversion of fuel into a mixture of hydrogen and CO_2) is carried out in a series of reactors: gasification for solid fuels or reforming for liquid/gas fuels, shift reactors (to convert CO in further hydrogen and CO_2) and separation units (pressure swing adsorption or methanation reactions). The mixture of CO_2 and H_2 contains a much higher concentration of carbon dioxide compared with the flue gas of a traditional power plant. For this reason, the associated energy penalty with CO_2 separation is less compared to post-combustion processes. Also in the pre-combustion route the application of membranes have been proposed: for instance, polymeric membranes (Ref. [5]) to separate CO_2 and H_2, and Pd/Ag alloy membranes in membrane reformers enabling enormous process intensification via integration of reaction (reforming) and hydrogen separation in the same unit [6, 7].

Oxyfuel combustion consists of burning a fuel with pure oxygen avoiding mixing of CO_2 and nitrogen, resulting in intrinsic CO_2 capture (Ref. [8, 9]). An air separation unit is used to produce pure oxygen, while for safety reasons the combustion is often carried out with a mixture of oxygen and CO_2 to decrease the temperature of combustion. One of the important challenges encountered with oxyfuel combustion is to minimize the energy efficiency loss associated with air separation. Air separation is typically carried out using cryogenic distillation and is therefore very energy intensive. The success of oxyfuel processes is thus connected to the improvements in oxygen separation. Membrane reactors (MRs) with oxygen permselective membranes have been proposed to perform combustion and oxygen separation in the same apparatus, but major improvements in the mechanical and chemical stability of the membranes are required [10].

In this chapter, the pre-combustion route is discussed with particular attention to novel reactors, namely the MR concepts, that integrate reforming reactions and separation in a single unit, thus resulting in both hydrogen production for power generation and CO_2 capture.

3.2 The Membrane Reactor Concept

A MR is a multi-functional reactor that integrates a reaction step and a separation step in a single unit, thus achieving a high degree of process intensification. Traditionally, MRs are studied for dehydrogenation reactions as in this case the reaction temperature and the working temperature of the membranes used (often Pd-based dense membranes) correspond so that reaction and separation can be integrated in the same unit.

As far as the pre-combustion capture is concerned, a fuel is converted into hydrogen and CO_2 (after gasification if the fuel is in solid form), then CO_2 is separated and hydrogen is used for power production at large scale or in fuel cells at smaller scales. It is therefore interesting to see how MRs can be used for hydrogen production.

Traditionally, hydrogen is produced via steam reforming (SR) of hydrocarbons such as methane, naphtha oil or from methanol/ethanol. On an industrial scale, however, most of the hydrogen (more than 80%) is currently produced by SR of natural gas carried out in huge multi-tubular fixed-bed reactors. In small-scale applications, two other main alternatives are generally considered along with SR: partial oxidation reactions, whose efficiency is, however, significantly lower than SR, and autothermal reforming, where the partial oxidation (exothermic reaction) and SR (endothermic reaction) are carried out in the same reactor.

The main drawbacks of SR, partial oxidation and autothermal conventional reactors is that all these reactions are equilibrium limited and (even in the case of complete fuel conversion) lead to a hydrogen-rich gas mixture also containing carbon oxides and other by-products. Consequently, in order to produce pure hydrogen, these chemical processes are carried out in a number of reaction units (typically high-temperature reformer, high- and low-temperature shift reactors) followed by separation units (mostly pressure swing adsorption units are used). The large number of different process steps decreases the system efficiency and makes scale down uneconomical. A typical reaction process scheme is reported in Figure 3.2 [11].

With this process, high hydrogen yields are achieved, but costly high-temperature heat exchangers and complex energy integration among different process units are required for obtaining the hydrogen at the desired purity.

Among different technologies related to production, separation and purification of H_2, membrane technologies seem to really play a fundamental role and membrane separation is nowadays considered as a good candidate for substituting the conventional systems. The specific thermodynamic limits affecting the traditional reactors can be circumvented by using innovative integrated systems, such as the so-called MRs, engineering systems in which both reaction and separation occur in the same device (see Figure 3.3).

If compared with a conventional configuration in which a reactor is followed by a separation unit, the use of MRs can bring various potential advantages such as reduced capital costs (due to the reduction of process unit), improved yields and selectivities (due to shift effect on the reaction) and reduced downstream separation costs (the separation is integrated).

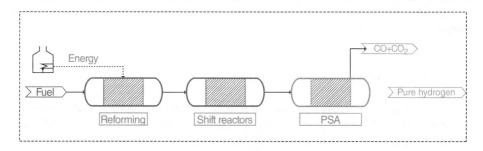

Figure 3.2 *Conventional steam reforming reaction scheme*

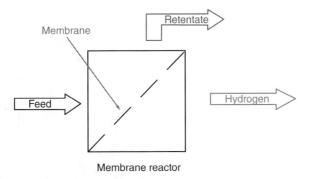

Figure 3.3   *Typical representation of a membrane reactor for hydrogen production*

The success of MRs for hydrogen production is basically connected to: i) the advances in the membrane production methods that allow to produce thin membranes with high hydrogen fluxes and high hydrogen permselectivities; ii) the design of innovative reactor concepts that allow the integration of separation and energy exchange, the reduction of mass and heat transfer resistances and simplifying the housing of the membranes.

The application of MRs for dehydrogenation reactions has been firstly proposed to the scientific community by Prof. Gryaznov in the late 1960s (see Ref. [12]). At that time, the application of the first thick membranes was just a way to shift the equilibrium reaction towards the product of interest (while hydrogen was considered by-product). MRs in dehydrogenation reactions were a scientific curiosity until around 1996, after that point mainly due to the production of more sophisticated membranes and the increasing discussion about the role of hydrogen as energy carrier in the future economy (in the last years), the attention towards MRs for hydrogen production increased sharply as indicated in Figure 3.4. Probably the echo produced by the two books "The Hydrogen Economy" (2002, 2004) [13, 14] was also reflected in the great increase of number of papers in the last 5 years. Accordingly, with the increase in number of papers, the patents awarded on hydrogen production in MRs also increased with time, and in the past years this number has been increasing rapidly. In fact, most of the patents have been awarded in the last 10 years [6, 7].

3.3   Types of Reactors

Different types of MRs for hydrogen production have been proposed in literature. Most of the previous work has been performed in packed bed membrane reactors (PBMRs); however, there is an increasing interest in novel configurations such as fluidized bed membrane reactors (FBMRs) and micro-membrane reactors (MMRs) especially because better heat management and decreased mass transfer limitations can be obtained in these reactor configurations. In the following, these reactor configurations along with membrane bio-reactors (MBRs) and catalytic membrane reactors (CMRs) for hydrogen production is discussed in detail.

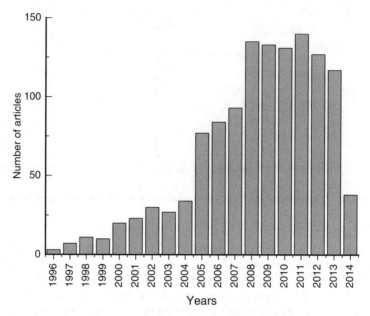

Figure 3.4 *Number of papers on hydrogen production in membrane reactors per year (Till May 2014). Database scopus (www.scopus.com)*

3.3.1 Packed Bed Membrane Reactors

The PBMR configuration was the first and most studied configuration for hydrogen production in MRs. This is because the first studies on MRs focussed on the effect of the hydrogen permeation through membranes on the reaction system. Thus, it was quite simple to compare two packed bed reactors (avoiding the complication of complex fluid dynamics such as in fluidized bed) in one of which a membrane was used.

In Table 3.2, the main investigators working with PBMR for hydrogen production are summarized (source Scopus – May 2014).

Table 3.2 *Major investigators on packed bed membrane reactors*

Investigator name	Institution	Number of papers
Rahimpour, M.R	Sharaz University (Iran)	61
Basile A.	Institute on Membrane Technology (Italy)	56
Tosti S.	ENEA (Italy)	35
Tsotsis T.T.	University of Southern California, Mork Family Department of Chemical Engineering and Materials Science, Los Angeles, United States	20
Lombardo E.A.	Instituto de Investigaciones en Catálisis y Petroquímica (Argentina)	14
Itoh N.	Utsunomiya University (Japan)	14

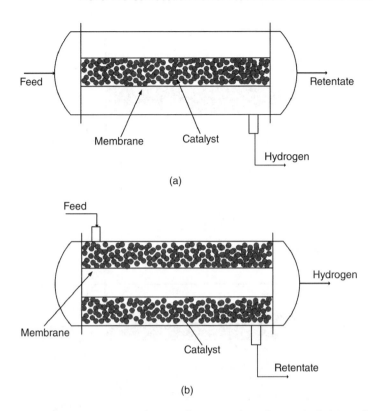

(a)

(b)

Figure 3.5 *Membrane reactor catalyst-in-tube- (a) and catalyst-in-shell (b) configurations*

PBMRs have been used for producing hydrogen via reforming of methane [15, 16], reforming of alcohols [17, 18], autothermal reforming [19], partial oxidation of methane (POM) [20], and so on.

In the packed bed, the catalyst is in fixed-bed configuration and in contact with a perms-elective membrane. The most used packed bed configuration is the tubular one where the catalyst may be packed either in the membrane tube (Figure 3.5a) or in the shell side (Figure 3.5b), while the permeation stream is collected on the other side of the membrane (in the case of hydrogen-selective membranes) or one reactant is fed on the other side of the membrane (in the case of oxygen-selective membrane) [21].

For multi-tubular MR configurations, the catalyst-in-tube configuration can be preferred especially for construction reason and for the extent of bed-to-wall mass and heat transfer limitations, which can be very detrimental in the catalyst in shell configuration.

Often, a sweep gas can be used in the permeation side of the membrane in order to keep the permeation hydrogen partial pressure as low as possible for minimizing the membrane area required for the hydrogen separation. This practice is, for example, welcome if hydrogen for ammonia plant is being produced, in which case an amount of nitrogen can be used for sweeping the permeation side producing a synthesis stream ($N_2/H_2 = 1/3$) ready for the final reaction step. If a sweep gas is used in the permeation side, then a PBMR can be used in both co-current (Figure 3.6a) and counter-current (Figure 3.6b) modes.

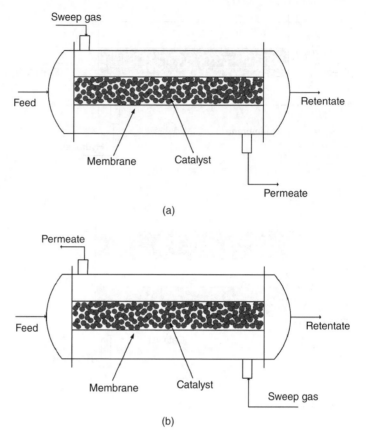

(a)

(b)

Figure 3.6   *Co-current configuration (a) and counter-current configuration (b) for membrane reactor*

Using a counter-current mode leads to completely different partial pressure profiles in reaction and permeation sides with respect to the co-current mode (independently on the reaction system considered), as demonstrated in Figure 3.7.

It is evident that in co-current mode hydrogen partial pressure in both the reaction side (here indicated as lumen) and permeation side (shell side) increases along the reactor. Moreover, the driving force for the hydrogen permeation related to the difference between the hydrogen partial pressure in the lumen side and in the shell side decreases by increasing the reactor length because the hydrogen partial pressure in the shell side tends to reach the hydrogen partial pressure in the lumen side. When the driving force tends to zero, the effect of membrane is not evident anymore. For counter-current mode, the hydrogen partial pressure in the lumen side increases in the first part of the reactor mainly due to the reaction, and afterwards it diminishes mainly due to the permeation through the membrane. The hydrogen partial pressure at the exit of the reaction side could also be as low as zero, but there is still a residue driving force for the permeation, the partial pressure in the permeation side being zero (inlet of fresh sweep gas). It is clear that in almost the whole reactor there is a positive driving force for the hydrogen permeation, and, consequently, by just using the

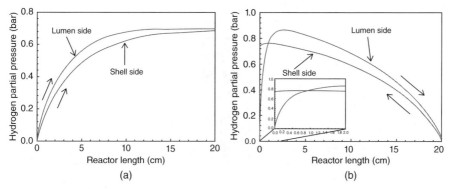

Figure 3.7 *Hydrogen partial pressure in co-current configuration (a) and counter-current configuration (b) for ethanol reforming in packed bed membrane reactors. (Source: Reprinted from Ref. 22 with permission of Elsevier)*

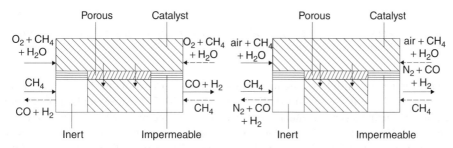

Figure 3.8 *RFCMR concepts with porous membranes [23]. (Source: Reprinted with permission of bepress)*

counter-current mode instead of the co-current one, it could be possible to recover 100% of the hydrogen produced in the reactor providing that enough membrane area is installed.

An interesting feature of PBMRs is the possibility to operate them in a reverse flow mode, integrating the reaction and separation with the recuperative heat exchange inside the reactor. This operational mode is quite interesting for POM, as indicated by Smit *et al.* [23]. In fact, as stated by the authors, in normal POM systems, air and CH_4 feed streams have to be preheated to the reaction temperature, while POM reaction being only slightly exothermic, the external heat transfer between feed and exhaust is very expensive. Therefore, recuperative heat exchange is preferably carried out inside the reactor. The authors studied POM in porous MR in a tube-in-shell configuration as reported in the following figure. By co-feeding the reactants in the two compartments in reverse flow mode (Figure 3.8), a good heat exchange can be obtained and a typical trapezoid temperature profile (high temperature at the centre of the reactor and low temperature at the inlet and outlet) can be achieved.

Although the tube-in-tube configuration is quite useful to work in laboratory scale and for proof of principle of MRs, for industrial scale some other configurations need to be used in order to increase the membrane area per volume of vessel used. In fact, the amount of hydrogen produced is directly related to the amount of membrane area installed in the reactor.

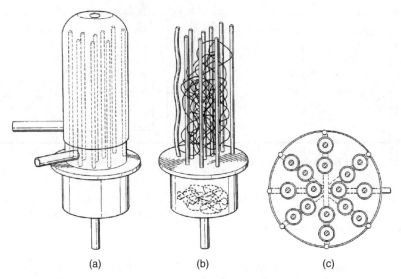

(a)   (b)   (c)

Figure 3.9   *Membrane housing (a), catalyst distribution (b) and membrane connectors (c) for a multi-tube membrane reactor [24]*

Starting for the tube-in-tube configuration, a straightforward way to increase the membrane area in packed bed is the tube-in-shell configuration [24, 25]. An example of multi-tube membrane housing has been patented by Buxbaum [24] and reported in Figure 3.9. In this case, the catalyst is loaded in the shell side of the reactor while the membrane tubes are connected to a collector for the pure hydrogen. In particular, in the figure, the possibility to use a catalyst in a separate chamber is shown. In the case of reforming reactions, this chamber acts as a pre-reforming zone where the greatest temperature profiles are concentrated. In this way, the membranes will work at an almost constant temperature.

The second way to increase the membrane area per volume of reactor is adopting the hollow fibre configuration. For example, in the case of perovskite membranes, the membrane flux is generally quite low and the hollow fibre configuration is quite interesting. The main investigators of hollow fibre MRs are summarized in Table 3.3.

Table 3.3   *Major investigators on hollow fibre membrane reactors*

Investigator name	Institution	Number of papers
Rittmann, B	Arizona State University, Center for Environmental Biotechnology, Tempe, United States	12
Li, K.	Imperial College London, UK	11
Nerenberg, R	University of Notre Dame, Department of Civil and Environmental Engineering and Earth Sciences, Notre	9
Xia, S.	Tongji University, College of Environmental Science and Engineering, Shanghai, China	9
Caro, J.	University of Hannover, Germany	7

Source: Scopus–May 2014

Kleinert *et al.* [26] studied, for example, POM in a hollow fibre MR. The perovskite membranes used by the authors were produced from Ba(Co,Fe,Zr)O$_{3-d}$ (BCFZ) powder via phase inversion spinning technique. A tube-in-tube configuration was used while the catalyst was packed in the shell side of the reactor.

In their paper, the authors show that the membrane was able to give quite interesting results with a methane conversion of 82% and a CO selectivity of 83%. Moreover, the membrane was quite stable under the reactive conditions investigated. Finally, the combination of SR and POM was studied by feeding steam along with methane in order to suppress the carbon formation. Even in these conditions, the MR showed good stability.

The membrane area required for the separation can be reduced by increasing membrane fluxes (by keeping the same high permselectivity). As already described earlier, membrane fluxes can be increased by decreasing the separation thickness (ultra-thin membranes). For example, for Pd-based membranes, defect-free separation layers as thin as 1–2 µm are now commonly produced in the laboratory [27]. The thinner is the separation layer, the higher is the hydrogen flux through the membrane, the lower the membrane area required for a given hydrogen recovery. However, the production of thin membranes brought under the spotlight one of the disadvantages of fixed-bed MRs: the influence of bed-to-wall mass transfer limitations on the membrane area required. Briefly, as long as the hydrogen flux through the membrane is a limiting step, the effect of external mass transfer limitations such as the limitations to hydrogen transport between the bulk of the catalytic bed (where hydrogen is produced) and the membrane wall (where hydrogen is recovered) can be neglected. However, by increasing the membrane flux, the external mass transfer limitations became limiting and determine the extent of membrane area. An example of the increase of extent of mass transfer limitations in MRs was given by Tiemersma *et al.* [28] and Gallucci *et al.* [29, 30] shown in Figure 3.10. In their paper, the authors claimed that with the membrane available in 2006 the effect of mass transfer limitations was quite limited. However, by increasing their experimental membrane flux (increase foreseen in a couple of years of membrane developments), the external mass transfer limitation could become a limiting step (as indicated in the figure).

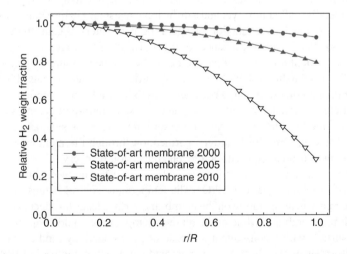

Figure 3.10 *Relative H$_2$ weight fraction profiles at changing membrane permeability*

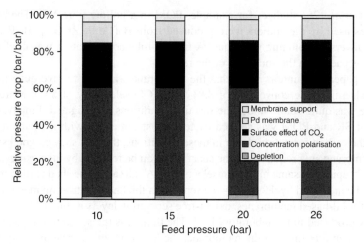

Figure 3.11   *Hydrogen pressure drops due to depletion, concentration polarization, transport in the palladium membrane and porous support compared to the total hydrogen partial pressure drop. (Source: Reproduced from Ref. [31])*

The conclusions drawn by Tiemersma and Gallucci were also confirmed by the work carried out at SINTEF by Bredesen and co-workers (see Figure 3.11 from Ref. [31]). The paper of Peters also reports the effect of poisoning by CO and depletion by other gases.

As indicated in Figure 3.11, it is evident that the extent of bed-to-wall mass transfer limitations (also called concentration polarization by membranologists) is the limiting factor for the permeation of hydrogen through the membrane, and, thus, it determines the amount of membrane area required for a given hydrogen recovery. These results have been also confirmed for self-supported membranes via detailed modelling by Caravella *et al.* [32]. Recently, Boon *et al.* [33] investigated via a 2D Navier–Stokes model, the main resistances to mass transfer such as concentration polarization in the retentate, hydrogen permeation through the metallic palladium layer and a diffusional resistance in the support layer, reporting the main resistance as a function of operating parameters. Again, the concentration polarization is often the main resistance for the transport of hydrogen.

A second important limitation of packed bed (MRs) is the pressure drop along the reactor, which dictates the dimension of the catalyst particle diameter to be used in the reactor. To decrease the pressure drop, large particles need to be used and this reflects in the extent of intra-particle mass transfer limitations. The increase of intra-particle mass transfer limitations lowers the local hydrogen partial pressure in the reaction side, and, consequently, it lowers the driving force for hydrogen permeation through the membrane, resulting in increased membrane area required for a given conversion/recovery.

Finally, the reactions for hydrogen production are often quite endothermic (reforming reactions) or exothermic (e.g. CPO). The temperature control is thus quite important because a decrease in temperature on the membrane surface leads to a decrease in hydrogen flux through the membrane while an increase in temperature could result in crack on the membrane surface with a consequent decrease of permselectivity (and thus deteriorating the MR performance). The heat management and temperature control in PBMRs is quite difficult, and temperature profiles along the membrane length are difficult to avoid in such a reactor.

The combination of these drawbacks has driven the research towards new reactor concepts such as MMRs or FBMRs, as discussed in the following sections.

3.3.2 Fluidized Bed Membrane Reactors

A typical fluidized membrane reactor (or membrane-assisted fluidized bed MAFBR) for hydrogen production consists of a bundle of hydrogen-selective membranes immersed in a catalytic bed operated in bubbling regime. The use of FBMRs makes possible the reduction of bed-to-wall mass transfer limitations but also allows operating the reactor at a virtually isothermal condition (due to the movement of catalyst). This possibility can be used for operating the autothermal reforming of hydrocarbons inside the MR. In fact, as indicated by Tiemersma *et al.* [28], the autothermal reforming of methane in a PBMR is quite difficult due to the hotspot at the reactor inlet, which can melt down the membrane. This problem is completely circumvented in FBMRs. In this case, both autothermal reforming and hydrogen recovery can be performed in a single reactor.

According to Deshmukh *et al.* [34], the main advantages of the MAFBR can be listed as follows:

- Negligible pressure drop, which allows using small particle sizes resulting in no internal mass and heat transfer limitations
- (Virtually) Isothermal operation
- Flexibility in membrane and heat transfer surface area and arrangement of the membrane bundles

  o Improved fluidization behaviour as a result of Compartmentalization, that is, reduced axial gas back-mixing
  o Reduced average bubble size due to enhanced bubble breakage, resulting in improved bubble to emulsion mass transfer

Different research groups applied membrane-assisted fluidized bed for different reaction systems (mostly involving hydrogen), as indicated in Table 3.4.

Table 3.4 *Major investigators on fluidized bed membrane reactors*

Investigator name	Institution	Number of papers
Grace, J. R. (John)	The University of British Columbia, Canada	20
Elnashaie, S. S. E. H. (Said)	Misr University for Science and Technology, Egypt	19
Van Sint Annaland, M. (Martin)	Eindhoven University, The Netherlands	14
Rahimpour, M. R. (Mohammad)	Shiraz University, Iran	10
Abashar, M. E. E. (Mohamed)	King Saud University, Saudi Arabia	6
Yan, Y. (Yibin)	Auburn University, USA	5

Source: Scopus – May 2014

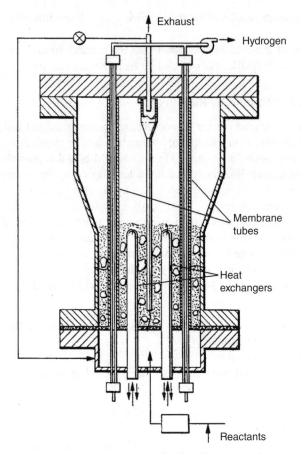

Figure 3.12 *Schematic representation of a fluidized bed membrane reactor for selective removal of hydrogen. (Source: After [45])*

Even though Rahimpour and co-workers often used FBMRs for distributive hydrogen feeding in methanol reactors [35–37], most of the literature has focussed on pure hydrogen production through Pd-based membranes (see among others [38–41]) and on autothermal reforming reactions (see a.o. Ref. [29, 42–44]).

A typical Pd-based membrane-assisted fluidized bed has been proposed by Adris and co-workers and is reported in Figure 3.12.

With this reactor, the bed-to-wall mass transfer limitations can be circumvented, while the heat required for the reforming reactions (often endothermic equilibrium reaction) is supplied through heat exchange surfaces inserted in the reactor system. In fact, fluidized bed reactors present higher heat exchange coefficients compared to fixed-bed reactors.

However, besides the high heat exchange coefficients, fluidized beds are also suitable for isothermal operations even if a highly exothermic reaction is occurring. This has been demonstrated a.o. by Deshmukh *et al.* [46, 47], who carried out oxidative dehydrogenation of methanol in laboratory-scale membrane fluidized bed reactors. The authors found virtually isothermal conditions even for very high methanol feed concentrations. This important aspect of fluidized bed reactors makes it possible to operate autothermal reforming of methane (and other hydrocarbons) by feeding oxygen directly in the MR preventing formation of hotspots and subsequent damage of the membranes.

Indeed, Roy *et al.* [48] firstly demonstrated that autothermal operation could be achieved by directly adding oxygen to a fluidized bed reformer to provide all of the required via partial oxidation reaction. The extension of this work to a membrane fluidized bed reactor has been performed by the same group few years later. In particular, the previous experiences with autothermal fluidized beds and membrane fluidized beds have been condensed in a modelling paper by Grace *et al.* [42]. The model chosen by Grace was an equilibrium model and this selection was made based on the previous experience on modelling and experiments on FBMRs which suggested that the overall composition of each species at each position inside the bed was found to be mainly controlled by two factors: i) the quantity of hydrogen that had permeated through the membranes (depending on the difference in hydrogen partial pressure and membrane characteristics); ii) chemical equilibrium at the given temperature and overall pressure on the reactor side (depending on the very high catalytic activity for methane reforming).

The model was applied in order to investigate the influence of various parameters on the performance of FBMR with oxygen addition. Although the results showed that autothermal operation can be achieved by using approximately $0.3\,O_2/CH_4$ feed ratio, the interaction between the different parameters is quite complex. For instance, in methane reformers an important parameter is the steam/carbon ratio. However, when feeding oxygen, the steam becomes also a product of the oxidation reaction and this makes the prediction of the reactor behaviour a bit more complicated. Furthermore, an important conclusion of the work is that oxygen addition reduces the coke formation and consequently the catalyst deactivation.

Based on this paper and on various other papers dealing with both theoretical and experimental works, recently the same group presented the experimental findings of pre-commercial MR prototypes for both methane reforming and ATR (with air addition) [49, 50]. The authors used optimized plane membranes, where a single two-side planar membrane panel consists of two Pd–Ag foils mounted on a porous stainless steel base using proprietary MRT sealing and protection techniques.

A scheme of the MR used is reported in Figure 3.13 (Ref. [51]).

Based on their experience with the fluidized bed, the authors suggested that the flux through membranes computed by Sieverts' equation must be multiplied by an effectiveness factor (< 1) for membranes immersed in fluidized bed reactors. The authors suggested that this effectiveness factor could possibly be due to fine dust generated in the bed or other mass transfer resistances. They demonstrated that by using ATR catalyst the effectiveness factor could be higher than the case of reforming catalyst and could in principle be higher than 0.9 (as high as 0.98 according to Ref. [49]). The experiments show that the maximum temperature difference along the reactor where membranes are immersed is always lower than 20 °C [50], which results in a higher membrane lifetime and avoids formation of pinholes. All these works confirm that the autothermal reforming of methane is possible in a FBMR with oxygen (or air) addition without hotspot formation.

It has to be pointed out that feeding air directly in the reactors brings two drawbacks: i) the exhaust gas mainly consists of nitrogen and CO_2, if CO_2 capture is required the separation is difficult (and costly); ii) in each case nitrogen dilutes the amount of hydrogen in the reactor, reducing the driving force for hydrogen permeation through the membranes

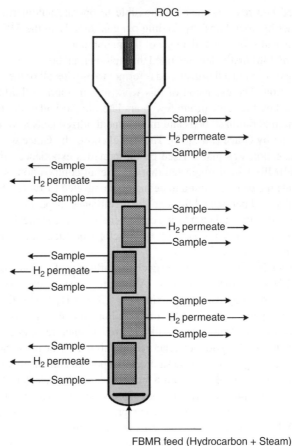

Figure 3.13 *Reactor configuration and Pd-based plane membrane details used by grace group. (Source: Reprinted with Permission from Ref. [51]. Copyright 2011 American Chemical Society)*

and resulting in higher membrane area required in the reactor. For these reasons, feeding oxygen in the FBMR for achieving autothermal reforming is more appropriate. However, an air separation unit (a costly process) outside the reactor is required to produce the oxygen for autothermal reforming.

This problem can be circumvented by using novel reactor configurations proposed by van Sint Annaland and co-workers [44, 52–54]. In particular, two configurations have been proposed to achieve autothermal methane reforming: methane combustion configuration and hydrogen combustion configuration.

The methane combustion configuration is sketched in Figure 3.14 and consists of two sections [44]. Hydrogen permselective membranes are integrated in a fluidized reforming/shift top section where ultra-pure H_2 is extracted and the energy required for the SR is supplied via in situ methane oxidation in a separate fluidized bottom section, where oxygen is selectively fed to the methane/steam feed via oxygen permselective membranes.

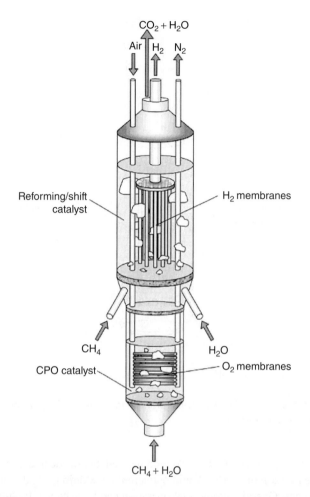

Figure 3.14   *Methane combustion configuration for pure hydrogen production through autothermal reforming of methane*

Two different sections are required because metallic Pd-based membranes for selective H_2 extraction can only be operated below typically 700 °C because of membrane stability, while acceptable O_2 fluxes through available perovskite-type O_2 permselective membranes can only be realized above 900–1000 °C. In this case, the air separation is achieved *in situ* resulting in lower costs and intrinsic CO_2 capture.

In the oxidation section, CH_4 is partially oxidized in order to achieve the high temperatures required for O_2 permeation through the perovskite membranes and to simultaneously preheat part of the CH_4/steam feed.

The preheated feed is mixed with additional CH_4 and steam and fed to the reforming/shift section, where CH_4 is completely converted into CO, CO_2 and H_2 because of the selective H_2 extraction through the Pd membranes which shifts methane SR. H_2 extraction can be achieved by using dead-end Pd membranes and applying a vacuum on the permeate side.

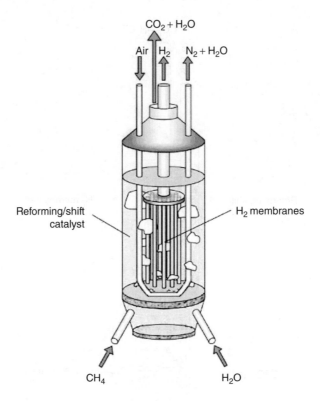

Figure 3.15 *Hydrogen combustion configuration for pure hydrogen production through autothermal reforming of methane*

Alternatively, a sweep gas (such as H_2O) could be used, but the decrease in membrane area due to the increased driving force should outweigh the additional costs for separating H_2 from the sweep gas. Overall autothermal operation can be achieved by tuning the overall CH_4, O_2 and steam fed to the reactor. The distinct advantage of this reactor concept is that the temperatures in both sections can be controlled independently by selecting the proper ratio of CH_4/H_2O fed at the oxidation and reforming/shift sections, while maintaining overall autothermal operation with optimal energy efficiency.

Figure 3.15 shows the hydrogen combustion configuration, where the energy for SR is delivered via burning part of the produced hydrogen [53]. This configuration consists of only one fluidized bed section, where two types of hydrogen permselective membranes are incorporated: dead-end Pd-based membranes to recover ultra-pure H_2 by applying a vacuum on the permeate side (similar to the ones used in the methane combustion configuration) and U-shaped Pd-based membranes with oxidative sweeping, by feeding air to the permeate side to burn the permeated hydrogen.

The hydrogen combustion configuration has the clear advantage that only one section is required, circumventing the need for a (costly) high-temperature bottom section. On the other hand, in the methane combustion, configuration steam is produced in situ, which enhances the CO conversion. Moreover, in the hydrogen combustion configuration, part of the expensive Pd-based membranes are used to burn part of the produced hydrogen, while for the methane combustion configuration further development of oxygen permselective membranes (especially the mechanical and chemical stability) is essential.

Patil *et al.* [44] and Gallucci *et al.* [53, 54] demonstrated from both experimental and theoretical points of view that autothermal reforming can be achieved in both reactor configurations.

Moreover, applying detailed models to both packed bed and FBMRs Gallucci *et al.* [30] demonstrated that for methane reforming the FBMR requires half membrane area compared to the PBMRs where external mass transfer limitations are prevailing. Although the system presented in Figure 3.15 is highly compact and integrated, to achieve hydrogen production for power generation with integrated CO_2 capture required a high amount of membranes while more than 25% of the costly Pd-based membranes are used "only" for heat supply to the reactor.

Recently, a new reactor has been proposed by combining the MR features with the chemical looping features for heat production with inherent CO_2 capture. The novel reactor concept called Membrane Assisted Chemical Looping Reforming (MA-CLR) was introduced by Medrano *et al.* [55]. In this system (Figure 3.16), a FBMR is located in the fuel reactor of a CLR system, where the incorporation of membranes substitutes the WGS and PSA steps of the traditional CLR process. The selective extraction of hydrogen provides a pure H_2 stream and also displaces the thermodynamic equilibria. Hence, reaction and

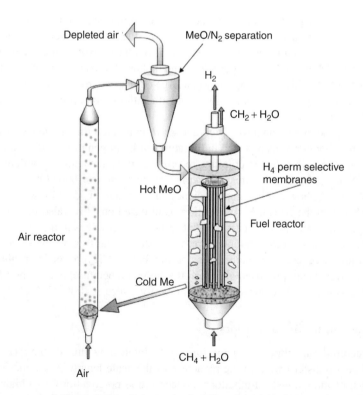

Figure 3.16 *Schematic drawing of the MA-CLR*

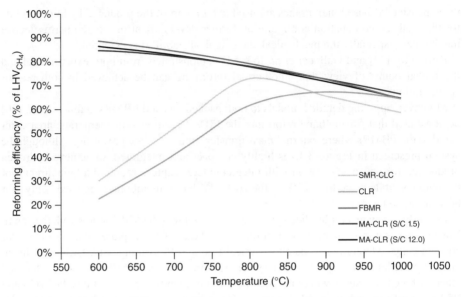

Figure 3.17   *Reforming efficiency profiles of the systems studied as a function of the reactor temperature*

separation are carried out in only one reactor. Autothermal operation is also achieved via the oxygen carrier, which is circulating between the fuel and air reactors in the chemical looping system. In the air reactor, oxygen carrier is oxidized (exothermic reaction), and the introduction of this hot oxygen carrier allows the fuel reactor to maintain elevated temperatures. With the MA-CLR, the complexity of a large number of process steps is reduced, and, hence, a high degree of process intensification with anticipated improved energy efficiency can be achieved.

With a thermodynamic analysis, Medrano *et al.* compared the efficiency of this novel concept with other concepts reported in literature for hydrogen production with CO_2 capture. A comprehensive comparison of the reforming efficiency obtained with the different technologies is depicted in Figure 3.17. At lower temperatures (600–700 °C), the FBMR and the MA-CLR show a reforming efficiency in the range of 75–90%, which is significantly higher than the typical H_2 production plant based on SMR (also in a combination with a mild system for the CO_2 capture) that shows typical values in the range of 70–75% [56]. Increasing the operating temperature, the systems tend to have the same reforming efficiency: this effect can be explained considering that the CH_4 conversion is almost 100% and the gas streams are produced almost at the same temperature, except the SMR-CLC, which is the system with the lowest reforming performance at 1000 °C (58%).

3.3.3   Membrane Micro-Reactors

Micro-structured (membrane) reactors are quite interesting due to their (i) improved mass and heat transfer owing to the reduction of the scale length in the micro-channels; (ii) removal of mass transfer limitations (concentration polarization); (iii) high degree of process intensification by integrating different process steps in a small-scale device.

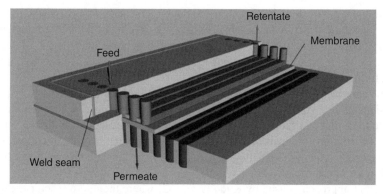

Figure 3.18 *Sketch of the micro-channel reactor configuration used by boeltken for Pd-based membrane tests. (Source: Reproduced from Ref. [59], with permission from Elsevier)*

Although there is a great interest in micro-reactors, the application of membrane micro-reactors is still limited. An interesting work is being carried out by Bredesen and co-workers [27, 57, 58] using thin, high selective and high permeable Pd-based membranes produced via magnetron spattering technique.

In particular, by comparing the performance of the same membrane in different configurations, the authors showed that in tubular configuration the extent of concentration polarization is the limiting step for hydrogen permeation (as also indicated above), while with the same membrane used in micro-channel configuration the concentration polarization effect can be completely neglected [27, 57].

A micro-channel reactor configuration proposed by Boeltken *et al.* is reported in Figure 3.18 [59].

The reactor consists of 6 micro-channels 13 mm long and a section 1 mm × 1 mm, while the membranes used are self-supported Pd-based membrane thinner than 3 μm. It has been shown that in this configuration a 1.4 μm thick membrane can withstand a differential pressure > 470 kPa.

In another work [58], the authors studied the influence of CO and CO_2 (components always present in reforming reactions) on the permeation of hydrogen. Although concentration polarization is not a problem, at low temperatures CO preferentially absorbs on the Pd surface (as found for all kind of configurations) depleting the hydrogen permeation rate.

Other studies on membrane micro-reactors were carried out for integrating different process steps in one micro-device for portable hydrogen production for fuel cell applications (see a.o. Ref. [60–62]).

In their work, Gervasio and co-workers designed and operated a membrane micro-reactor for generating pure hydrogen at ambient temperature. The system can generate hydrogen for a small fuel cell system. The interesting idea is based on the efficiency of the macro-reactor system (integrating feeding, reaction and hydrogen separation) coupled with an efficient hydrogen carrier. The carrier is a liquid aqueous alkaline borohydride solution, which reacts (completely) at room temperature over a Ru-based catalyst packed in the micro-reactor. One wall of the reactor is made of a Celgard membrane, which is impermeable for the liquid solution but permeable for gaseous hydrogen.

Another integration study has been carried out by Kim and co-workers who prepared an integrated catalytic structured mini-channel network covered by a Pd-based membrane for hydrogen recovery. This first step work led to an hydrogen separator made of defect-free Pd layer. However, this is a step further towards the use of such a system as integrated micro-reformer and separator.

A same approach has been used by Karnik and co-workers who successfully produced Pd-based membrane via micro-fabrication and used it for hydrogen separation. The authors were able to design and fabricate a micro-reactor that was used for hydrogen separation; however, the support (or a part of it) for the micro-membrane is made of copper, which is also active catalyst for WGS reaction. The reactor can be thus used for on-board hydrogen production for micro-fuel cell applications.

A different approach is used by Kudo *et al.* [62] where the micro-reactor applied consists of an array of up to 15 stages and the membrane is used as a catalytic membrane in order to achieve hydrogen production through methanol reforming reaction. The particular gas profiles in the micro-channels due to the laminar flow are responsible, according to the authors, for the higher methanol conversion and higher hydrogen yield compared to a traditional tubular reactor system.

Finally, Alfadhel and Kothare [63] modelled a membrane micro-reactor for WGS reaction with a simplified (1D, isothermal) model, suggesting the limitation of the simple model in simulating a membrane micro-reactor.

As mentioned, membrane micro-reactors are very interesting systems to be studied in case external mass transfer limitations could not be ignored (such as for high-flux Pd-based membranes) and when different steps need to be coupled (coupling exothermic and endothermic reaction is an example). However, more research is required before being able to design an optimum membrane micro-reactor.

3.4   Conclusions

MRs, integrating reaction and product separation in a single step, can be effectively used for conversion of fossil fuels into hydrogen with CO_2 capture. As such, these reactors can be used for pre-combustion CO_2 capture. In this chapter, several MR configurations were explained ranging from micro-reactors to hollow fibre reactors to fluidized bed reactors. Among the different reactor configurations, the fluidized bed reactors seem to have most advantages over packed beds (in terms of reduced mass transfer resistances and uniform temperature profiles) and over micro-reactors (allowing enough catalyst and a simpler construction for higher throughput). Even combination of MRs in fluidized bed configuration with chemical looping process can improve the hydrogen production and help the CO_2 capture process while decreasing the amount of membranes required.

It should be noted that the amount of membranes required for CO_2 capture is the most important parameter. Indeed, in our recent paper [64], we reported a power plant of "only" 1 GWe net electricity production by utilizing membranes with the best reported performance (but not integrated in the reactor), a relatively large ($\sim$0.7%) amount of palladium is required compared to the total world supply. This means that the amount of membranes should be reduced to the minimum or a different membrane material using much less Pd should be produced and ways to completely recover the Pd should be implemented, to make the full-scale pre-combustion capture via MRs possible.

3.5   Note for the reader

This chapter is based on previous published papers from which the information was just updated to May 2014. An interested reader can find even more information in the following papers:

Gallucci *et al.* [6, 7], Gallucci *et al.* [65], Medrano *et al.* [55]

References

1. IPCC – Chapter 2. http://www.ipcc.ch/pdf/special-reports/srccs/srccs_chapter2.pdf
2. IPCC – Chapter 3. http://www.ipcc.ch/pdf/special-reports/srccs/srccs_chapter3.pdf
3. IPCC report http://www.ipcc.ch/
4. CATO2 project report http://www.co$_2$-cato.nl/
5. Merkel T.C., Zhou M., Thomas S., Lin H. and Serbanescu A. (2010), Novel Polymer Membrane Process for Pre-Combustion CO_2 Capture From Coal-Fired Syngas, AIChE spring meeting 2010, San Antonio TX, paper 60b
6. Gallucci, F., Basile, A., Iulianelli, A. and Kuipers, J.A.M. (2009a) A review on patents for hydrogen production using membrane reactors. *Recent Patents on Chemical Engineering*, **2** (3), 207–222.
7. Gallucci, F., Van Sint Annaland, M. and Kuipers, J.A.M. (2009b) Autothermal reforming of methane with integrated CO_2 capture in novel fluidized bed membrane reactors. *Asia-Pacific Journal of Chemical Engineering*, **4** (3), 334–344.
8. Buhre, B.J.P., Elliott, L.K., Sheng, C.D. *et al.* (2005) Oxy-fuel combustion technology for coal-fired power generation. *Progress in Energy and Combustion Science*, **31**, 283–307.
9. Anderson, K. and Johnsson, F. (2006) Process evaluation of an 865 MWe lignite fired O2/CO_2 power plant. *Energy Conversion and Management*, **47**, 3487–3498.
10. Tan, X., Li, K., Thursfield, A. and Metcalfe, I.S. (2008) Oxyfuel combustion using a catalytic ceramic membrane reactor. *Catalysis Today*, **131** (1–4), 292–304.
11. Basile, A. and Gallucci, F. (2009) Ultra-pure hydrogen production in membrane reactors, in *Chapter 3 in Handbook of Exergy, Hydrogen Energy and Hydropower Research* (eds G. Pélissier and A. Calvet), Nova Science Pub, New York. ISBN: 978-1-60741-715-6.
12. Gryaznov, V.M., Polyakova, V.P., Savitskii, E.M. *et al.* (1970) Influence of the nature and amount of the second component of binary-palladium alloys on their catalytic activity with respect to the dehydrogenation of cyclohexane. *Bulletin of the Academy of Sciences of the USSR, Division of Chemical Science*, **19** (11), 2368–2371.
13. National Academy Of Engineering (2004) *The Hydrogen Economy Opportunities, Costs, Barriers*, And R&D Needs, The National Academies Press, Washington DC.
14. Rifkin, J. (2002) *The Hydrogen Economy*, Jeremy P. Tarcher. ISBN: 1-58542-193-6.

15. Gallucci, F., Comite, A., Capannelli, G. and Basile, A. (2006) Steam reforming of methane in a membrane reactor: An industrial case study. *Industrial & Engineering Chemistry Research*, **45** (9), 2994–3000.

16. Matsumura, Y. and Tong, J. (2008) Methane steam reforming in hydrogen-permeable membrane reactor for pure hydrogen production. *Topics in Catalysis*, **51** (1–4), 123–132.

17. Kikuchi, E., Kawabe, S. and Matsukata, M. (2008) Steam reforming of methanol on Ni/Al2O3 catalyst in a pd-membrane reactor. *Journal of the Japan Petroleum Institute*, **46** (3), 93–98.

18. Tosti, S., Basile, A., Borelli, R. *et al.* (2009) Ethanol steam reforming kinetics of a Pd-Ag membrane reactor. *International Journal of Hydrogen Energy*, **34** (11), 4747–4754.

19. Simakov, D.S.A. and Sheintuch, M. (2009) Demonstration of a scaled-down autothermal membrane methane reformer for hydrogen generation. *International Journal of Hydrogen Energy*, **34**, 8866–8876.

20. Tan, X. and Li, K. (2009) Design of mixed conducting ceramic membranes/reactors for the partial oxidation of methane to syngas. *AIChE Journal*, **55** (10), 2675–2685.

21. Jin, W., Gu, X., Li, S. *et al.* (2000) Experimental and simulation study on a catalyst packed tubular dense membrane reactor for partial oxidation of methane to syngas. *Chemical Engineering and Science*, **55** (14), 2617–2625.

22. Gallucci, F., De Falco, M., Tosti, S., Marrelli, L. and Basile, A. (2008) Co-current and counter-current configurations for ethanol steam reforming in a dense Pd–Ag membrane reactor, *International Journal of Hydrogen Energy*, **33**, 21, 6165–6171, ISSN 0360-3199.

23. Smit, J., Bekink, G.J., Van Sint Annaland, M. and Kuipers, J.A.M. (2005) A reverse flow catalytic membrane reactor for the production of syngas: An experimental study. *International Journal of Chemical Reactor Engineering*, **3**, A12, 1–11.

24. Buxbaum, R.E. (2002): Patent US20026461408.

25. Tosti, S., Basile, A., Bettinali, L. *et al.* (2008) Design and process study of Pd membrane reactors. *International Journal of Hydrogen Energy*, **33** (19), 5098–5105.

26. Kleinert, A., Feldhoff, A., Schiestel, T. and Caro, J. (2006) Novel hollow fibre membrane reactor for the partial oxidation of methane. *Catalysis Today*, **118**, 44–51.

27. Mejdell, A.L., Jøndahl, M., Peters, T.A. *et al.* (2009a) Experimental investigation of a microchannel membrane configuration with a 1.4 µm Pd/Ag23 wt.% membrane-Effects of flow and pressure. *Journal of Membrane Science*, **327** (1–2), 6–10.

28. Tiemersma, T.P., Patil, C.S., Sint Annaland, M.V. and Kuipers, J.A.M. (2006) Modelling of packed bed membrane reactors for autothermal production of ultrapure hydrogen. *Chemical Engineering and Science*, **61** (5), 1602–1616.

29. Gallucci, F., Van Sint Annaland, M. and Kuipers, J.A.M. (2010a) Pure hydrogen production via autothermal reforming of ethanol in a fluidized bed membrane reactor. *International Journal of Hydrogen Energy*, **35**, 1659–1668.

30. Gallucci, F., Van Sint Annaland, M. and Kuipers, J.A.M. (2010b) Theoretical comparison of packed bed and fluidized bed membrane reactors for methane reforming. *International Journal of Hydrogen Energy*, **35**, 7142–7150.

31. Peters, T.A., Stange, M., Klette, H. and Bredesen, R. (2008) High pressure performance of thin Pd-23%Ag/stainless steel composite membranes in water gas shift gas

mixtures; influence of dilution, mass transfer and surface effects on the hydrogen flux. *Journal of Membrane Science*, **316** (1–2), 119–127.

32. Caravella, A., Barbieri, G. and Drioli, E. (2009) Concentration polarization analysis in self-supported Pd-based membranes. *Separation and Purification Technology*, **66** (3), 613–624.

33. Boon, J., Pieterse, J.A.Z., Dijkstra, J.W. and van Sint Annaland, M. (2012) Modelling and systematic experimental investigation of mass transfer in supported palladium-based membrane separators. *International Journal of Greenhouse Gas Control*, **11** (SUPPL), 122–129.

34. Deshmukh, S.A.R.K., Heinrich, S., Mörl, L. *et al.* (2007) Membrane assisted fluidized bed reactors: Potentials and hurdles. *Chemical Engineering and Science*, **62**, 416–436.

35. Rahimpour, M.R. and Elekaei, H. (2009) Enhancement of methanol production in a novel fluidized-bed hydrogen-permselective membrane reactor in the presence of catalyst deactivation. *International Journal of Hydrogen Energy*, **34** (5), 2208–2223.

36. Rahimpour, M.R. and Lotfinejad, M. (2008) Co-current and countercurrent configurations for a membrane dual type methanol reactor. *Chemical Engineering & Technology*, **31** (1), 38–57.

37. Rahimpour, M.R., Bayat, M. and Rahmani, F. (2010) Enhancement of methanol production in a novel cascading fluidized-bed hydrogen permselective membrane methanol reactor. *Chemical Engineering Journal*, **157** (2–3), 520–529.

38. Adris, A.M., Elnashaie, S.S.E.H. and Hughes, R. (1991) A fluidized bed reactor for steam reforming of methane. *Canadian Journal of Chemical Engineering*, **69**, 1061.

39. Prasad, P. and Elnashaie, S.S.E.H. (2002) Novel circulating fluidized-bed membrane reformer for the efficient production of ultraclean fuels from hydrocarbons. *Industrial & Engineering Chemistry Research*, **41**, 6518–6527.

40. Chen, Z., Yan, Y. and Elnashaie, S.S.E.H. (2003) Novel circulating fast fluidized bed membrane reformer for efficient production of hydrogen from steam reforming of methane. *Chemical Engineering and Science*, **58** (19), 4335–4349.

41. Khademi, M.H., Jahanmiri, A. and Rahimpour, M.R. (2009) A novel configuration for hydrogen production from coupling of methanol and benzene synthesis in a hydrogen-permselective membrane reactor. *International Journal of Hydrogen Energy*, **34** (12), 5091–5107.

42. Grace, J.R., Li, X. and Lim, C.J. (2001) Equilibrium modelling of catalytic steam reforming of methane in membrane reactors with oxygen addition. *Catalysis Today*, **64**, 141–149.

43. Prasad, P. and Elnashaie, S.S.E.H. (2003) Coupled steam and oxidative reforming for hydrogen production in a novel membrane circulating fluidized-bed reformer. *Industrial & Engineering Chemistry Research*, **42**, 4715–4722.

44. Patil, C.S., van Sint Annaland, M. and Kuipers, J.A.M. (2005) Design of a novel autothermal membrane-assisted fluidized-bed reactor for the production of ultrapure hydrogen from methane. *Industrial & Engineering Chemistry Research*, **44** (12), 9502–9512.

45. Adris, A.M., Grace, J.R., Lim, C.J., Elnashaie, S.S., (1994). Fluidized bed reaction system for steam/hydrocarbon gas reforming to produce hydrogen. US Patent number 5326550.

46. Deshmukh, S.A.R.K., Laverman, J.A., Cents, A.H.G. *et al.* (2005a) Development of a membrane assisted fluidized bed reactor. 1: Gas phase back-mixing and bubble to emulsion mass transfer using tracer injection and ultrasound. *Industrial & Engineering Chemistry Research*, **44** (16), 5955–5965.
47. Deshmukh, S.A.R.K., Laverman, J.A., van Sint Annaland, M. and Kuipers, J.A.M. (2005b) Development of a membrane assisted fluidized bed reactor. 2. Experimental demonstration and modeling for partial oxidation of methanol. *Industrial & Engineering Chemistry Research*, **44** (16), 5966–5976.
48. Roy, S., Pruden, B.B., Adris, A.M. *et al.* (1999) Fluidized bed steam methane reforming with oxygen input. *Chemical Engineering and Science*, **54**, 2095–2102.
49. Chen, Z., Grace, J.R., Lim, C.J. and Li, A. (2007) Experimental studies of pure hydrogen production in a commercialized fluidized-bed membrane reactor with SMR and ATR catalysts. *International Journal of Hydrogen Energy*, **32**, 2359–2366.
50. Mahecha-Botero, A., Boyd, T., Gulamhusein, A. *et al.* (2008) Pure hydrogen generation in a fluidized-bed membrane reactor: Experimental findings. *Chemical Engineering and Science*, **63**, 2752–2762.
51. Rakib, M.A., Grace, J.R., Lim, C.J. and Elnashaie, S.S.E.H. (2011) Modeling of a fluidized bed membrane reactor for hydrogen production by steam reforming of hydrocarbons. *Industrial & Engineering Chemistry Research*, **50** (6), 3110–3129.
52. Patil, C.S., van Sint Annaland, M. and Kuipers, J.A.M. (2007) Fluidised bed membrane reactor for ultrapure hydrogen production via methane steam reforming: Experimental demonstration and model validation. *Chemical Engineering and Science*, **62** (11), 2989–3007.
53. Gallucci, F., Van Sint Annaland, M. and Kuipers, J.A.M. (2008a) Autothermal reforming of methane in a novel fluidized bed membrane reactor. Part 1: Experimental demonstration. *Topics in Catalysis*, **51**, 133–145.
54. Gallucci, F., Van Sint Annaland, M. and Kuipers, J.A.M. (2008b) Autothermal reforming of methane in a novel fluidized bed membrane reactor. Part 2: Comparison of reactor configurations. *Topics in Catalysis*, **51**, 146–157.
55. Medrano, J.A., Spallina, V., van Sint Annaland, M. and Gallucci, F. (2014) Thermodynamic analysis of a membrane-assisted chemical looping reforming reactor concept for combined H2 production and CO_2 capture. *International Journal of Hydrogen Energy*, **39**, 4725–4738.
56. DOE/NETL (2010). Assessment of Hydrogen Production with CO_2 Capture Volume 1: Baseline State-of-the-Art Plants.
57. Mejdell, A.L., Peters, T.A., Stange, M. *et al.* (2009b) Performance and application of thin Pd-alloy hydrogen separation membranes in different configurations. *Journal of the Taiwan Institute of Chemical Engineers*, **40**, 253–259.
58. Mejdell, A.L., Jøndahl, M., Peters, T.A. *et al.* (2009c) Effects of CO and CO_2 on hydrogen permeation through a 3 μm Pd/Ag 23 wt.% membrane employed in a microchannel membrane configuration. *Separation and Purification Technology*, **68**, 178–184.
59. Boeltken, T., Belimov, M., Pfeifer, P. *et al.* (2013) Fabrication and testing of a planar microstructured concept module with integrated palladium membranes. *Chemical Engineering and Processing: Process Intensification*, **67**, 136–147.
60. Gervasio, D., Tasic, S. and Zenhausern, F. (2005) Room temperature micro-hydrogengenerator. *Journal of Power Sources*, **149** (1–2), 15–21.

61. Kim, D., Kellogg, A., Livaich, E. and Wilhite, B.A. (2009) Towards an integrated ceramic micro-membrane network: Electroless-plated palladium membranes in cordierite supports. *Journal of Membrane Science*, **340**, 109–116.
62. Kudo, S., Maki, T., Kitao, N. and Mae, K. (2009) Efficient hydrogen production from methanol by combining micro channel with carbon membrane catalyst loaded with Cu/Zn. *Journal of Chemical Engineering of Japan*, **42** (9), 680–686.
63. Alfadhel, K. and Kothare, M.V. (2005) Modeling of multicomponent concentration profiles in membrane microreactors. *Industrial & Engineering Chemistry Research*, **44** (26), 9794–9804.
64. Helmi, A., Gallucci, F. and Van Sint Annaland, M. (2014) Resource scarcity in palladium membrane applications for carbon capture in integrated gasification combined cycle units. *International Journal of Hydrogen Energy*, **39** (20), 10498–10506.
65. Gallucci, F., Fernandez, E., Corengia, P. and van Sint Annaland, M. (2013) Recent advances on membranes and membrane reactors for hydrogen production. *Chemical Engineering and Science*, **92** (5), 40–66.

4

Oxy Fuel Combustion Power Production Using High Temperature O₂ Membranes

Vesna Middelkoop and Bart Michielsen
Flemish Institute for Technological Research, Materials Technology, Mol, Belgium

4.1 Introduction

Since the turn of the millennium, there has been growing interest in oxygen permeation membranes due to their potential for use in industrial processes and commercial applications such as gas separation, commercial chemical reactions and solid oxide fuel cells. This chapter provides a brief review of the current state of research and development of oxygen permeation membranes that could play an important role in more efficient oxy-combustion processes.

Oxygen permeation membranes (hereinafter abbreviated to OPM), also referred to as ion transport membranes (ITM), oxygen transport membranes (OTM) or oxygen-ion conducting membranes, are oxygen-permeable ceramic membranes that are made from mixed ionic and electronic conducting (MIEC) metal oxides. It is a versatile group of materials that, when fabricated into membranes, can be successfully employed in oxy-fuel combustion power production and thereby facilitate large-scale oxygen production from air – an energy and cost-saving alternative to the conventional cryogenic air separation. When they are subjected to sufficiently high temperatures and oxygen chemical potential gradient, these materials exhibit unique properties that can be exploited to continuously supply pure oxygen. In light of pressing energy and environmental concerns, their main advantage lies in providing pure membrane-supplied oxygen in a single step with no additional energy required.

Process Intensification for Sustainable Energy Conversion, First Edition.
Edited by Fausto Gallucci and Martin van Sint Annaland.
© 2015 John Wiley & Sons, Ltd. Published 2015 by John Wiley & Sons, Ltd.

Thus far, attempts at scaling up to full industrial commercialisation have been hindered by chemical and mechanical stability issues, membrane life span, high material manufacturing costs and compatibility and integration issues with large-scale components. All of these key aspects will be presented herein, focusing on material and membrane properties, their fabrication, characterisation and optimisation including application and integration with power generation units and energy conversion processes.

The fossil fuel power generation industry – a large source of greenhouse gas emissions – faces one of the most challenging environmental and energy issues [1]. To improve CO_2 capture at power plants, various separation techniques and innovative concepts are being pursued. There are three main technology streams for capturing CO_2 from fossil fuel combustion: post-combustion, pre-combustion and oxy-combustion. The latter's drawback is its cost. Up to 20% of the total value of power station production is used by cryogenic air separation units (ASU) that separate the oxygen from air [2]. The term *oxy-fuel* combustion was coined to describe oxygen-fired pulverised coal combustion. The oxy-combustion process involves the use of pure oxygen instead of air for burning fuel. One of the main benefits of oxy-fuel combustion is that it takes place in a virtually nitrogen-free environment and, therefore, the combustion of the input fuel results in a flue gas stream with high CO_2 concentration (80–98%) and steam. After the steam has been removed by condensation, an almost pure CO_2 exit gas is ready for compression and storage. However, as the combustion with pure oxygen generates very high temperatures, an additional limitation in the current design of the oxy-fuel combustion is the requirement to cool and recycle a portion of the captured CO_2 back into the system in order to maintain the temperature in the oxy-fuel combustion system. Emerging oxygen-permeable membrane technologies are expected to meet this challenge by directly integrating high-temperature ceramic membranes with the combustion system, delivering a solution to both of these issues: replace cryogenic air separation and improve process efficiency [3–5]. The block diagram in Figure 4.1 shows the basic principle of operation of an oxygen separation membrane process that is integrated with an oxy-fuel power process [6].

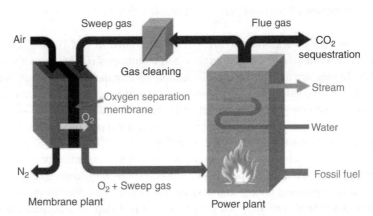

Figure 4.1   *Block diagrams illustrating a membrane-based oxy-fuel combustion system for CO_2 capture with ceramic membranes for oxygen separation from air integrated with the combustion process. (Source: Reproduced with permission from Ref. [6]. Copyright © 2011, The Royal Society of Chemistry)*

4.2   MIEC Perovskites as Oxygen Separation Membrane Materials for the Oxy-fuel Combustion Power Production

MIEC metal oxides for high-temperature ceramic membranes are a very versatile group of materials. By far the largest and most studied families of these compounds are the cubic perovskites of general formula ABO_3 [7]. In addition, a wide range of perovskite-like mixed metal oxides, known as the A_2BO_4 Ruddlesden–Popper family, have attracted considerable attention as membrane materials. Figure 4.2 illustrates both the structure of an ideal cubic ABO_3 perovskite and its layered structural analogue, an A_2BO_4 structure. These simplified formulas do not directly reflect the whole diversity of compositions that can be generated, from almost all the elements of the periodic table. MIEC metal oxides are usually formed from alkaline earth metals and lantanoides on the A site of the ABO_3 structure and first- and second-row transition metals on the B site of the ABO_3 structure. By doping with other suitable metal ions (A' or B') on one or both of their sites, these solid oxide perovskite structures may become richer in defects; as a result their oxygen transport behaviour will be improved. The general ABO_3 formula of the doped systems can be rewritten as $A_xA'_{1-x}B_yB'_{1-y}O_{3-\delta}$, wherein δ denotes the presence of oxygen vacancies within the oxygen sub-lattice. These complex formulas are commonly abbreviated to the first letters of the elements in the compound, often followed by the first significant figures of the stoichiometric composition. An example is $La_xSr_{1-x}Co_yFe_{1-y}O_{3-\delta}$, which allows a large variation in cation substitution, such as $La_{0.6}Sr_{0.4}Co_{0.2}Fe_{0.8}O_{3-\delta}$ (or LSCF6428), which is further discussed in this chapter.

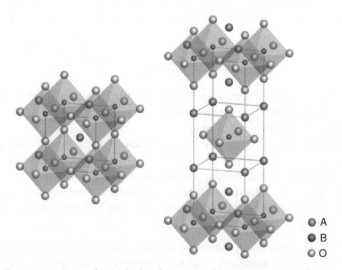

○ A
● B
○ O

Figure 4.2 *Representations of an ideal cubic perovskite structure, ABO_3 (a), showing an A-centred unit cell with corner-sharing BO_6 octahedra and a perovskite-related crystal structure, A_2BO_4 (b), with alternating stacking of perovskite-type layers consisting of BO_6 octahedra and an AO rock salt-like structure; in both representations the A cations are shown in mid grey, B cations in dark grey and BO_6 octahedra in light grey with oxygen atoms in light grey*

Potential applications of perovskite materials for oxygen separation membranes rely on their intrinsic ability to simultaneously and efficiently conduct oxygen ions and electrons (excluding other gas-phase species) through their crystal structure when exposed to elevated temperatures (typically above 700 °C) and different oxygen partial pressures (different gas streams) on either side of the membrane without the necessity of using electrodes and external electric circuits. Oxygen transport through the symmetric (dense) membranes is governed by two processes: oxygen bulk diffusion and surface exchange processes (at both feed side and permeate side) [8, 9]. The slower of these two affects the kinetics of the overall permeation and will make the other process less rate limiting. Generally, in order to improve the oxygen permeation flux, it is necessary to identify which of these steps is rate controlling and to tune membrane thickness. Figure 4.3 illustrates the oxygen transport mechanism (a) and permeation flux changes with membrane thickness (b). Regarding the membrane thickness, there are two situations where the critical thickness is typically in the range of several 100 μm to less than several 10 μm:

1. When the membrane thickness is greater than the critical membrane thickness, the oxygen bulk diffusion is rate limiting, in which case Wagner's transport equation [11] is applicable, and the oxygen flux through the membrane can be calculated as follows:

$$j(O_2) = \frac{V}{RT_V S} \times \frac{p(O_2)_{out} - p(O_2)_{in}}{1 - \frac{p(O_2)_{in}}{P}} \tag{4.1a}$$

where $j(O_2)$ is the oxygen permeation flux, V is the total gas flow rate, T_V is the temperature at a given gas flow rate, S is the membrane surface area, $p(O_2)_{in}$ and $p(O_2)_{out}$ are the inlet and outlet oxygen partial pressures at the high and low pO_2 side, respectively, P is the total pressure of the system and R is the universal gas constant. It should be noted that Eq. (4.1a) can also be derived in other ways such as Eq. (4.1b):

$$j(O_2) = -\frac{RT}{16F^2 L} \int_{\ln p(O_2)_{in}}^{\ln p(O_2)_{out}} t_{el} \sigma_{ion} d \ln P \tag{4.1b}$$

showing that the oxygen permeation flux varies linearly with the inverse of L, the membrane thickness, where t_{el} is the electronic transference number, σ_{ion} the oxygen ionic conductivity, F is the Faraday constant.

2. When the membrane thickness is less than the critical membrane thickness, the surface exchange kinetics are rate determining [12]. This can be improved by a surface modification technique such as coating the membrane surface with a porous MIEC layer (and so increasing the effective surface area of the membrane) or depositing a layer of catalytic material on the membrane surface, all of which, in turn, have the desired effect of increasing the oxygen permeation flux.

It should be noted that mixed rate control is often reported in the literature on MIEC for oxygen separation in three cases: when the membrane thickness is very close to its critical value, when the oxygen permeation flux values no longer seem to progress in a linear manner and when the oxygen permeation is co-controlled by both bulk diffusion and surface exchange processes (as shown in the illustration in Figure 4.3) [10]. To better

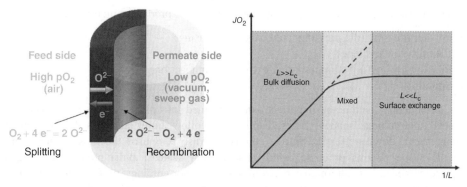

Figure 4.3 *Schematic of the counter-current oxygen transport mechanism of a symmetric (dense) MIEC membrane exploited for oxygen separation applications (a) and the change in oxygen permeation flux with the membrane thickness (b). (Source: Reproduced from Ref. [10], with permission from Elsevier)*

illustrate the oxygen transport across the membrane, the mechanism is described by the following three steps [13]:

(i)   Surface exchange reaction at the high $p(O_2)$ interface (at the outer membrane surface)
(ii) Simultaneous bulk diffusion of anionic oxygen (oxygen vacancy) and electrons (electron holes) in the bulk phase
(iii) Surface exchange reaction at the low $p(O_2)$ interface (at the inner membrane surface)

4.3   MIEC Membrane Fabrication

The different methods available to produce mixed-oxide ceramic powders are solid-state reactions and chemical synthesis methods such as coprecipitation, sol–gel processing, hydrothermal synthesis, alkoxide route, spray pyrolysis and spray and freeze drying. These powders can be produced on both laboratory and industrial scale. Among the powder preparation methods the most common and direct route for producing mixed oxides is by the solid-state reaction method from a mixture of metal oxides, hydroxides, carbonates or salts in the solid state at high temperatures [14]. This method typically results in the formation of agglomerates, a broad particle size distribution and poor homogeneity and purity. On the other hand, spray pyrolysis, as a less laborious and less energy-intensive up-to-date alternative to the conventional solid-state reaction method allows for the simple and fast production of homogeneous powders of high chemical purity with a better control of both the overall particle size distribution and degree of agglomeration, although still incurring high production costs [15, 16]. The preparation of the starting powders with the desirable characteristics is crucial to the subsequent production of as well as to the properties and performance of the membranes.

Most membranes fabrication routes generally involve the following steps: powder preparation, shape forming, calcination and high-temperature sintering and sometimes a back-end membrane processing step (such as surface modification).

The following are suitable shape-forming methods for producing membranes:

- Powder pressing (die pressing, cold isostatic pressing)
- Casting (slip and tape)
- Plastic forming (injection molding, extrusion, compaction)
- Spinning

Consequently, membrane materials can be formed into planar shapes (discs and flat sheets) and tubular shapes (tubes, capillaries, hollow fibres and possibly U-shaped geometries). Powder pressing into a compacted disc is one of the most popular and simple shape-forming processes used on laboratory scale. Tubular membrane designs have emerged as an alternative to overcome the shortcomings of the planar models associated with joining the components, pressure resistance and membrane longevity. Their surface-area-to-volume ratios and wall thicknesses are still insufficient, making them impracticable on a pilot plant scale or in large-scale industrial processes.

Since large-scale gas separation applications require high membrane surface/volume ratios, the reported oxygen permeation flux results suggest that capillary or hollow fibre geometry offers a distinct advantage over tubular and flat-sheet membranes. The production parameters (including the spinnerets design, the orifice dimensions, bore liquid flow rate and content, coagulation bath) during the spinning process can be altered to control both the diameter and wall thickness. The final membrane product is a capillary with a diameter normally in the range of 2–5 mm or a hollow fibre membrane with a diameter of normally up to 2.0 mm. The combined spinning and phase inversion followed by sintering is a well-established method of developing both symmetric and asymmetric MIEC ceramic capillary and hollow fibre membranes [17–20]. Furthermore, by fine-tuning the spinning suspension composition and the aforementioned spinning parameters for asymmetric hollow fibres, a range of novel morphologies (porous sponge-like structure and finger-like macrovoids) can be obtained [21, 22].

Figure 4.4 shows six consecutive steps in the preparation of capillaries and hollow fibres. The spinning precursor is prepared from a polymer solution (dope) containing a polymer binder, a solvent and an additive mixed with the perovskite powder that has been milled to achieve a fine particle size. In the starting polymer suspension, polyether sulphone (PESF), polysulfone (PSF), polyethene (PE), polyetherimide (PEI), polyethylene terephthalate (PET), polyurethane (PU) and polyetherimide cellulose acetate (CA) are commonly used as a polymer binder; *N*-methyl-2-pyrrolidone (NMP) or dimethyl sulfoxide (DMSO) as a solvent; water as a non-solvent and up to 1%, w/w polyvinyl pyrrolidinone (PVP) or glycerol as an additive to adjust the viscosity of the dope mixture. While the powder mixture is spun into a capillary or a hollow fibre, the spinning process also relies on the phase separation of the polymer solution and the mass transfer. Thereafter, the membranes are allowed to dry at room temperature. With the binder removed before the temperature reaches $\sim 300\,°C$, the calcined membrane is further sintered in the temperature range from $500\,°C$ up to $1350\,°C$. By altering the calcination and sintering parameters for different powders, it is possible to have a high degree of control over the final membrane microstructure (i.e. its density and grain size) and mechanical strength.

The effects of all the membrane preparation steps (from the powder particle morphology and size distribution, starting composition and membrane shape-forming process and the subsequent thermal treatment) on the structure and permeation behaviour of the resulting

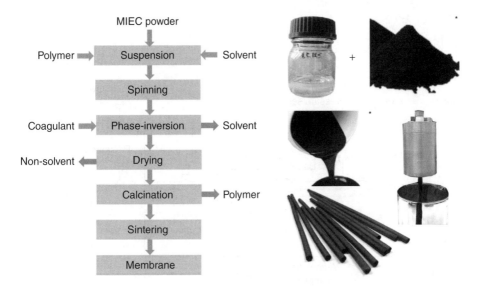

Figure 4.4 *Process diagram and illustration showing various sequences involved in the fabrication of MIEC capillary or hollow fibre membranes. (Source: Reproduced with permission from www.fuelcellmaterials.com)*

membranes have been extensively studied in the literature [23–26]. With all the essential characterisation tools for investigating structural, morphological and mechanical properties, pre-operation and post-operation (XRD, SEM/FESEM-EDS, EDX, XPS, dilatometry, etc.), the measurements of transmembrane oxygen fluxes are the key to studying the membrane performance.

4.4 High-temperature ceramic oxygen separation membrane system on laboratory scale

4.4.1 Oxygen permeation measurements and sealing dense MIEC ceramic membranes

Oxygen permeation measurements are performed in an experimental set-up that typically consists of a shell-and-tube arrangement with a membrane inside, a tubular furnace, air and inert gas (argon, helium) supply, stainless steel gas inlets/outlets and Swagelok fittings, gas flow controllers, gas flow metres and an online mass spectrometer or gas chromatographer. The membrane capillary (hollow fibre) is placed and sealed in a pair of gas-tight ceramic tubes through which the inert gas is supplied to the core (also called the lumen or permeate) side of the membrane. One of the tubes houses a thermocouple that can be positioned along the length of the membrane to provide an accurate temperature reading through the measurement. This module is enclosed in a quartz (or steel or mullite) tube in which the shell (feed) side of the membrane is exposed to air. The desired gas flow rates (usually in the range from 0 to 200 ml/min) are regulated by a mass flow controller before entering the

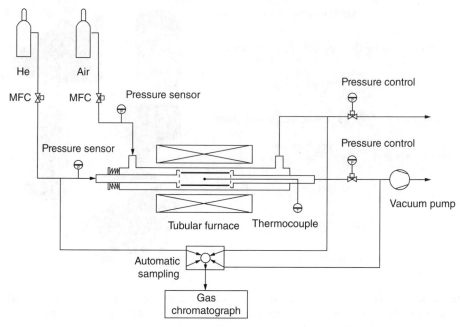

Figure 4.5 *Schematic diagram of a typical experimental apparatus to study oxygen permeation through hollow fibre membranes.* (Source: *Adapted from Ref. [27], with permission from Elsevier*)

lumen of the membrane. The total flow rate at the outlet of the membrane is measured by a gas flow metre, and the product gases that exit the tube are studied by the gas analysis instrument. A diagram of a typical apparatus layout is illustrated in Figure 4.5.

An important aspect to consider when setting up an oxygen permeation experiment is employing a reliable method of hermetically joining the membrane sample under study to gas supply tubes. Details of sealant use and application are, however, seldom reported in the literature. In general, two approaches are used for sealing membrane modules depending on where the joining points are situated in the heating zone: "cold" sealing (outside of the furnace) and high-temperature sealing (in the central isothermal hot zone of the furnace). The former requires longer membrane lengths, and it should be taken into account that oxygen fluxes vary along the length of the membranes as does the temperature profile along the furnace. In contrast, the latter approach ensures isothermal operation in which the seal needs to be susceptible to fracture and match the thermal expansion coefficient of the membrane and tubes in order to withstand high temperatures, numerous thermal cycles and relatively long duration of experiments. During the course of the experiment, both the sealant and the membrane can be susceptible to such effects, both non-equilibrium thermal and chemical stresses (due to oxidation and reduction) leading to premature failure of the material, fracture of the module and leaks. By far the most commonly used high-temperature resistant, gas-tight sealants are glass and glass–ceramic powders with a variety of compositions. The most common "cold" sealants are silicone sealants. The second most commonly used sealing materials are gold and silver. There is an alternative novel method that has also been devised for ceramic-to-ceramic and ceramic-to-metal joining and sealing of solid oxide fuel cell systems, called reactive active metal brazing, which utilises a molten oxide and a noble metal–based braze filler (e.g. Ag–CuO braze system) [28]. This method has recently been successfully applied on MIEC membrane systems operating at high temperatures [29]. On the whole, long-life sealing of a membrane is still a technical challenge, and almost none of the available sealants yet fully meet the long-term requirements of the membrane module design.

4.4.2 $Ba_xSr_{1-x}Co_{1-x}Fe_yO_{3-\delta}$ and $La_xSr_{1-x}Co_{1-y}Fe_yO_{3-\delta}$ Membranes

In the last ten years, much of the research focus has been directed towards the $Ba_xSr_{1-x}Co_{1-x}Fe_yO_{3-\delta}$ and $La_xSr_{1-x}Co_{1-y}Fe_yO_{3-\delta}$ perovskite families. They have been identified as the most promising among oxygen-permeable MIEC materials. The first part of this section outlines some of the main characteristics of the $Ba_xSr_{1-x}Co_{1-x}FeyO_{3-\delta}$ and $La_xSr_{1-x}Co_{1-y}Fe_yO_{3-\delta}$ type of membranes, focusing on how their oxygen permeation and stability may be suitable for integration into existing power plants. The second part of this section addresses the potential of A_2BO_4-based membranes, including a discussion of the materials' carbon dioxide tolerance if intended for use in an industrially relevant environment.

Following on from previous work on the parent perovskite-type $SrCoO_{3-\delta}$, a significant number of work has been carried out on the doped $A_xSr_{1-x}B_yCo_{1-y}O_{3-\delta}$ systems that were suggested as OTM [30, 31]. One of the early landmark studies on these materials was carried out by Teraoka *et al.* specifically on a series of $La_{1-x}Sr_xCo_{1-y}Fe_yO_{3-\delta}$ oxides made into discs of 1 mm thickness. The highest oxygen permeation rates of 2.4 and 3.38 ml(STP)/cm^2/min at the lowest onset temperature (~850 °C) were observed for the value of $x=1$ in $SrCo_{0.4}Fe_{0.6}O_{3-\delta}$ and $SrCo_{0.8}Fe_{0.2}O_{3-\delta}$, respectively [32, 33]. These higher permeation rates were explained by the higher content of Sr and in particular Co. Because all the La^{3+} sites were being substituted by Sr^{+2} and additional oxygen vacancies were formed, there was an increase in the oxygen flux (compared to other studied $La_{1-x}Sr_xCo_{1-y}Fe_yO_{3-\delta}$). Moreover, the Co^{3+} ion, having a smaller ionic radius and smaller bonding energy to other oxide ions compared to Fe^{3+}, has better oxygen permeating properties than Fe, although the presence of the latter is still necessary to preserve the perovskite-type structure at high Sr contents. However, the main disadvantage of $SrCo_{0.8}Fe_{0.2}O_{3-\delta}$ was its phase stability and mechanical properties, due to the small Sr^{+2} radius.

4.4.2.1 $Ba_xSr_{1-x}Co_{1-x}Fe_yO_{3-\delta}$ Membranes

The next landmark set of experiments in the development of the $A_xSr_{1-x}B_yCo_{1-y}O_{3-\delta}$ systems took place when Shao *et al.* replaced a fraction of Sr ions with divalent Ba ion in the parent compound $SrCo_{0.8}Fe_{0.2}O_{3-\delta}$, which yielded a material with higher structural stability [34]. Shao *et al.* proposed that the optimum Ba doping level for the $Ba_xSr_{1-x}Co_{0.8}Fe_{0.2}O_{3-\delta}$ series is between $0.3 \leq x \leq 0.5$. With the Ba doping strategy, the oxygen permeation of BSCF discs has also shown a slight increase in the permeation rates [35]. The highest oxygen permeation rates reported for $Ba_{0.5}Sr_{0.5}Co_{0.8}Fe_{0.2}O_{3-\delta}$ (BSCF5582) discs are similar to those obtained for a dense tubular BSCF5582 membrane (about 3.0 ml/cm^2/min) at a temperature of 900 °C [36, 37]. The latter suggested the tubular membranes prepared by the plastic extrusion method were suitable for industrial applications with sufficient oxygen permeation fluxes at different oxygen partial pressures in the shell side and stability during 150 h of operation.

Liu *et al.* reported the first oxygen permeation through MIEC hollow fibre membranes. Their permeation measurements were undertaken on dense and gas-tight BSCF5582 hollow fibre membranes with a thickness of 0.22 mm and an effective length of 7 cm at a temperature range of 850–950 °C, by using helium as the sweep gas supplied at 18.8 to 217 ml/min through the hollow fibre lumen side. As noted by other researchers (see for example Wang *et al.* [37]), the oxygen flux increased with the increasing helium sweep rate due to the decrease in the oxygen pressure at the lumen side. The maximum oxygen flux measured at a flow of 217 ml/min and at 950 °C was 5.1 ml/min/cm^2 [38].

This was followed by several other studies carried out on BSCF5582 hollow fibres by different research groups in an attempt to produce hollow fibres with an improved performance and long-term operation. Leo *et al.* investigated the effect of sulphur-free binders on the oxygen permeation performance of BSCF5582 hollow fibres. In all previous studies, sulphur-containing binders (mainly PESF or PSF) were used in the production of hollow fibres to retain better control of the spinning process [17, 39, 40]. One obvious disadvantage of the use of sulphur-containing polymers is that they act as a source of sulphur contaminants, which affect the oxygen ionic diffusion by forming non-ionic domains in the form of barium and strontium sulphates and small amount of cobalt oxide. Leo *et al.* demonstrated that BSCF5582 hollow fibres prepared with sulphur-free binders (PU and PEI in their case) exhibited an increase of more than 100% in oxygen permeation fluxes compared to the previous values obtained for the BSCF-PESF hollow fibres by Liu *et al.* [38]. At 950 °C, a flux of 9.5 ml/min/cm^2 was observed for the BSCF-PEI–based hollow fibre. Flux values of the BSCF-PEI-Pd hollow fibre, which were catalytically modified with palladium, increased even further reaching 14.5 ml/min/cm^2 at a flow rate of 150 ml/min and with a thickness of ~0.32 mm. Furthermore, the BSCF-PEI-Pd membrane showed a significant enhancement in oxygen permeations at 700 °C with a flux of 4.1 ml/min/cm^2, which in effect indicates that the operating temperatures can be lowered (by appropriately 250 °C) and energy demand reduced (Figure 4.6) [41].

In the most recent study on BSCF5582, Han *et al.* reported the development of a novel membrane morphology based on PEI polymer binder and consisting of one dense layer and one porous layer that has provided markedly higher rates of oxygen permeation (and

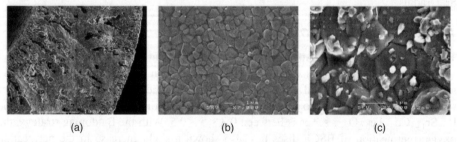

(a)   (b)   (c)

Figure 4.6 *SEM micrographs of BSCF hollow fibre membranes prepared using sulphur-free binder (PEI) (a) cross section showing a wall thickness of 0.32 mm, (b) outer surface showing BSCF grains and (c) outer surface modified with Pd nanoparticles. (Source: Reproduced from Ref. [41], with permission from Elseiver)*

thereby the highest reported oxygen permeation fluxes for an unmodified hollow fibre in the literature to date) of 11.46 ml/min/cm^2 at $950\,^\circ$C, at a helium sweep of 200 ml/min and through a wall thickness of 0.3 mm [42].

Relatively high oxygen permeation fluxes for BSCF5582 membranes have recently been achieved by several research groups showing the effect of an increase in sweep gas flow rate and surface modification [43–45]. The permeation flux increases with the increase in the sweep gas flow rate, which effectively reduces flow-dependent concentration polarisation and increases the pressure driving force across the membrane (and thereby the surface exchange rates). A recent study by Baumann *et al.* highlights the noteworthy influence of these factors on the permeation performance if the membrane geometry is planar. The dense BSCF5582 membrane layer and support were prepared by a tape casting process. A thin porous BSCF activation layer was deposited on the top membrane surface by the screen printing process. When the most effective parameters were applied, namely, pure oxygen on the feed side and an argon sweep rate of 400 min/cm^2 on the permeate side at a temperature of $1000\,^\circ$C, the highest oxygen flux obtained (ever reported to date) was 67.7 ml/cm^2/min through the coated and supported BSCF thin film (in the form of a $70\,\mu$m thick disc) and 31.8 ml/cm^2/min through the non-coated membrane (of the same thickness) (Figure 4.10). When only synthetic air was supplied to the feed side, the coated and supported BSCF thin film showed a flux of 12.2 ml/min/cm^2 at $1000\,^\circ$C. Although the optimisation issues remain pertinent, these results point out the potential of thin-film-supported perovskite membranes for use in the oxy-fuel processes [46].

The second recently developed membrane type with relatively high oxygen permeability (compared to that of BSCF5582 membranes) is its zirconia-containing analogue $BaCo_xFe_yZr_zO_{3-\delta}$ (BCFZ). Some of the first materials of the BCFZ perovskite family that were investigated, ZrO_2- or YSZ-doped $SrCo_{0.4}Fe_{0.6}O_{3-\delta}$ (i.e. $BaCo_{0.4}Fe_{0.6-x}Zr_xO_{3-\delta}$, $x = 0–0.4$), have shown oxygen permeation flux up to 0.9 ml/min/cm^2 through 1.0 mm thick discs, stable lattice structure and high mechanical strength [47, 48].

Schiestel *et al.* developed and reported a BCFZ hollow fibre membrane of an outer diameter of 0.88 mm, a thickness of 0.175 mm and an effective fibre length of 11.9 mm with an oxygen flux of 7.6 ml/min/cm^2 at $900\,^\circ$C and at a fixed air feed flow rate of 150 ml/min and a sweep flow rate of 30 ml/min of helium–neon gas mixture (29.5 and 0.5 ml/min, respectively) [49]. This was about twice the flux value reported by Liu and Gavalas [38]. The BCFZ fibre (the precise composition of which has not been disclosed by the manufacturer due to proprietary information) had a similar microstructure to that developed by Liu and Gavalas [38], a slightly reduced wall thickness ($\sim45\,\mu$m), and it was prepared in a single-step, while the preparation of the BSCF5582 hollow fibre of Liu and Gavalas [38] involved both spinning and coating. In the subsequent study, Tablet *et al.* with Caro *et al.* established that the same BCFZ hollow fibre membrane can deliver stable operation at $850\,^\circ$C over the course of 5 days and an extremely high flux of 6 ml/min/cm^2 for the same wall thickness and helium flow rate of 30 ml/min with no observable degradation, which confirms the stable phase structure of the hollow fibre membranes [50, 51]. In additional long-time tests in different gas atmospheres, the oxygen flux remained stable and the microstructure studies indicated the role of large grain size in increasing oxygen permeability. It was also suggested by Tablet *et al.* [50] that by increasing the air pressure on the feed side, oxygen permeation fluxes of BCFZ would be highly suitable for industrial use.

4.4.2.2 $La_xSr_{1-x}Co_{1-y}Fe_yO_{3-\delta}$ Membranes

In parallel to the development of the $Ba_xSr_{1-x}Co_{1-x}Fe_yO_{3-\delta}$ based membranes, which have been confirmed to have superior oxygen permeation performance, a significant research interest has been maintained in an ever-increasing number of La_xSr_{1-x} $Co_{1-y}Fe_yO_{3-\delta}$-based membranes for oxygen permeation, and, in particular, in $La_{0.6}Sr_{0.4}Co_{0.2}Fe_{0.8}O_{3-\delta}$ (LSCF6428) membranes that offer a high level of chemical and mechanical stability [52]. As one of the front runners in the development of LSCF6428 membranes, the group of Professor Kang Li at Imperial College had carried out extensive characterisation of hollow fibres in the past decade. The LSCF hollow fibres were first manufactured and tested for oxygen permeation by Tan *et al.* [20, 53]. The oxygen permeation fluxes obtained from these LSCF6428 hollow fibres, compared to that of conventional LSCF6428 disc [54] and tubular [55] membranes, was much higher, up to $0.8 \, ml/min/cm^2$ [56].

An SEM image of typical spun LSCF6428 hollow fibres (produced by Tan *et al.* [20]) shown in Figure 4.7 illustrates the effect of sintering at high temperatures (1100–1280 °C) on the hollow fibre membranes with an asymmetric structure. In the aforementioned measurements, Tan *et al.* [20] employed the gas-tight hollow fibres sintered at 1280 °C for 4 h.

Thursfield *et al.* [57] tested a LSCF6428 membrane reactor module consisting of four gas-tight hollow fibres (prepared by the group of Professor Kang Li at Imperial College), which were catalytically modified by applying a porous layer of platinum (with a thickness of ca. 0.2 μm) on the outer surface of each hollow fibre. The maximum oxygen flux achieved was $1 \, ml/min/cm^2$ (mass and non-mass transfer limited value) at 1000 °C at a constant helium sweep of 100 ml/min, which remained stable. The starting composition was retained although post-operation characterisation revealed localised changes in the perovskite cation stoichiometry with higher proportions of strontium and sulphur detected only on the membrane air-side surface. This indicates the presence of strontium sulphate, which may have played a role in the observed drop in oxygen flux in the long-term measurements (performed at 1000 and 750 °C), in particular after 4 days of operation. One surprising observation was that for a test hollow fibre, the outer surface of which was not coated, sulphur was detected at low levels on the air-side, on the permeate-side and within the membrane bulk. Possible sources of the sulphur contamination have been discussed. Tan *et al.* [56, 58] later reported on the improved performance of a similar, unmodified (non-coated) hollow fibre, a porous LSCF-modified membrane and an Ag-coated hollow fibre having flux rates of up to 0.81, 1.48 and $1.85 \, ml/min/cm^2$, respectively, in the temperature range of 700–1000 °C (see Table 4.1 for comparison).

There has been a number of oxygen permeation studies carried out on LSCF6428 [58, 60–62, 64–67], one of which by Han *et al.* [60] stands out as a cornerstone of recent research activity on this particular material. They compared the respective oxygen permeation fluxes from five different LSCF membranes: disc, conventional hollow fibre, modified hollow fibre and Ag- or Pt-deposited hollow fibre. The membranes have consecutively increasing fluxes of 0.10, 0.33, 0.84, 1.42 and $2.62 \, ml/min/cm^2$, respectively. These results demonstrate that a change from the planar design to a catalytically modified hollow fibre (including both thickness reduction and micro-structural difference) leads to a reduced bulk diffusion and surface reaction resistance enhancing the flux by 250%.

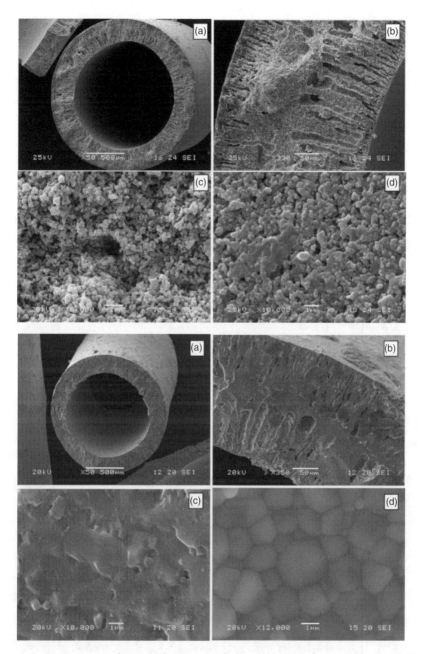

Figure 4.7 *SEM images of a typical LSCF6428 hollow fibre membrane sintered at 1100°C (top set of four images) and 1280°C (bottom set of four mages) displaying the sponge-like structure at the centre sandwiched between the finger-like structures. The latter set illustrates the more integrated membrane body with fainter porous structures after full densification and gas-tightness have been achieved at 1280°C. (Source: Reproduced with permission from Ref. [20]. Copyright © 2005, American Chemical Society)*

Table 4.1 Comparison of oxygen permeation performance of BSCF5582 membranes in different geometries with other most studied $A_xSr_{1-x}B_yCo_{1-y}O_{3-\delta}$ based perovskite membranes. All flux values and sweep gas flow rates are quoted at standard temperature and pressure (STP)

MIEC membrane material	Membrane type	Membrane wall thickness (mm)	Temperature (°C)	Flux jO_2 (ml(STP)/cm²/min)	Sweep gas flow rate (cm³/min)	Reference
SCF: $SrCo_{0.8}Fe_{0.2}O_{3-\delta}$ <	Flat sheet (5 cm × 5 cm)	0.2	1000	13	He, 600 ml/min	[59]
BSCF: $Sr_{0.5}Ba_{0.5}Co_{0.8}Fe_{0.2}O_{3-\delta}$	Flat disc	1.1	900	3.07	He, 100 ml/min	[36]
BSCF: $Ba_{0.5}Sr_{0.5}Co_{0.8}Fe_{0.2}O_{3-\delta}$	Coated supported BSCF thin film	0.07	1000	12.2	Ar, 400 ml/min, air feed	[46]
	Non-coated supported BSCF thin film			31.8	Ar 400 ml/minO₂ feed	
	Coated supported BSCF thin film			67.7	Ar 400 ml/minO₂ feed	
	Tube	1.7	900	3.0	He, 60 ml/min	[37]
	PESf-based hollow fibre	0.22	950	5.1	He, 217 ml/min	[38]
	PEI hollow fibre	320	950	9.5	Ar, 150 ml/min	[41]
	Pd-coated PEI hollow fibre			14.5		
	(Unmodified) PEI hollow fibre	300	950	11.5	He, 200 ml/min	[60]
BCFZ: $BaCo_{0.4}Fe_{0.4}Zr_{0.2}O_{3-\delta}$/ $BaCo_{0.4}Fe_{0.5}Zr_{0.1}O_{3-\delta}$/	Dense disc	1	950	0.9	He 30 cm³/min	[48]
BCFZ: $BaCo_xFe_yZr_{1-x-y}O_{3-\delta}$	Hollow fibre	0.175	900	7.6	He, 29.5 ml/ min + Ne, 0.5 ml/min	[49]

	Thickness	Temperature	O_2 flux	Sweep	Reference
BSCZF: $Ba_{0.5}Sr_{0.5}(Co_{0.0.3}xZr_{0.5})Fe_{0.2}O_{3-\delta}$					
Hollow fibre	0.3	950	4	He, 215 ml/min	[68]
LSCF: $La_{0.6}Sr_{0.4}Co_{0.2}Fe_{0.8}O_{3-\delta}$					
Tube	1.5	900	0.21	He, 43.5 ml/min	[55]
Hollow fibre	0.25	1000	0.65	He, 180 ml/min	[57]
Pt-coated hollow fibre			1.0		
Hollow fibre	0.3	1000	0.81	He, 50 ml/min	[56, 58]
Porous modified hollow fibre			1.48		
Ag-coated hollow fibre			1.85		
Hollow fibre	0.088 μm dense 0.299 μm porous	1000	2.19	He, 51.9 ml/min	[61]
Disc	1	900	0.10	He 100 ml/min	[60]
Hollow fibre	0.3		0.33		
Modified hollow fibre			0.84		
Ag-deposited modified hollow fibre			1.42		
Pt-deposited modified hollow fibre			2.62		
LSCF: $La_{0.2}Sr_{0.8}Co_{0.4}Fe_{0.6}O_{3-\delta}$					
Flat disc	1	870	0.7	He, 30 cm^3/min	[32]
LSCF: $La_{0.1}Sr_{0.9}Co_{0.9}Fe_{0.1}O_{3-\delta}$					
Flat dense disc	0.8	900	3.3	He, 200 cm^3/min	[62]
SCF: $SrCo_{0.4}Fe_{0.6}O_{3-\delta}$					
Flat disc	1	870	2.4	He, 30 cm^3/min	[32]
LSF: $La_{0.7}Sr_{0.3}FeO_{3-\delta}$					
Hollow fibre	0.22	1000	1.6	He, 100 cm^3/min	[63]

Table 4.1 provides a comparison overview of the maximal oxygen flux values as a function of temperature and sweep gas rate and membrane thickness for the most widely studied perovskite structure types for oxygen permeation membranes. The primary conclusion is that BSCF5582 can yield the highest oxygen flux throughout the temperature range. In addition, the table also includes results obtained from a few other well-performing perovskite materials belonging to the families of $Ba_xSr_{1-x}Co_{1-x}Fe_yO_{3-\delta}$ (such as $SrCo_{0.8}Fe_{0.2}O_{3-\delta}$ (SCF)) [59] and $Ba_{0.5}Sr_{0.5}(Co_{0.8x}Zr_x)Fe_{0.2}O_{3-\delta}$ (BSCZF) [68] and of $La_xSr_{1-x}Co_{1-y}Fe_yO_{3-\delta}$ (such as $La_{0.7}Sr_{0.3}FeO_{3-\delta}$ (LSF)) [63] and Nb-substituted $SrCoO_x$ oxides [69])

4.4.3   Chemical Stability of Perovskite Membranes Under Flue-Gas Conditions

Membrane materials for the two possible integration designs, which are discussed in more detail in Section 4.5.1, have to comply with several requirements regarding chemical stability. In the three-end design, the membrane material is only exposed to air. Therefore, many different materials can be used, and the two main design considerations are oxygen flux and cost of the materials. In the four-end design, the membrane comes in direct contact with flue gas. The composition of the flue gas is around 25–30 vol% water vapour, 70–75 vol% CO_2, 1–3 vol% O_2 and SO_2 content of about 400 ppm [70, 71]. Consequently, the membrane material has to be able to produce acceptable oxygen flux in the presence of these gases.

A material for which the chemical stability in the presence of CO_2 and SO_2 has been extensively studied is BSCF, as it is one of the membrane materials with the highest flux. The presence of CO_2 causes a reversible decline in oxygen flux. Even at concentrations of 500 ppm CO_2 (ambient air), a slight decrease in the oxygen permeation rate is observed [34]. This decrease is caused by the formation of carbonates of alkaline earth metals on the exposed surface of the membrane. The carbonate formation of perovskite materials $ABO_{3-\delta}$ in the presence of CO_2 can be expressed as follows [72]:

$$2ABO_{3-\delta} + 2CO_2 \leftrightarrow 2ACO_3 + B_2O_3 + ((1-\delta)/2)O_2$$

A is an alkaline earth metal. The formation of the carbonate occurs mainly through the bonding between the negatively charged oxide ions $O^{\delta-}$ of the perovskite and the positively charged carbon atom $C^{\delta+}$ of CO_2 [72, 73]. This process only affects the exposed surface up to a depth of 40–50 μm and is reversible. The flux can be easily recovered by sweeping with a non-poisoning sweep gas such as helium or argon.

The carbonate formation proceeds more rapidly as the CO_2 concentration increases, resulting in a declined oxygen flux. Engels *et al.* [70] studied the permeation flux through a BSCF tubular membrane as a function of the CO_2 concentration at 850 °C. They found that at a CO_2 concentration of 5% a decrease in flux of only 6% is observed. This decrease amounted to 20% at a CO_2 concentration of 10%. At a CO_2 concentration of 15%, a dramatic decrease of 80% was observed for the oxygen flux. When pure CO_2 was used as a sweep gas (shown in Figure 4.8), no appreciable oxygen flux was observed.

Temperature has an effect on the formation of carbonates too. For example, in Figure 4.8, a low oxygen flux can be observed for a BSCF membrane at 900 °C. The full recovery of the membrane material by the reversed reaction is faster at higher operating temperatures.

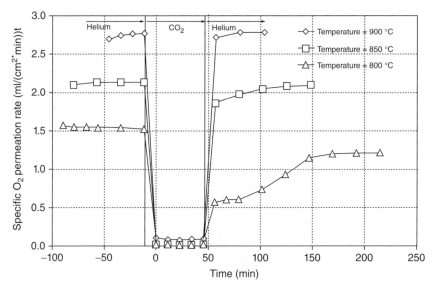

Figure 4.8 *Effect of CO$_2$ on the oxygen permeation rate of a BSCF membrane at 10 bar feed pressure and temperatures between 800 and 900°C. (Source: Reproduced from Ref. [70], with permission from Elsevier)*

At lower temperatures, the recovery is much slower and not always complete. At 800°C, the flux of the regenerated membrane lies significantly below that of the initial membrane.

The Ellingham diagram can be used to roughly estimate the performance of a perovskite membrane material in a CO$_2$-containing atmosphere at a certain temperature, as shown in Figure 4.9 [74]. In this diagram, the thermodynamic stability of the resulting carbonates is given for a certain temperature and CO$_2$ partial pressure. The dashed lines represent the chemical potential of CO$_2$ at various partial pressures. The solid lines represent the chemical potential of CO$_2$ during the decomposition of the corresponding carbonate. If the chemical potential (the solid line) is higher than the corresponding partial pressure (the dashed line) at a certain temperature, the carbonate is thermodynamically unstable. As can be seen in the Ellington diagram, BaCO$_3$ is much more stable than LaCO$_3$. This would indicate that, if the Barium in BSCF is replaced by Lanthanum, the material will be less likely to form carbonates and will show a better performance in a CO$_2$-containing environment.

Tan *et al.* investigated the effect of CO$_2$ in the sweep gas on the oxygen permeation behaviour of La$_{0.6}$Sr$_{0.4}$Co$_{0.8}$Fe$_{0.2}$O$_3$ hollow fibre membranes. The resulting fluxes at different temperatures and CO$_2$ concentrations are shown in Figure 4.10. The oxygen flux of the membranes exposed to CO$_2$ decreases but not as rapidly as the BSCF above it. At CO$_2$ concentrations of 20%, the flux is still half of that of the unexposed membrane. Even if pure CO$_2$ is used as flue gas, the oxygen flux does not vanish. So, LSCF is a better membrane material to operate under flue gas conditions than BSCF [75].

Several groups have tried to extend the chemical stability of MIEC membrane materials against CO$_2$ by using different approaches, some of which are discussed in the following sections.

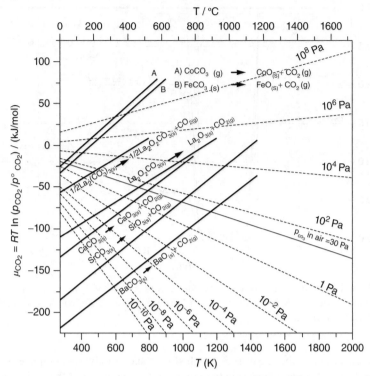

Figure 4.9   *The ellingham diagram for the decomposition of carbonates under different partial pressures. The dashed lines represent the chemical potential of CO_2 in the surrounding atmosphere for different partial pressures. $p°(CO_2) = 101.3\,kPa$ refers to standard conditions. (Source: Reproduced from Ref. [74], with permission from Elsevier)*

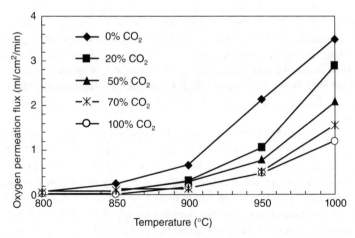

Figure 4.10   *Oxygen permeation behaviour of $La_{0.6}Sr_{0.4}Co_{0.8}Fe_{0.2}O_3$ hollow fibre membranes with highly concentrated CO_2 exposure. (Source: Reproduced from Ref. [75], with permission from Elsevier)*

Table 4.2   *Comparison of oxygen permeation performance of CO$_2$-tolerant MIEC materials*

MIEC membrane material	Membrane type	Flux (ml/cm^2/min)	Temperature (°C)	Reference
CTF: CaTi$_{1-x}$Fe$_x$O$_{3-\delta}$	Asymmetric sheet	0.45	1000	[80]
	Disc	0.02	900	[81]
Sr$_{0.5}$Ca$_{0.5}$Mn$_{0.8}$Fe$_{0.2}$O$_{3-\delta}$	Tube	0.13	900	[70]
La$_2$NiO$_{4+\delta}$	Flat sheet	0.46	950	[82]
Ba$_{0.6}$Sr$_{0.4}$Co$_{0.88}$Ti$_{0.12}$O$_{3-\delta}$	Disc	2.12	900	[79]
Ba$_{0.6}$Sr$_{0.4}$Co$_{0.7}$Ti$_{0.3}$O$_{3-\delta}$	Disc	1.46	900	[79]
BaCo$_{0.7}$Fe$_{0.2}$Ta$_{0.1}$O$_{3-\delta}$	Disc	2	900	[78]
(Pr$_{0.9}$La$_{0.1}$)$_2$ (Ni$_{0.74}$Cu$_{0.21}$Ga$_{0.05}$)O$_{4+\delta}$	Hollow fibre	0.45	900	[83]

4.4.4   CO$_2$-Tolerant MIEC Membranes

Most studies on the stability of MIEC membrane materials show that perovskites with a high-alkaline earth metal and cobalt content are more susceptible to CO$_2$ poisoning. Therefore, the development of CO$_2$-resistant membranes focuses on partly replacing this element by doping the material with Zr, La, Ti, Cr, Ga [76–79]. Some of these materials with their highest observed flux are listed in Table 4.2. The development of these chemically stable perovskites is rapidly speeding up, and a complete review of these materials is beyond the scope of this chapter. Instead, we discuss some distinct examples of both ABO$_3$ and A$_2$BO$_4$ structures.

4.4.4.1   CaTi$_{1-x}$Fe$_x$O$_{3-\delta}$ (CTF)

CaTi$_{1-x}$Fe$_x$O$_{3-\delta}$ materials are of great interest for membrane applications for a number of reasons. They have great chemical stability and an almost composition and temperature independent thermal expansion coefficient of 12×10^{-6} 1/K. In addition, its cost is much lower compared to that of BSCF and LSCF. However, a serious drawback is its poor oxygen transport capability. Its oxygen flux is about 0.02 ml/cm^2/min, around two orders of magnitude lower than that of BSCF.

To improve the oxygen flux without changing the material composition, there are two possible strategies: Either the surface of the membrane can be tuned to enhance the surface exchange or the membrane thickness can be decreased to avoid bulk diffusion. Both strategies have been applied to CaTi$_{0.8}$Fe$_{0.2}$O$_{3-\delta}$, the material composition with the highest flux value.

Figueiredo *et al.* improved the surface exchange of CaTi$_{0.8}$Fe$_{0.2}$O$_{3-\delta}$ membrane discs by increasing the surface with porous layers of the same material using screen printing. This led to a 55% increase in oxygen flux if the porous layer was situated on the permeate side and to a 70% increase when both sides of the membrane had this porous layer applied to it. They further improved the surface exchange properties by incorporating a small amount of Ag, a material with a higher exchange rate. This resulted in an improvement factor of 1.86 compared to the standard membrane [81].

Fontaine *et al.* prepared asymmetric CaTi$_{0.9}$Fe$_{0.1}$O$_{3-\delta}$ membranes by tape casting. They printed a 30 µm dense membrane layer on a 500 µm porous support. Different porosities of the support were obtained by using different filler materials or concentrations. It was found that these porosities had a large influence on the oxygen flux. The maximum flux achieved this way amounted to 0.45 ml/cm^2/min. This is a significant improvement compared to the flux of the 1 mm disc membranes, which amounts to 0.02 ml/cm^2/min [80].

4.4.4.2   K$_2$NiF$_4$ Materials

A relatively new class of materials are based on the K$_2$NiF$_4$ perovskite-related structure. Most of these materials with a Ruddlesden–Popper crystal structure show higher thermo-chemical stability than perovskites structures do. The performance of La$_2$NiO$_{4+\delta}$ and its doped versions for the separation of oxygen has been investigated by several groups [84]. The oxygen flux of this material is about 0.39 ml/cm^2/min at 900 °C when helium is used as a pure sweep gas. Upon switching the sweep gas to pure CO$_2$, the flux declines to a value of 0.22 ml/cm^2/min. The lower flux value is due to different oxygen surface exchange rates in different gas atmospheres. In addition, CO$_2$ can hinder the release of oxygen from the membrane surface as CO$_2$ molecules tend to absorb on this surface [82].

Engels *et al.* [70] investigated the performance of this membrane for more than 100 h by using CO$_2$ as sweep gas. The resulting flux data are shown in Figure 4.11. It can be seen that the flux declines only slightly over this long time period. In addition, after operation for this long period of time, no formation of carbonates is detected.

Another material with the same structure that has recently been developed is (Pr$_{0.9}$La$_{0.1}$)$_2$(Ni$_{0.74}$Cu$_{0.21}$Ga$_{0.05}$)O$_{4+\delta}$ (PLNCG) [71, 83, 85]. The flux of this material when pure helium is used as a sweep gas is similar to that of La$_2$NiO$_{4+\delta}$. But the decrease

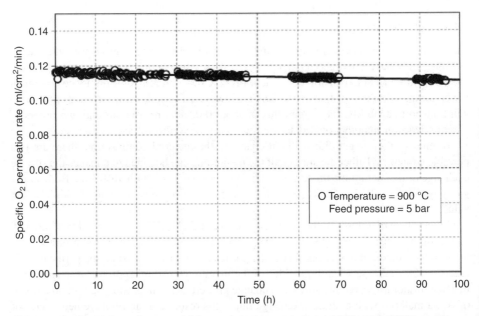

Figure 4.11   *La$_2$NiO$_{4+\delta}$ decrease of the permeation rate about 100 h under CO$_2$ sweep gas at 5 bar feed pressure and temperature of 900 °C. (Source: Reproduced from Ref. [70], with permission from Elsevier)*

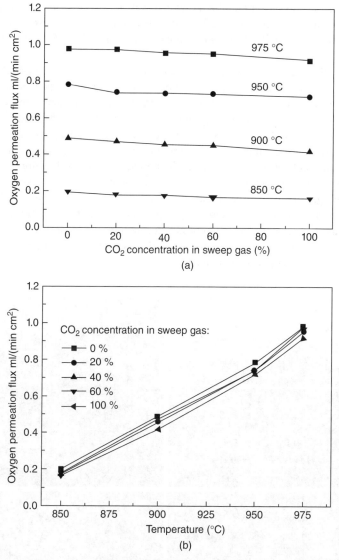

Figure 4.12 *Effects of CO$_2$ concentration in the sweep gas on the oxygen permeation fluxes through a PLNCG hollow fibre membrane at different temperatures. (Source: Reproduced from Ref. [85], with permission from Elsevier)*

in the flux upon changing the sweep gas to pure CO$_2$ is only 6%, which is much better. The effect of different operating temperatures and CO$_2$ concentrations in the sweep gas is given in Figure 4.12. It can be seen that the CO$_2$ concentration has almost no effect on the flux, which is only temperature dependant. Wei *et al.* further enhanced the oxygen permeation of a PLNCG hollow fibre membrane by applying a porous catalytic coating of La$_{0.8}$Sr$_{0.2}$CoO$_{3-\delta}$, which has better oxygen exchange properties. This coating was only applied to the outside of the membrane, which was not exposed to CO_2. It prevents the degradation of the catalytic coating. The coating significantly improved the oxygen flux of the membrane, especially at the lower temperature end. At 800 °C the flux improved by 3.5 times.

CO_2 is not the only problematic component in the flue gas of an oxy-fuel combustion power plant. If coals are used, SO_2 will also be present. Therefore, the membrane material are required to remain stable with at least 400 ppm SO_2. There are a number of recent studies into the effect of SO_2 on $La_2NiO_{4+\delta}$ and PLNCG. Engels *et al.* [70] studied the effect of 360 ppm SO_2 on $La_2NiO_{4+\delta}$. He observed a drastic decrease in the permeation rate and the formation of sulphur-containing reaction products on the surface. Wei *et al.* investigated the effect of different concentrations of SO_2 in the sweep gas on hollow fibre PLNCG membranes (Figure 4.13). A decline in the oxygen flux to almost zero was observed at 206 ppm of SO_2 in the gas stream. When it was exposed for only a short time, the flux could be completely recovered. After exposure, sulphates were found on the surface of the PLNCG membrane and the development of a porous morphology was observed on the membrane side that was exposed to SO_2. The formation of these sulphates can be reversed, but the formation of the porous part is permanent.

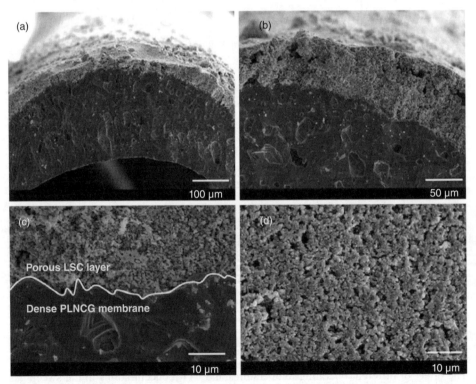

Figure 4.13 *SEM images of the PLNCG hollow fibre membranes after surface modification. (a–c) cross section of the LSC-coated membrane; (d) LSC porous layer coated on the outer surface of the PLNCG hollow fibre membrane. (Source: Reproduced from Ref. [85], with permission from Elsevier)*

4.5   Integration of High-Temperature O$_2$ Transport Membranes into Oxy-Fuel Process: Real World and Economic Feasibility

This section reviews the relevant literature on high-temperature oxygen-permeable membranes and what it suggests about their future role in oxy-fuel processes. Current projects looking into membrane integration into oxy-fuel combustion and CO$_2$ capture draw upon laboratory-scale research and development with the aim of identifying appropriate geometries and conditions to migrate to large-scale systems. As a result of numerous advantages over flat-sheet and tubular membranes, hollow fibre membranes continue to gather momentum.

4.5.1   Four-End and Three-End Integration Modes

There are two different designs for the integration of OTM membranes into the oxy-fuel combustion process: the three-end design and the four-end design. A schematic of each of these is shown in Figure 4.14. Each uses a different approach to obtain the necessary pressure difference across the membrane. In the four-end integration mode, flue gases comprising mainly CO$_2$ and H$_2$O are used to sweep away the permeated oxygen [87]. This results in low pO_2 pressure at the permeate side. In the three-end integration mode, a vacuum pump is used to extract the pure oxygen at the permeate side. In this design, contact between the membrane and flue gases is avoided. In contrast to conventional power plants,

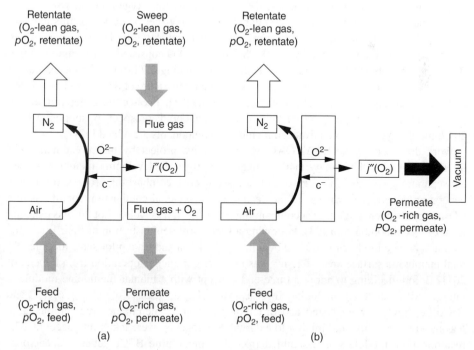

Figure 4.14   *Schematic of the four-end (a) and three-end designs (b). (Source: Reproduced from Ref. [86], with permission from Elsevier)*

the three-end design uses additional energy for the vacuum pump. In the four-end design, additional energy is used for the recirculation fan [86].

There are several studies that compare the efficiency of these designs [27, 88]. Stadler *et al.* [86] compared the three-end and four-end designs, assuming 90% CO_2 capture, with a conventional power plant. They found that the four-end design has the best overall energy efficiency at 40.7% compared to 40.1% for the three-end design. Also, in the four-end design, oxygen separation requires less membrane area, thus reducing capital investment. However, the four-end design requires membranes that can provide a sufficiently large oxygen flux under flue gas conditions. At the moment, a limited amount of materials are capable of operating under these conditions, and in most cases their oxygen conducting capacity is low. Sections 4.4.2–4.4.4 discussed the development of membrane materials with high enough oxygen flux under flue gas conditions, including additional information on some of their properties. As only the implementation of the three-end design is technically possible at this stage, this design is likely to be the one used initially. The existing membrane materials with a high oxygen flux such as BSCF and BCFZ would be the most suitable and readily available materials to use for this purpose.

4.5.2   Pilot-Scale Membrane Systems

Kriegel *et al.* from the Fraunhofer Institute for Ceramic Technologies and Systems (IKTS) are among the first researchers who have tried to take the OTM systems beyond laboratory tests. The IKTS group has been working within the Helmholtz alliance Mem-Brain project on developing dense ceramic MIEC membranes for oxygen/nitrogen separation that operate at 800–1000 °C and are suitable for the oxy-fuel process. For this application, BSCF5582 is considered the best reference material [89, 90]. The first stand-alone demonstration unit constructed by the IKTS group in 2009 consisted of 19 single-channel BSCF5582 membrane tubes and was capable of producing 2.7 L/min of pure oxygen. It worked in the "three-end" mode (using vacuum) at a temperature of 850 °C and completed 27 heating/cooling cycles in 1700 h of operation time [91]. The second-generation device (shown in Figure 4.15) produces 4 L/min of oxygen but holds prospects of supplying 15 to 20 L/min of oxygen if more advanced membrane components are added [92].

Work done within the OXYCOAL-AC cooperative project by the researchers from the RWTH Aachen University marks a major step towards developing membrane-based air separation systems for a zero CO_2-emission coal combustion process for power generation. Kneer *et al.* discussed the design under oxy-coal combustion conditions in a CO_2/O_2 atmosphere including accompanying burner design as well as hot flue gas cleaning from oxy-coal combustion [93]. Following a successful demonstration of their first 1-m² laboratory-scale module in 2008 (shown in Figure 4.16a and b), a pilot-scale module of a total membrane surface area of 15 m² was ready to commence operation from the end of 2011. Before deciding to adopt a three-end concept with a tubular membrane module, a range of MIEC materials including $Sr_{0.5}Ca_{0.5}Mn_{1-x}Fe_xO_{3-\delta}$, $Ba_{0.5}Sr_{0.5}Co_{1-x}Fe_xO_{3-\delta}$ and $BaCo_xFe_yZrzO_{3-\delta}$ were tested and different parameters (membrane area/volume ratio, leakage-free operation and joining to external components) were taken into account. The resulting pilot module was assembled from 570 monolithic BSCF tubes and enclosed into a vertical furnace with a combustion chamber of 2100 mm long and 400 mm inner diameter to run at 850 °C, air pressure of 15–20 bar on the feed site and a low vacuum of

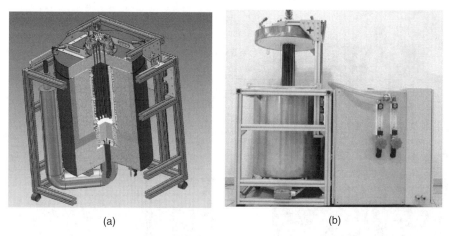

Figure 4.15 *(a) 3D CAD model and (b) a photo of a demonstration unit equipped with MIEC membranes for oxygen separation. (Source: Reproduced with permission from Ref. [92]. Copyright © 2012, Ralf Kriegel and Ingolf Voigt)*

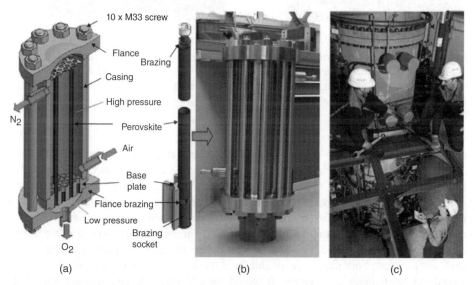

Figure 4.16 *120 kW oxy-coal pilot plant (amounting to 150 kg HT-steel) constructed by RWTH aachen university; 1 m² membrane module consisting of 42 perovskite tubes (with a length of 500 mm and diameter of 15 mm) housed in a heating chamber operating at $T_{max} = 850\,°C$ and $P_{max} = 20\,bar$. (Source: Reproduced from Ref. [94], with permission from Elsevier)*

about 0.8 bar on the permeate site. The pilot module was estimated to have the capacity to produce more than 300,000 L of pure oxygen per day as part of a 120 kW designed pilot plant (see Figure 4.16b) [94, 95].

As far as other MIEC membrane materials are concerned, Tan *et al.* developed a pilot-scale membrane system (shown in Figure 4.17) comprising 889 LSCF6428

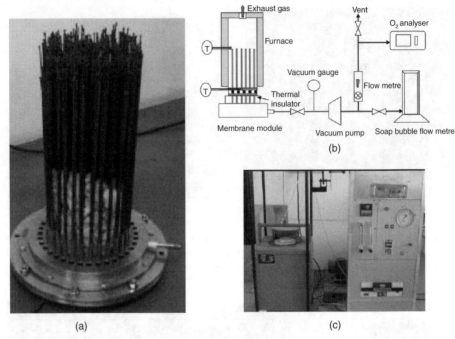

Figure 4.17 *Photos of a hollow fibre membrane module consisting of 129 bundles (889 fibres) propped up on a vacuum supply plate holder (a) and integrated in the system for oxygen production; flow chart (b) and photo of the system (c). (Source: Reproduced from Ref. [96], with permission from Elsevier)*

perovskite hollow fibres (28–32 cm in length and 0.3 mm thickness). The system produced over 3 L(STP)/min oxygen gas with a purity of 99.8% at the maximum operating temperature of 1070 °C. A Co–Ag layer was coated on the outer surface of all hollow fibres to enhance their permeation performance. The system was operated continuously for a period of 1067 h, at 960 °C and 97–98 kPa applied vacuum. The oxygen production rate reached 0.86 L/min with 99.4% oxygen purity. Through this investigation, the design and performance (separation, stability) as well as energy efficiency parameters have been analysed to assess both their experimental and theoretical effect on scaling up. For this module to be commercially viable, it would have to integrate heat exchangers within it to recover the heat energy in exhaust gas and oxygen products. To keep the overall production costs down (and bearing in mind that LSCF6428 produces modest fluxes in comparison to BSCF and BCFZ), either the total membrane surface area or the operating temperature has to be increased and the oxygen recovery limited to 20–40% [96].

4.5.3 Further Scale-Up of O_2 Production Systems

Besides the aforementioned projects, other key players worldwide are working on the possibilities of integrating inorganic membranes in power generation processes under the umbrella of different government-funded programmes such as the EC Seventh Framework

Programme, US DOE's Vision 21 R&D Program, the Australian COAL21, the Japanese NEDO programmes, the National High Technology Research and Development Programs ("863" program), the National Basic Research Program and National Basic Research Program of China.

The US Department of Energy has been working on improving ITM systems, which could be pursued as a potential large-scale, high-efficiency and low-cost oxygen production route. As a collaborative effort, under the auspices of the National Energy Technology Laboratory (NETL) of the Department of Energy (DOE), Air Products and Chemicals, Inc., along with their partners, have made impressive progress in ITM Oxygen technology and its integration with both conventional industrial processes and advanced power generation.

At the core of this technology are planar membrane wafers, made by ceramic tape casting and then stacked and joined into multi-wafer modules. These modules are then placed in serial and parallel arrays, inside a terminating end cap with a ceramic oxygen product pipe. In the latest commercial-scale version (described as Phase III), these modules are scaled up to produce up to 1 ton per day (TPD) of oxygen. All module components are constructed of the same ceramic material to ensure uniform thermal expansion and stress distribution. Large ITM Oxygen modules are arranged in the so-called ITM Oxygen vessel shown in Figure 4.18 [97]. It has already been demonstrated that Air Products and Chemicals' commercial-scale modules can produce up to 5 TPD for a number of operating conditions in a prototype facility. An ITM Oxygen system for the production of 100 TPD of oxygen integrated with a hot gas expander for the co-production of power has already entered the actual phase of development. This pilot unit, called the intermediate-scale test unit (ISTU) (a schematic of which is shown in Figure 4.19) started commissioning in early 2014 reaching full operating conditions with the initial loading of ITM modules under way in late 2014 [98]. In attempt to commercialise ITM Oxygen, Air Products and Chemicals and NETL are taking tangible steps towards a test platform for the design of larger

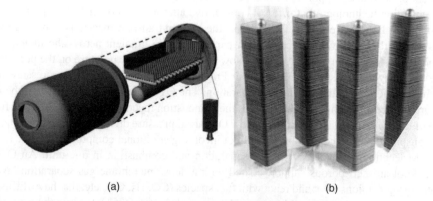

(a)   (b)

Figure 4.18   *Schematic of air products commercial ITM oxygen vessel (a) ...... multi-wafer modules (ITM module array shown only partially for clar...... assembled in a common flow duct and connected to an oxygen he...... manifold tubes. Four 1-TPD ITM oxygen modules capable of pro...... shown separately (b). The size of each of these modules can be ...... roughly the size of a tennis racket. (Source: Reproduced with ...... and Chemicals, Inc. All rights reserved)*

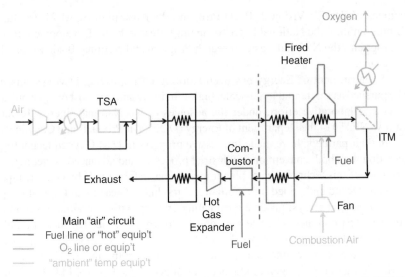

Figure 4.19 *Simplified block flow diagram of the intermediate-scale test unit (ISTU) using ITM oxygen technology with a nominal capacity of 100 TPD of oxygen production. Main "air" circuit is shown in black, fuel line or "hot" equipment is shown in dark grey, O_2 line or equipment is shown in mid grey and "ambient" temperature equipment is shown in light grey.* (Source: Reproduced with permission from Air Products and Chemicals, Inc. All rights reserved)

(pre-commercial) scale plants of a 2000 TPD capacity for the deployment of future energy applications to produce thousands of TPD of oxygen from 2018 onwards.

Alongside the work of Air Products, Praxair has been carrying out a research and development project, also supported by DOE/NETL, to utilise durable and cost-effective OTM air separation technology that offers a competitive alternative to other CO_2 capture technologies for large power plants [99]. Unlike the oxygen transport mechanism in Air Products' ITMs that is driven by the oxygen partial pressure gradient across the membrane (typically 13–20 bar on the feed side and low to sub-atmospheric pressure on the permeate side), the oxygen transport mechanism in Praxair's OTMs is driven not by pressure but by the chemical potential for the O_2 separation. The OTM oxy-combustion system being integrated directly with the boiler enables the combustion reaction to take place on the fuel side of the membrane. This produces a low O_2 partial pressure driving force and maintains O_2 transport through the membrane without involving additional compressors.

OTM technology involves both O_2 separation and combustion in one unit. An OTM is made of an inert porous support coated with a dense membrane gas separation layer, where oxygen anions meet and react with fuel species (CO, H_2, CH_4 etc. that have diffused through the porous support) and form oxidation products (H_2O, CO_2), while the electrons travel in the opposite way back through the membrane layer [100].

The first-generation OTM modules were assembled and tested. The reactor for the pilot-scale syngas plant was delivered and installed. Until 2015, the OTM-based oxy-combustion project is expected to have delivered the second generation of improved OTM modules that will be tested for their performance and fully integrated on a development scale of a 1 megawatt thermal (MWth) oxy-combustion system (Figure 4.20) [103].

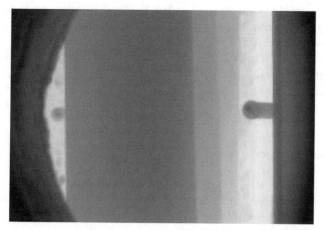

Figure 4.20 *Praxair's OTM tubes during combustion testing seen through multi-tube reactor port. (Source: Reproduced with permission from DOE/NETL 2010) [101, 102]*

References

1. Figueroa, J.D., Fout, T., Plasynski, S. *et al.* (2008) Advances in CO_2 capture technology—The U.S. Department of Energy's Carbon Sequestration Program. *International journal of greenhouse gas control*, **2**, 9–20.
2. Soundararajan, R. (2011) *Efficiency loss analysis for oxy-combustion CO_2 capture process*, Norwegian University of Science and Technology – NTNU, Department of Energy and Process Engineering, Trondheim, Norway.
3. Bernardo, P., Drioli, E. and Golemme, G. (2009) Membrane gas separation: A review/state of the art. *Industrial & Engineering Chemistry Research*, **48**, 4638–4663.
4. Bredesen, R., Jordal, K. and Bolland, O. (2004) High-temperature membranes in power generation with CO_2 capture. *Chemical Engineering and Processing*, **43**, 1129–1158.
5. Leo, A., Liu, S. and Diniz da Costa, J.C. (2009) Development of mixed conducting membranes for clean coal energy delivery. *International Journal of Greenhouse Gas Control*, **3**, 357–367.
6. Dong, X., Jin, W., Xu, N. and Li, K. (2011) Dense ceramic catalytic membranes and membrane reactors for energy and environmental applications. *Chemical Communications*, **47**, 10886.
7. Baran, E.J. (1990) Structural chemistry and physicochemical properties of perovskite-like materials. *Catalysis Today*, **8** (2), 133–276.
8. Virkar, A.V. (2005) Theoretical analysis of the role of interfaces in transport through oxygen ion and electron conducting membranes. *Journal of Power Sources*, **147** (1–2), 8–31.
9. Kim, S., Yang, Y.L., Jacobson, A.J. and Abeles, B. (1998) Diffusion and surface exchange coefficients in mixed ionic electronic conducting oxides from the pressure dependence of oxygen permeation. *Solid State Ionics*, **106** (3–4), 189–195.

10. Wu, Z., Thursfield, A., Metcalfe, I. and Li, K. (2012) Effects of separation layer thickness on oxygen permeation and mechanical strength of DL-HFMR-ScSZ. *Journal of Membrane Science*, **415–416**, 229–236.

11. Wagner, C. and Schottky, W. (1930) Beitrag zur theorie des anlaufvorganges. *Zeitschrift fuer Physiologische Chemie*, **B11**, 25.

12. Bouwmeester, H.J.M., Kruidhof, H. and Burggraaf, A.J. (1994) Importance of the surface exchange kinetics as rate limiting step in oxygen permeation through mixed-conducting oxides. *Solid State Ionics*, **72**, 185.

13. Sunarso, J., Baumann, S., Serra, J.M. *et al.* (2008) Mixed ionic–electronic conducting (MIEC) ceramic-based membranes for oxygen separation. *Journal of Membrane Science*, **320**, 13–41.

14. Bhalla, A.S., Guo, R. and Roy, R. (2000) The perovskite structure—a review of its role in ceramic science and technology. *Materials Research Innovations*, **4**, 3.

15. Patil, P.S. (1999) Versatility of chemical spray pyrolysis technique. *Materials Chemistry and Physics*, **59** (3), 185–198.

16. Li, M. and Guo, L. (2011) Spray Pyrolysis in Membranes for Membrane Reactors: Preparation, Optimization and Selection in Basile, A. and Gallucci, F. *Membranes for Membrane Reactors: Preparation, Optimization and Selection, Wiley*.

17. Luyten, J., Buekenhoudt, A., Adriansens, W. *et al.* (2000) Preparation of LaSrCoFeO$_{3-x}$ membranes. *Solid State Ionics*, **135**, 637.

18. Tan, X. and Li, K. (2004) LSCF hollow fibre membranes for air separation and oxidative coupling reactions, Proceedings of the Sixth International Conference on Catalysis in Membrane Reactors, Lahnstein, Germany, 70.

19. Tablet, C., Grubert, G., Wang, H., Caro, J., Schiestel, T., Schroeder, M. and Hederer, H. (2004) Oxygen permeation study of perovskite hollow fibre membranes for the partial oxidation of methane to syngas, Proceedings of the Sixth International Conference on Catalysis in Membrane Reactors, Lahnstein, Germany, 211.

20. Tan, X., Lui, Y. and Li, K. (2005) Preparation of LSCF ceramic hollow fiber membranes for oxygen production by a phase-inversion/sintering technique. *Industrial & Engineering Chemistry Research*, **44**, 61.

21. Kingsbury, B.F.K. (2010) A morphological study of ceramic hollow fibre membranes: a perspective on multifunctional catalytic membrane reactors, Imperial College London PhD thesis.

22. Tan, X., Liu, N., Mengand, B. and Liu, S. (2011) Morphology control of the perovskite hollow fibre membranes for oxygen separation using different bore fluids. *Journal of Membrane Science*, **378**, 308–318.

23. Alaee, M.A. and Mohammadi, T. (2009) Preparation and characterization of Ba$_x$Sr$_{1-x}$Co$_{0.8}$Fe$_{0.2}$O$_{3-\delta}$ perovskite-type membranes: Part I. *Membrane Technology*, **6** (2), 6–12.

24. Alaee, M.A. and Mohammadi, T. (2009) Preparation and characterization of Ba$_x$Sr$_{1-x}$Co$_{0.8}$Fe$_{0.2}$O$_{3-\delta}$ perovskite type membranes: Part II. *Membrane Technology*, **6** (3), 7–11.

25. Baumann, S., Schulze-Küppers, F., Roitsch, S. *et al.* (2010) Influence of sintering conditions on microstructure and oxygen permeation of Ba$_{0.5}$Sr$_{0.5}$Co$_{0.8}$Fe$_{0.2}$O$_{3-\delta}$ (BSCF) oxygen transport membranes. *Journal of Membrane Science*, **359**, 102–109.

26. Zou, Y., Zhou, W., Liu, S. and Shao, Z. (2011) Sintering and oxygen permeation studies of La$_{0.6}$Sr$_{0.4}$Co$_{0.2}$Fe$_{0.8}$O$_{3-\delta}$ ceramic membranes with improved purity. *Journal of the European Ceramic Society*, **31** (15), 2931–2938.

27. Engels, S., Beggel, F., Modigell, M. and Stadler, H. (2010) Simulation of a membrane unit for oxyfuel power plants under consideration of realistic BSCF membrane properties. *Journal of Membrane Science*, **359**, 93–101.

28. Scott Weilz, K., Yong Kim, J. and Hardy, J.S. (2005) Reactive air brazing: A novel method of sealing SOFCs and other solid-state electrochemical devices. *Electrochemical and Solid-State Letters*, **8** (2), 133–136.

29. Chen, H., Li, L., Kemps, R., Michielsen, B., Jacobs, M., Snijkers, F., Middelkoop, V. (2015) Reactive air brazing for sealing mixed ionic electronic conducting hollow fibre membranes, Acta Materialia, 10.1016/j.actamat.2015.01.029 Accepted manuscript.

30. Tenelshof, J.E., Bouwmeester, H.J.M. and Verweij, H. (1995) Oxygen transport through La$_{1-x}$Sr$_x$FeO$_{3-\delta}$ membranes. I. Permeation in air/He gradients original research article. *Solid State Ionics*, **81**, 97.

31. Kharton, V.V., Yaremchenko, A.A., Kovalevsky, A.V. *et al.* (1999) Perovskite-type oxides for high-temperature oxygen separation membranes. *Journal of Membrane Science*, **163** (2), 307–317.

32. Teraoka, Y., Zhang, H.M., Furukawa, S. and Yamazoe, N. (1985) Oxygen permeation through perovskite-type oxides. *Chemistry Letters*, **14**, 1743–1746.

33. Santos, L.C., Moraes, C. and Hughes, R. (2011) Characterization of hollow fibre membranes for oxygen permeation and partial oxidation reactions. *Brazilian Journal of Petroleum and Gas*, **5**, 45–54.

34. Shao, Z.P., Yang, W.S., Cong, Y. *et al.* (2000) Investigation of the permeation behavior and stability of a Ba$_{0.5}$Sr$_{0.5}$Co$_{0.8}$Fe$_{0.2}$O$_{3-\delta}$ oxygen membrane. *Journal of Membrane Science*, **172**, 177–188.

35. Shao, Z., Xiong, G., Tong, J. *et al.* (2001) Ba effect in doped Sr(Co$_{0.8}$Fe$_{0.2}$)O$_{3-\delta}$ on the phase structure and oxygen permeation properties of the dense ceramic membranes. *Separation and Purification Technology*, **25**, 419–429.

36. Chen, Z., Ran, R., Zhou, W. *et al.* (2007) Assessment of Ba$_{0.5}$Sr$_{0.5}$Co$_{1-y}$Fe$_y$O$_{3-\text{[delta]}}$ ($y = 0.0$ 1.0) for prospective application as cathode for ITSOFCs or oxygen permeating membrane. *Electrochimica Acta*, **52** (25), 7343–7351.

37. Wang, H., Cong, Y. and Yang, W. (2002) Oxygen permeation study in a tubular Ba$_{0.5}$Sr$_{0.5}$Co$_{0.8}$Fe$_{0.2}$O$_{3-\delta}$ oxygen permeable membrane. *Journal of Membrane Science*, **210**, 259–271.

38. Liu, S. and Gavalas, G.R. (2005) Oxygen selective ceramic hollow fiber membranes. *Journal of Membrane Science*, **246**, 103.

39. Li, K., Tan, X. and Liu, Y. (2006) Single-step fabrication of ceramic hollow fibers for oxygen permeation. *Journal of Membrane Science*, **272**, 1–5.

40. Wang, H., Tablet, C. and Caro, J. (2008) Oxygen production at low temperature using dense perovskite hollow fiber membranes. *Journal of Membrane Science*, **322**, 214–217.

41. Leo, A., Smart, S., Liu, S. and Diniz da Costa, J.C. (2011) High performance perovskite hollow fibres for oxygen separation. *Journal of Membrane Science*, **368**, 64–68.

42. Han, D., Tan, X., Yana, Z. *et al.* (2013) New morphological $Ba_{0.5}Sr_{0.5}Co_{0.8}Fe_{0.2}O_{3-\delta}$ a hollow fibre membranes with high oxygen permeation fluxes. *Ceramics International*, **39** (1), 431–437.
43. Leo, A., Liu, S. and Diniz da Costa, J.C. (2009) The enhancement of oxygen flux on $Ba_{0.5}Sr_{0.5}Co_{0.8}Fe_{0.2}O_{3-\delta}$ (BSCF) hollow fibers using silver surface modification. *Journal of Membrane Science*, **340**, 148–153.
44. Chen, Z., Ran, R., Shao, Z. *et al.* (2009) Further performance improvement of $Ba_{0.5}Sr_{0.5}Co_{0.8}Fe_{0.2}O_{3-\delta}$ perovskite membranes for air separation. *Ceramics International*, **35**, 2455–2461.
45. Van Noyen, J., Middelkoop, V., Buysse, C. *et al.* (2012) Fabrication of perovskite capillary membranes for high temperature gas separation. *Catalysis Today*, **193** (15), 172.
46. Baumann, S., Serra, J.M., Lobera, M.P. *et al.* (2011) Ultrahigh oxygen permeation flux through supported $Ba_{0.5}Sr_{0.5}Co_{0.8}Fe_{0.2}O_{3-\delta}$ membranes. *Journal of Membrane Science*, **377** (1–2), 198–205.
47. Li, S.G., Jin, W.Q., Huang, P. *et al.* (1999) Perovskite-related ZrO_2-doped $SrCo_{0.4}Fe_{0.6}O_{3-\delta}$ membrane for oxygen permeation. *AIChE Journal*, **45** (2), 276.
48. Tong, J., Yang, W., Zhu, B. and Cai, R. (2002) Investigation of ideal zirconium-doped perovskite-type ceramic membrane materials for oxygen separation. *Journal of Membrane Science*, **203**, 175.
49. Schiestel, T., Kilgus, M., Peter, S. *et al.* (2005) Hollow fibre perovskite membranes for oxygen separation. *Journal of Membrane Science*, **258**, 1–4.
50. Tablet, C., Grubert, G., Wang, H. *et al.* (2005) Oxygen permeation study of perovskite hollow fiber membranes. *Catalysis Today*, **104**, 126–130.
51. Caro, J., Wang, H.H., Tablet, C. *et al.* (2006) Evaluation of perovskites in hollow fibre and disk geometry in catalytic membrane reactors and in oxygen separators. *Catalysis Today*, **118** (1–2), 128–135.
52. Jin, W., Li, S., Huang, P. *et al.* (2001) Preparation of an asymmetric perovskite-type membrane and its oxygen permeability. *Journal of Membrane Science*, **185**, 237–243.
53. Tan, X., Lui, Y. and Li, K. (2005b) Mixed conducting ceramic hollow fiber membranes for air separation. *AIChE Journal*, **51**, 1991.
54. Lane, J.A., Benson, S.J., Waller, D. and Kilner, J.A. (1999) Oxygen transport in $La_{0.6}Sr_{0.4}Co_{0.2}Fe_{0.8}O_{3-\delta}$. *Solid State Ionics*, **121**, 201–208.
55. Li, S., Jin, W., Huang, P. *et al.* (2000) Tubular lanthanum cobaltite perovskite type membrane for oxygen permeation. *Journal of Membrane Science*, **166** (1), 51.
56. Tan, X., Wang, Z., Liu, H. and Liu, S. (2008) Enhancement of oxygen permeation through $La_{0.6}Sr_{0.4}Co_{0.2}Fe_{0.8}O_{3-\delta}$ hollow fibre membranes by surface modifications. *Journal of Membrane Science*, **324**, 128–135.
57. Thursfield, A. and Metcalfe, I.S. (2007) Air separation using a catalytically modified mixed conducting ceramic hollow fibre membrane module. *Journal of Membrane Science*, **288**, 175–187.
58. Tan, X., Li, K., Thursfield, A. and Metcalfe, I.S. (2008) Oxyfuel combustion using a catalytic ceramic membrane reactor. *Catalysis Today*, **131**, 292–304.
59. Vente, J.F., Haije, W.G. and Rak, Z.S. (2006) Performance of functional perovskite membranes for oxygen production. *Journal of Membrane Science*, **276**, 178.

60. Han, D., Sunarso, J., Tan, X. *et al.* (2012) Optimizing oxygen transport through La$_{0.6}$Sr$_{0.4}$Co$_{0.2}$Fe$_{0.8}$O$_{3-\delta}$ hollow fiber by microstructure modification and Ag/Pt catalyst deposition. *Energy & fuels*, **26**, 4728.

61. Wang, Z., Yang, N., Meng, B. and Tan, X. (2009) Preparation and oxygen permeation properties of highly asymmetric La$_{0.6}$Sr$_{0.4}$Co$_{0.2}$Fe$_{0.8}$O$_{3-\delta}$ perovskite hollow-fiber membranes. *Industrial & Engineering Chemistry Research*, **48**, 510–516.

62. Middelkoop, V., Chen, H., Michielsen, B., *et al.* (2014) Development and characterisation of dense lanthanum-based perovskite oxygen-separation capillary membranes for high-temperature applications. *Journal of Membrane Science*, **468**, 250–258.

63. Tan, X., Shi, L., Hao, G. *et al.* (2012) La$_{0.7}$Sr$_{0.3}$FeO$_{3-\delta}$ a perovskite hollow fiber membranes for oxygen permeation and methane conversion. *Separation and Purification Technology*, **96**, 89–97.

64. Büchler, O., Serra, J.M., Meulenberg, W.A. *et al.* (2007) Preparation and properties of thin La$_{1-x}$Sr$_x$Co$_{1-y}$Fe$_y$O$_{3-\delta}$ perovskitic membranes supported on tailored ceramic substrates. *Solid State Ionics*, **178**, 91–99.

65. Zydorczak, B., Wu, Z. and Li, K. (2009) Fabrication of ultrathin La$_{0.6}$Sr$_{0.4}$Co$_{0.2}$Fe$_{0.8}$O$_{3-\delta}$ hollow fibre membranes for oxygen permeation. *Chemical Engineering Science*, **64**, 4383–4388.

66. Tan, X., Wang, Z. and Li, K. (2010) Effects of sintering on the properties of La$_{0.6}$Sr$_{0.4}$Co$_{0.2}$Fe$_{0.8}$O$_{3-\delta}$ perovskite hollow fiber membranes. *Industrial & Engineering Chemistry Research*, **49**, 2895–2901.

67. Yacou, C., Sunarso, J., Lin, C.X.C. *et al.* (2011) Palladium surface modified La$_{0.6}$Sr$_{0.4}$Co$_{0.2}$Fe$_{0.8}$O$_{3-i}$ hollow fibres for oxygen separation. *Journal of Membrane Science*, **380**, 223–231.

68. Meng, X., Yang, N., Meng, B. *et al.* (2011) Zirconium stabilized Ba$_{0.5}$Sr$_{0.5}$(Co$_{0.8-x}$Zr$_x$)Fe$_{0.2}$O$_{3-\alpha}$ perovskite hollow fibre membranes for oxygen separation. *Ceramics International*, **37**, 2701–2709.

69. Zhang, K., Ran, R., Shao, Z. *et al.* (2010) Effects of niobium doping site and concentration on the phase structure and oxygen permeability of Nb-substituted SrCoO$_x$ oxides. *Ceramics International*, **36**, 635.

70. Engels, S., Markus, T., Modigell, M. and Singheiser, L. (2011) Oxygen permeation and stability investigations on MIEC membrane materials under operating conditions for power plant processes. *Journal of Membrane Science*, **370**, 58–769.

71. Wei, Y., Liao, Q., Xue, J. *et al.* (2013) Influence of SO$_2$ on the phase structure, oxygen permeation and microstructure of K$_2$NiF$_4$-type hollow fiber membranes. *Chemical Engineering Journal*, **217**, 34–40.

72. Zeng, Q., Zuo, Y., Fan, C. and Chen, C. (2009) CO$_2$-tolerant oxygen separation membranes targeting CO$_2$ capture application. *Journal of Membrane Science*, **335**, 140–144.

73. Freund, H.J. and Roberts, M.W. (1996) Surface chemistry of carbon dioxide. *Surface Science Reports*, **25**, 225–273.

74. Efimov, K., Klande, T., Juditzki, N. and Feldhoff, A. (2012) Ca-containing CO$_2$-tolerant perovskite materials for oxygen separation. *Journal of Membrane Science*, **389**, 205–215.

75. Tan, X., Liu, N., Meng, B. *et al.* (2012) Oxygen permeation behavior of $La_{0.6}Sr_{0.4}Co_{0.8}Fe_{0.2}O_3$ hollow fibre membranes with highly concentrated CO_2 exposure. *Journal of Membrane Science*, **389**, 216–222.

76. Kathiraser, Y., Wang, Z.G., Yang, N.T. *et al.* (2013) Oxygen permeation and stability study of $La_{0.6}Sr_{0.4}Co_{0.8}Ga_{0.2}O_{3-delta}$ (LSCG) hollow fiber membrane with exposure to CO_2, CH_4 and He. *Journal of Membrane Science*, **427**, 240–249.

77. Liang, F.Y., Partovi, K., Jiang, H.Q. *et al.* (2013) B-site La-doped $BaFe_{0.95-x}La_x Zr_{0.05}O_{3-delta}$ perovskite-type membranes for oxygen separation. *Journal of Materials Chemistry A*, **1**, 746–751.

78. Luo, H.X., Wei, Y.Y., Jiang, H.Q. *et al.* (2010) Performance of a ceramic membrane reactor with high oxygen flux Ta-containing perovskite for the partial oxidation of methane to syngas. *Journal of Membrane Science*, **350**, 154–160.

79. Xu, N.S., Zhao, H.L., Shen, Y.N. *et al.* (2012) Structure, electrical conductivity and oxygen permeability of $Ba_{0.6}Sr_{0.4}Co_{1-x}Ti_xO_{3-delta}$ ceramic membranes. *Separation and Purification Technology*, **89**, 16–21.

80. Fontaine, M.L., Smith, J.B., Larring, Y. and Bredesen, R. (2009) On the preparation of asymmetric $CaTi_{0.9}Fe_{0.1}O_{3-\delta}$ membranes by tape-casting and co-sintering process. *Journal of Membrane Science*, **326**, 310–315.

81. Figueiredo, F.M., Kharton, V.V., Viskup, A.P. and Frade, J.R. (2004) Surface enhanced oxygen permeation in $CaTi_{1-x}Fe_xO_{3-\delta}$ ceramic membranes. *Journal of Membrane Science*, **236**, 73–80.

82. Klande, T., Efimov, K., Cusenza, S. *et al.* (2011) Effect of doping, microstructure, and CO_2 on $La_2NiO_{4+delta}$-based oxygen-transporting materials. *Journal of Solid State Chemistry*, **184**, 3310–3318.

83. Wei, Y., Ravkina, O., Klande, T. *et al.* (2013) Effect of CO_2 and SO_2 on oxygen permeation and microstructure of $(Pr_{0.9}La_{0.1})_2(Ni_{0.74}Cu_{0.21}Ga_{0.05})O_{4+\delta}$ membranes. *Journal of Membrane Science*, **429**, 147.

84. Tsipis, E., Naumovich, E., Shaula, A. *et al.* (2008) Oxygen nonstoichiometry and ionic transport in $La_2Ni(Fe)O_{4+delta}$. *Solid State Ionics*, **179**, 57.

85. Wei, Y., Liao, Q., Li, Z. and Wang, H. (2013) Enhancement of oxygen permeation through U-shaped K_2NiF_4-type oxide hollow fiber membranes by surface modifications. *Separation and Purification Technology*, **110**, 74–80.

86. Stadler, H., Beggel, F., Habermehl, M. *et al.* (2011) Oxyfuel coal combustion by efficient integration of oxygen transport membranes. *International Journal of Greenhouse Gas Control*, **5**, 7.

87. Buysse, C., Michielsen, B., Middelkoop, V. *et al.* (2013) Modeling of the performance of BSCF capillary membranes in four-end and three-end integration mode. *Ceramics International*, **39**, 4113.

88. Wang, H., Wang, R., Liang, D.T. and Yang, W. (2004) Experimental and modeling studies on $Ba_{0.5}Sr_{0.5}Co_{0.8}Fe_{0.2}O_{3-\delta}$ (BSCF) tubular membranes for air separation. *Journal of Membrane Science*, **243**, 405–415.

89. Czyperek, M., Zapp, P., Bouwmeester, H.J.M. *et al.* (2010) Gas separation membranes for zero-emission fossil power plants: MEM-BRAIN. *Journal of Membrane Science*, **359** (1–2), 149–159.

90. Liang, F., Jiang, H., Luo, H. *et al.* (2012) High-purity oxygen production by a dead-end $Ba_{0.5}Sr_{0.5}Co_{0.8}Fe_{0.2}O_{3-\delta}$ tube membrane. *Catalysis Today*, **193**, 95.

91. Kriegel, R. (2010) Fraunhofer Institute for Ceramic Technologies and Systems – IKTS, Hermsdorf, Germany http://www.ikts.fraunhofer.de/content/dam/ikts/en/documents/Demonstrator_fuerdieO2-Membranseparation2010tcm244-61774.pdf

92. Kriegel, R., Schulz, M., Ritter, K. *et al.* (2012) *Ceramic Membranes for Oxygen Production*, Fraunhofer IKTS, presented at Achema.

93. Kneer, R., Toporov, D., Förster, M. *et al.* (2010) OXYCOAL-AC: Towards an integrated coal-fired power plant process with ion transport membrane-based oxygen supply. *Energy & Environmental Science*, **3**, 198–207.

94. Dong, X. and Jin, W. (2012) Mixed conducting ceramic membranes for high efficiency power generation with CO_2 capture. *Current Opinion in Chemical Engineering*, **1**, 163–170.

95. Pfaff, E.M., Kaletsch, A. and Broeckmann, C. (2012) Design of a mixed ionic/electronic conducting oxygen transport membrane pilot module. *Chemical Engineering & Technology*, **35** (3), 455–463.

96. Tan, X., Wang, Z., Meng, B. *et al.* (2010) Pilot-scale production of oxygen from air using perovskite hollow fibre membranes. *Journal of Membrane Science*, **352**, 189–196.

97. Repasky, J.M., Anderson, L.L., Stein, V.E.E., Armstrong, P.A. and Foster, E.P. (2012) ITM Oxygen technology: scale-up toward clean energy applications, International Pittsburgh Coal Conference.

98. Woods, C.M., Maxson, A. (2014), ITM Update, Gasification Technologies Conference, Washington, DC, http://www.gasification.org//uploads/eventLibrary/2014_13.2_Air_Products_Charles_Woods.pdf

99. NETL Oxy-combustion: Oxygen Transport Membrane Development http://www.netl.doe.gov/publications/factsheets/project/NT43088.pdf

100. DOE/NETL Advanced Carbon Dioxide Capture R&D Program: Technology Update May 2011 http://www.netl.doe.gov/technologies/coalpower/ewr/pubs/CO$_2$Handbook/CO$_2$-Capture-Tech-Update-2011_Front-End%20Report.pdf

101. http://www.netl.doe.gov/publications/factsheets/project/Proj470.pdf

102. http://www.netl.doe.gov/publications/factsheets/project/NT43088.pdf SAME AS 97

103. Grants - Award Summary, Praxair, Inc, http://www.recovery.gov/Transparency/RecoveryData/pages/RecipientProjectSummary508.aspx?AwardIdSur=119460

5

Chemical Looping Combustion for Power Production

V. Spallina, H. P. Hamers, F. Gallucci and M. van Sint Annaland
Eindhoven University of Technology, Chemical Process Intensification, Department
of Chemical Engineering and Chemistry, Eindhoven, The Netherlands

5.1   Introduction

Climate change due to the increasing greenhouse gas emissions due to human activities is amply recognized as one of the most important challenges for the power industries in the mid-long term. Different strategies may be implemented to comply with this requirement. A moderate CO_2 emission reduction may be achieved by improving the conversion efficiency or a fuel switching to lower carbon content (such as natural gas), but CO_2 capture and storage (CCS) technology implementation is required to reduce greenhouse gas emissions of about 19% [1].

The processes aiming at CO_2 removal in power plants may conceptually be split into three different great families:

- CO_2 separation from flue gases (post-combustion removal).
- CO_2 separation from fuel processing including CO conversion of synthesis gas, carbon separation from high-hydrogen content fuels (pre-combustion removal).
- CO_2 separation from concentrated exhaust gases (e.g. oxy-fuel combustion).

The post-combustion technologies generally consist of chemical absorption systems that are operated at almost atmospheric pressure and with amine-based solvent. The advantage of this technique is the possibility to retrofit existing power plants, but high efficiency

Process Intensification for Sustainable Energy Conversion, First Edition.
Edited by Fausto Gallucci and Martin van Sint Annaland.
© 2015 John Wiley & Sons, Ltd. Published 2015 by John Wiley & Sons, Ltd.

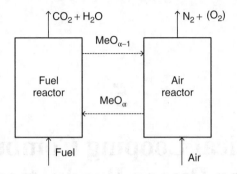

Figure 5.1   *Schematic of a CLC process*

penalties are associated with the steam consumptions for the desorption column that is used to regenerate the solvent and release the CO_2-rich stream, and the CO_2 compression up to the condition of the final storage.

In the case of pre-combustion plants, the H_2-rich gas is produced after coal or natural gas conversion into H_2/CO-rich syngas, and, subsequently, CO is shifted to H_2 and CO_2. The CO_2 is separated in an absorption unit (chemical or physical process may be used). In this case, the energy cost associated with the steam consumption for the WGS and the lower cold gas efficiency of the processes are partly compensated by full steam cycle integration with all the units that are used in the plant [2].

In the case of oxy-combustion CCS technology the CO_2 is easily separated from exhaust gases that are obtained in oxy-fired boiler but the cost to produce the O_2 from cryogenic system leads to 13–30 percentage points efficiency decay [3].

Different alternative technologies have been discussed in the last decades. One of the most promising technologies is chemical looping combustion (CLC): the conversion is performed by using a solid metal (called oxygen carrier), which is alternatively oxidized and reduced by sequential contact with an air and a fuel stream (Figure 5.1). The technology is based on the use of a fuel reactor and an air reactor that are operated in a loop: in the first reactor the fuel is reacting with a metal oxide, which is in an oxidative state and the oxygen is transferred from the solid to the fuel phase so that CO_2 and H_2O are produced; in the air reactor, an oxidant stream (i.e. air) is fed and the metal oxide is oxidized to the original form. The oxidation reaction is an exothermic reaction and the O_2-depleted air is produced at high temperature. The reduction reaction can be both exothermic and endothermic depending on the fuel and the oxygen carriers (OCs).

The OCs react with the oxygen in the air reactor, while in the fuel reactor different types of fuel can be adopted. The generic reactions that occur during the CLC are listed as follows (Eqs. 5.1–5.4).

$$MeO_{\alpha-1} + \frac{1}{2}O_2 \rightarrow MeO_\alpha \qquad (5.1)$$

$$MeO_\alpha + H_2 \rightarrow MeO_{\alpha-1} + H_2O \qquad (5.2)$$

$$MeO_\alpha + CO \rightarrow MeO_{\alpha-1} + CO_2 \tag{5.3}$$

$$\left(2x + \frac{y}{2}\right) MeO_\alpha + C_xH_y \rightarrow \left(2x + \frac{y}{2}\right) MeO_{\alpha-1} + xCO_2 + \frac{y}{2}H_2O \tag{5.4}$$

In the last years, a lot of research has been devoted to the study of the direct solid CLC in order to extend the technology of interconnected fluidized bed reactors to the use of solid fuels (such as coal, petcoke or biomass). Coal can be used in a CLC process in two ways: in the first case the coal is first gasified in a CO/H_2-rich stream and then is combusted with OC in the fuel reactor, or it is possible to convert the pulverized coal directly in the fuel reactor with the OCs. In the second case, two different options have been proposed (Figure 5.2): the *in situ* gasification chemical looping combustion (isG-CLC) and chemical looping with oxygen uncoupling (CLOU). The first process consists of feeding coal with H_2O and CO_2 (which act as fluidization agents) [4, 5]: the solid fuel is first devolatilized and char gets gasified producing H_2 and CO; the volatile matter and the syngas are then oxidized, reacting with the OCs such as in the CLC with gaseous fuels. The CLOU mechanism is based on the use of an OC, which is first reduced by releasing O_2 in the gas phase, and then the gaseous oxygen reacts with the fuel [6–8]: using the CLOU mechanism it is possible to overcome the low reactivity, which is associated with the char gasification stage because the char can directly react with $O_{2(g)}$.

In this chapter, the CLC combustion technology is discussed. In the first part, the description mainly concerns the different OCs that have been studied and the main properties of the solid materials; in the second part, the reactor concepts and the main operating conditions that have been studied and tested are presented; in the third part, the integration of

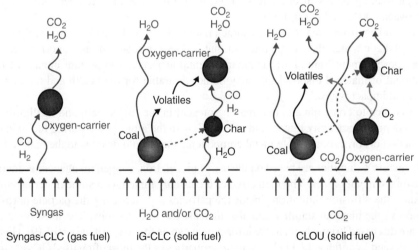

Figure 5.2 *Different mechanisms to convert coal with a CLC process.* (Source: Reproduced from Ref. [9], with permission from Elsevier)

reactors loop in large-scale power plant with CO_2 capture is discussed and compared with the reference technologies with CO_2 capture.

5.2   Oxygen carriers

The CLC technology can be successfully operated if the OCs are selected and designed properly to achieve high performance for the fuel conversion. Several factors have to be considered in order to achieve the desirable process:

1. High oxygen capacity in order to reduce the total amount of solid that has to be circulated and used in the system. The oxygen capacity R_0 is defined as the amount of O_2 that is transferred from the solid to the gaseous streams. It is calculated as follows:

$$R_0 \left[\frac{\text{kg}_{\text{OC,reduced}}}{\text{kg}_{\text{OC,oxidized}}} \right] = \frac{\dot{m}_{\text{OC,ox}} - \dot{m}_{\text{OC,red}}}{\dot{m}_{\text{OC,ox}}} \tag{5.5}$$

2. High selectivity to H_2O and CO_2 in order to achieve a complete fuel conversion in the reduction reactions. In the case of low reactivity of the OCs with the fuel, the system needs additional units to process the unconverted fuel and the CO_2 capture rate is lower. The high reaction rates for reduction and oxidation reactions are also desirable to reduce the size of the reactors and the possibility to operate in a wide range of operating conditions.
3. Good material stability including the long-term recyclability and durability in order to minimize the material losses and the costs associated. Hence, it is important that the particles have a low attrition rate and that they do not agglomerate, otherwise the particles defluidize.
4. High melting points in order to be used in a process at high temperature (typically between $800\,°C < T < 1200\,°C$)
5. Cost and environmental friendly characteristics: the cost of the OCs, especially in case of synthetic materials, can be relevant, but it does not represent an important technology development limitation; the environmental aspects are important in case of large amount of material disposal that may be not desirable for the health and the environmental impact issues.
6. Resistance to contaminants that may be present in the fuel stream and negligible carbon deposition activity that would release CO_2 in the atmosphere during the oxidation reaction or form some stable metal components that would deactivate the material.

The main OC families are based on the use of nickel, iron, copper, manganese as metal or some mixed metal oxides. The metal oxides are usually mixed with some material support to reduce the diffusion limitations inside the particles (by increasing the particle porosity), to improve the material stability and also increase the heat capacity. The most often used OCs are discussed briefly here. More information can be found in Adanez *et al.* [9], Hossain *et al.* [10] and Lyngfelt *et al.* [11]. The main properties of the most often used OCs are listed in Table 5.1 [9, 11, 12].

Table 5.1 *List of material properties used for CLC.*

Material type	Weight content in the particle (%wt)	Support type	Fuel type	Process	Temperature range tested (°C) [11]	Selectivity to CO_2/H_2O* (800–1000°C) [12, 20]	Oxygen Carrier pair considered	Melting points °C, [12]	Oxygen ratio, R_0 (not considering support)	Reaction enthalpy at 1000°C**(kJ/mol reactant gas) [19, 12]					Metal cost ($/ton metal) [9]
										CO	H_2	CH_4	C	O_2	
Ni based	18–100%	α-Al_2O_3, γ-Al_2O_3, Al_2O_3, $NiAl_2O_4$, $NiAl_2O_4$–MgO, $MgAl_2O_4$, Bentonite, ZrO_2–MgO	CH_4, C_2H_6, C_3H_8, H_2, CO, syngas, CH_4+H_2S	CLC/CLR	450–1200	>98.9% (Ni)	NiO/Ni	1455	0.214	−47	−15	134	75	−468	15,000
Cu based	12–15%	α-Al_2O_3,γ-Al_2O_3, $MgAl_2O_4$	CH_4, H_2, CO, syngas, C_xH_y, CH_4+H_2S	CLC	300–1000	100% (Cu)	CuO/Cu	1085	0.201	−134	−101	−212	−99	−296	7000
Cu based	15–80%	Al_2O_3,γ-Al_2O_3, Sepiolite, $MgAl_2O_4$, Bentonite, ZrO_2, TiO_2, SiO_2	CH_4, coke, char, N_2, CO_2	CLOU	850–985	100% (Cu)	CuO/Cu_2O	1235	0.112	−151	−119	−283	−135	−260	7000
Fe based	20–100%	Al_2O_3, Bentonite	CH_4, PSA-off-gas, biomass	CLC	430–1000	100% (Fe_3O_4) 54–78% (FeO)	Fe_2O_3/Fe_3O_4	1565	0.033	−42	−10	154	84	−479	200
Mn based	40%	ZrO_2–MgO	syngas	CLC	810–1000	100% (MnO)	Mn_2O_3/MnO	1347	0.101	−102	−70	−85	−36	−359	<200
Mn based	80%	SiO_2	CH_4	CLOU	800–1000	100% (MnO)	Mn_2O_3/Mn_3O_4	1347	0.034	−192	−160	−446	−217	−179	<200
Ilmenite ($FeTiO_3$)	100%	-	Coal, petcoke, syngas	CLC	813–1030	almost 100%	Fe_2O_3/FeO***	1565	0.100	−4.7	27.5	304	158	−554	<200

*The species between brackets are the reduced components.

**The dependency of the reaction enthalpy on the temperature is small.

***Different components containing iron, titanium and oxide can be considered, like Fe_2TiO_5, $FeTiO_3$. In that case, the reaction enthalpy is a bit different.

5.2.1   Nickel-based OCs

The most studied OCs are based on nickel [9]. It is a fast reacting material, withstands a temperature of at least 1185 °C [13] and NiO has a high oxygen content. However, the selectivity is in the range of 98–100%, which means that some of the fuel is not converted to CO_2 and H_2O. The selectivity decreases with increasing temperature. At 1000 °C, the selectivity from CO to CO_2 is 98.9% [12]. A high methane conversion has been obtained with nickel oxide; due to the thermodynamic limitations, the fuel cannot be completely converted [9]. The reaction with methane and carbon is endothermic, while syngas is converted exothermically. Despite its relatively good properties, there are two important drawbacks: the cost of nickel is relatively high and the material is considered to be toxic.

Low reaction rates were obtained with pure NiO, because of low porosity [14]; this is the case of all the OCs. If alumina is used as support material, $NiAl_2O_4$-spinel can be formed, which causes a decrease in reactivity during reduction. This spinel has been used successfully as support material with NiO as active material with a capacity of $120\,kW_{th}$ [15, 16]. This support is not fully inert, but it becomes inert after adding Mg or Ca [17]. NiO/Al_2O_4-particles with a small amount of MgO were demonstrated as a good OC for the conversion of CH_4 during a test of 611 hours [18].

5.2.2   Iron-based OCs

A cheaper and non-toxic alternative to Ni is iron. Iron can be present in several phases: Fe_2O_3, Fe_3O_4, FeO, Fe. Hematite, Fe_2O_3, is the most oxidized form below 1300 °C. To which form it is reduced depends on the oxygen content of the system. If it is further reduced than Fe_3O_4, the selectivity becomes an issue. For that reason, only the properties of the Fe_2O_3/Fe_3O_4-pair are listed in Table 5.1. If only the Fe_2O_3/Fe_3O_4-pair is used, the oxygen capacity of the OC is limited. It is possible to further reduce Fe_3O_4 and obtain full gas conversion by applying a countercurrent solids flow [19] or adding alumina or titanium to the carrier [20–22].

The latter is the case if ilmenite is used as OC [20]. Ilmenite is a natural material that consists of different species containing iron, titanium and oxide, like Fe_2TiO_5. It is a natural material, which is obtained from pits in Norway, South Africa and Australia. Before ilmenite can be used for CLC, it has to be activated. During the activation period, the porosity and reactivity is increased [23]. Activated ilmenite has been demonstrated as good OC for conversion of syngas [24] and also for solids in dual fluidized bed systems [25].

5.2.3   Copper-based OCs

Copper is a fast reacting OC and also the selectivity is high. The material is cheaper than nickel and all the reactions are exothermic. The critical issue for the copper-based OC is the melting temperature of 1085 °C, which makes it not applicable at high temperatures. Successful operation was achieved with 15 wt% CuO/Al_2O_3 in a continuous $10\,kW_{th}$ CLC circulating fluidized bed system with CH_4 as fuel for 120 h; a high conversion was achieved and no problems were observed with the particles [26].

Copper can be present as CuO, Cu_2O and Cu. Above 1000 °C, it can only be oxidized to Cu_2O at atmospheric pressure [6]. Because of this thermodynamic equilibrium, this carrier is applicable for the CLOU concept. In this concept, CuO is formed in the air reactor and in the fuel reactor, which has a higher temperature than the air reactor, oxygen is released and Cu_2O is formed. The fuel is subsequently combusted with the released oxygen. For that reason, relatively fast conversion of solid materials can be achieved.

5.2.4   Manganese-based OCs

Manganese is also a cheap and environmental friendly alternative. Manganese can be present as Mn_2O_3, Mn_3O_4, MnO. The thermodynamic equilibrium between Mn_2O_3 and Mn_3O_4 is at 850°C with air at atmospheric pressure [6]. Hence, also the Mn_2O_3/Mn_3O_4-couple can be used as OC for the CLOU process.

Manganese can also be used for the regular CLC process. In that case, it can be reduced to MnO and the most oxidized form depends on the operation temperature during oxidation; above 900 °C the most oxidized form is only Mn_3O_4. For that reason, the OC capacity of this material is relatively low. Tests with manganese in a 300 W_{th} continuous CLC unit demonstrated that high syngas conversion can be obtained, but the reactivity with CH_4 is much lower [27]. Natural manganese ore can also be used as OC, but poor mechanical stability and fluidization properties were observed [28].

5.2.5   Other Oxygen Carriers

Along with the four main OCs described above, some other materials have been studied, such as cobalt and $CaSO_4$.

Cobalt is present as Co_3O_4, CoO and Co; the thermodynamic equilibrium for Co_3O_4 and CoO is favoured at 900 °C, and. therefore, this material can also be used for the CLOU process. The reason that cobalt is not often used is that its costs are high (even higher than that of nickel), it is toxic and the selectivity is low (87–97% for CO) in the temperature range of 800–1200 °C [9].

From natural anhydrite, the $CaSO_4$/CaS-oxygen carrier can be obtained, which is a low-cost material with a high oxygen capacity. Low reaction rates were measured and the thermodynamic equilibrium caused that always a small amount of CO and H_2 remains unconverted [9]. Also the side reaction could occur, in which CaO and SO_2 are formed.

As alternative mixes of OCs have been used to evaluate if advantages of OCs could be combined and the drawbacks be reduced [9]. For example, the reactivity and the fuel conversion could be increased, the amount of expensive or toxic material decreased and the strength of the particles increased. Studies have been conducted with many different combinations (Cu/Fe, Cu/Ni, Fe/Ni, Co/Ni, Fe/Mn), but have not resulted in an application on a larger scale until now [9]. Perovskite structures have been proposed as well, which might also be applicable for the CLOU process in case perovskite contains Mn–Ca–O [29].

5.2.6   Sulfur Tolerance

If the fuel contains sulfur, SO_2 is formed during reduction or sulfur reacts with the OC, forming sulfides or sulfates (i.e. MeS and $MeSO_4$). If sulfides or sulfates are formed, SO_2 could be formed during oxidation (and the flue gas needs to be desulfurized) or sulfur remains on the carrier, which might be deactivated. The formation of sulfides and sulfates during reduction is enhanced at high sulfur concentrations, low temperature, high pressure and low oxygen content (with the OC or by adding H_2O or CO_2) [12, 30].

The formation of sulfides is especially the case for nickel-based OCs [31]. Continuous operation experiments in a circulating fluidized bed system with CH_4 and nickel-based carriers demonstrated that fuels with a sulfur content below $100\,ppm_v$ H_2S can be tolerated in an industrial plant [31]. At higher H_2S concentrations, a desulfurization step might be required before the fuel is fed to the CLC system. Copper has been demonstrated to be more tolerant of sulfur than nickel. During continuous operation, most of the sulfur was released in the fuel reactor as H_2S and SO_2 [32].

If coal is directly fed to the fuel reactor, copper- and iron-based OCs are preferred, because they are harmless if they are mixed with residual ashes [9]. In Ref. [25] it is demonstrated that in case the CLC system is fuelled with petroleum coke and ilmenite is used as OC, all the sulfur is released as H_2S and SO_2 in the fuel reactor. Thus, with ilmenite and petroleum coke, no sulfur poisoning problems are expected.

5.3   Reactor Concepts

The chemical looping concept can be accomplished with two main configurations: in the first configuration, the solid material is circulating between the fuel reactor and the air reactor, while in the second case the solid material is stationary and the gases are alternatively fed to the reactor by using a gas switching system.

Most of the research about chemical looping has been focused on the first configuration, and different reactor concepts have been developed in recent years to make the process feasible as discussed in Ref. [33]. The utilization of solid fuel has been also investigated in recent years and some pilot plants based on interconnected fluidized beds have been built at different sizes [34]. The chemical looping combustor units have been built at different sizes and tested with several OCs. In the second case, the research is at the early stage.

5.3.1   Interconnected Fluidized Bed Reactors

The first design has been proposed in Ref. [35]. Two interconnected fluidized beds are used as fuel and air reactors, respectively. The fuel reactor is a low-velocity bubbling fluidized bed where the fuel is converted by using the OCs. The size of the fuel reactor is determined by the terminal gas velocity, minimum fluidization velocity, solid circulation and high residence time that is preferable to achieve a complete solid conversion. The high velocity riser as air reactor provides the required driving force for the solid material circulation and allows converting the OC in the original form.

The inventory of the system includes the total amount of material that is in the fuel and air reactors (expressed in kg of solid or in kg_s/MW_{th}) in the loop seals and in the piping units that connect the different components. The solid circulation is selected to ensure the complete fuel conversion. In terms of mass balance of the system, it is possible to identify two important parameters: (i) the solid circulation flow rate and (ii) the solid conversion.

Assuming $1\,MW_{th}$ of the fuel entering the system and full conversion of the gas, the solid circulation flow rate is calculated as follows:

$$\dot{m}_{OC}\left[\frac{kg_{OC}}{MJ_{LHV}}\right] = \frac{1}{R_{OC}\Delta Xs}\frac{2\xi MM_O}{LHV_{fuel}}\tag{5.6}$$

where ξ is the stoichiometric coefficient of O_2 in the oxidation reaction with reduced OCs, MM_O is the molecular weight of the atomic oxygen (16 kg/kmol), LHV_{fuel} is the lower heating value of the fuel used in the fuel reactor (expressed as MJ/kg) and R_{OC} is the oxygen transport capacity, which is calculated as follows:

$$R_{OC}\left[\frac{kg_{OC,active}}{kg_{OC,oxidized}}\right] = y_{OC,act}R_0 \tag{5.7}$$

($y_{OC,act}$ is the weight fraction of the active weight material in the OCs.)

And ΔXs is the solid conversion, which is calculated as the difference in solid mass flow rate at the inlet and the outlet of the fuel and air reactors divided per the total solid mass flow rate difference that is achieved in case of complete solid conversion:

$$\Delta Xs\left[\frac{kg_{OC,converted}}{kg_{OC,available}}\right] = \frac{\dot{m}_{OC,i} - \dot{m}_{OC,red}}{\dot{m}_{OC,ox} - \dot{m}_{OC,red}} = \frac{\dot{m}_{OC,i} - \dot{m}_{OC,red}}{\dot{m}_{OC,ox} \cdot R_{OC}} \tag{5.8}$$

The combination of bubbling fluidized bed as fuel reactor and riser as air reactor has been also presented for a 10 kW$_{th}$ unit in other works [36–38]. The O_2-depleted air and the OCs are separated in a HT cyclone to avoid the particle going to the other plant components. Two particle seals are also included to prevent gas mixing between the two reactors: the first one is placed between the fuel reactor and the air reactor and it can be fluidized with air or steam, the second one at the bottom of the downcomer that connects the cyclone and the fuel reactor.

The present facility has been used recently for CLC with solid fuels [39] by changing the fuel feed configuration in order to increase the residence time in the fuel reactor. A carbon stripper is also included to separate the unconverted char and the OC that is transferred in the air reactor. The setup has been also used to scale up a new experimental facility of 100 kW$_{th}$, which is operated at atmospheric pressure with solid fuels (Figure 5.3).

A similar facility has been proposed in Ref. [40] for 50 kW$_{th}$ combustor with a bubbling fuel reactor and high-velocity riser as air reactor (Figure 5.4). In this case, the solid level in the fuel reactor is controlled by a valve that is connected with an horizontal pipe at the bottom of the reactor where air stream is passing and moving the solids to the air reactor.

A different prototype of 10 kW$_{th}$ have been built and operated continuously for 200 h at 800 °C [26] using copper-based OCs: two bubbling fluidized bed reactors are used as fuel and air reactors; the solid movement is ensured by using a gravity system where the solid is going from the fuel reactor to the air reactor. At the lowest level and after O_2 regeneration, the solid is taken up by a pneumatic system that uses air as fluid vector (Figure 5.5). The same configuration has been adopted in 1.5 kW$_{th}$ experiments of CLC with bituminous coal with a Cu-based OC in order to demonstrate the CLOU process [41] and also for Fe-based materials to study the performance of the isG-CLC of 0.5 kW$_{th}$ [42, 43].

Another experimental facility has been presented in Ref. [44]. The interconnected fluidized bed configuration is designed in an annular shape circulating fluidizing bed reactor of 1 kW$_{th}$ (Figure 5.6). It is composed of two bubbling bed zones in the core and in the annular section and two risers where the OCs are circulated through each section. The use of a bubbling flow reactor as air reactor in the core section is required to achieve a complete conversion of the solid after the riser. The thermal integration is also optimized to the annular shape where the exothermic reaction in the air section is transferred to the fuel reactor, which performs an endothermic reaction.

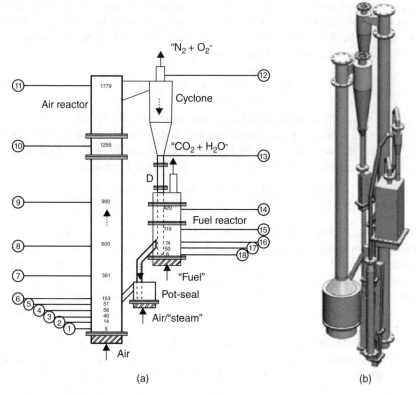

Figure 5.3   *10 kW$_{th}$ unit (a) (Source: Reproduced from Ref. [38], with permission from Elsevier) and the 100 kW$_{th}$ unit (b) (Source: Reproduced from Ref. [34], with permission from Elsevier) at chalmers university*

Another configuration based on circulating interconnected fluidized beds has been proposed in Refs. [16, 45]. The dual circulating fluidized bed reactor system is based on a very high global solid circulation rate between the air and fuel reactors and very high gas–solid contact. In this respect, turbulent fluidization (or fast fluidization) is also desirable in the fuel reactor with the perspective of system scale up (Figure 5.7). The air reactor is operated as a fast fluidized bed and the solids are transferred to a cyclone separator and a loop seal where the solid moves to the fuel reactor by using steam. The fuel reactor is fluidized with the gaseous fuel and the solid is separated in the cyclone and recirculated back to the fuel reactor. In this configuration, the solid holdup is stabilized by the direct hydraulic link between the two reactors. Moreover, the solid circulation rate is only dependent on the air flow. A 120–140 kW$_{th}$ unit has been built and successfully tested with ilmenite and NiO for more than 390 h using CH_4 and natural gas for CLC and also CLR. A new reactor of 10 MW$_{th}$ fuelled with natural gas is now under construction to be fully integrated in a steam generation unit (100 bar) for steam-assisted gravity drainage process to enhance oil recovery from bitumen [46].

A 65 kW$_{th}$ unit based on circulating fluidized bed reactors has been successfully tested from Alstom [47] operated with $CaSO_4$ as OC, and the same configuration has been scaled

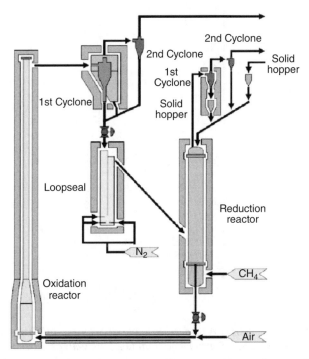

Figure 5.4   *50 kW$_{th}$ unit at the Korean Institute of Energy (KIER) and research as proposed in Ref. [40]. (Source: Reproduced from Ref. [40], with permission from Elsevier)*

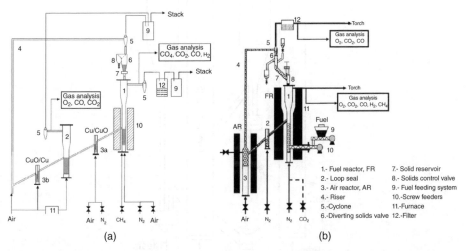

Figure 5.5   *10 kW$_{th}$ unit at Instituto de Carboquimica (CSIC) in zaragoza as proposed in Ref. [26] (a) and 1.5 kW$_{th}$ used in Refs. [41] (b) and [43]. (Source: Reproduced from Ref. [26], with permission from Elsevier)*

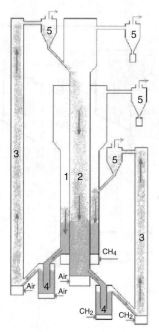

Figure 5.6 *1 kW$_{th}$ unit at Korea Advanced Institute of Science and Technology (KAIST) as proposed in Ref. [44]. (Source: Reproduced from Ref. [44], with permission from Elsevier)*

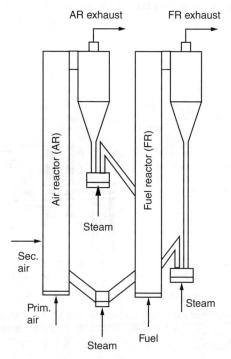

Figure 5.7 *120 kW$_{th}$ unit at TU wien as proposed in Ref. [45]. (Source: Reproduced from Ref. [45], with permission from Elsevier)*

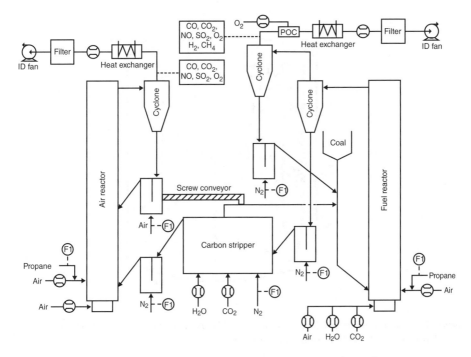

Figure 5.8   *1 MW$_{th}$ unit at TU darmstadt as proposed in Ref. [49]*

up to 3 MW$_{th}$ unit [48]. A 1 MW$_{th}$ unit (auto-thermal prototype in Figure 5.8) has been built at Technical University of Darmstadt (Germany) and the operations are carried out using ilmenite as OC for the isG-CLC mechanism for the same configuration [49].

A similar reactor design has been proposed in Refs. [50, 51] as 150 kW$_{th}$ interconnected fluidized bed combustors (Figure 5.9). The system is designed to achieve high solid circulation and high flexibility in the operation. The use of a compact system with high fuel conversion is also very relevant if the unit is pressurized. In this case, the air reactor is meant to operate in a fast fluidization regime, while the FR in both turbulent and fast fluidization regimes. The use of double loop seals is considered in order to have flexibility in both the reactors to lead the solid entrained by one reactor into the other or recirculate back. In this respect, high solid circulation is achieved with a relative low solid inventory.

The use of pressurized fluidized bed reactors has been investigated in Ref. [52], and an experimental facility of 50 kW has been built and continuously tested with bituminous coal at a pressure in the range of 1–5 bar with iron-based material for 19 h (Figure 5.10). The reactors loop consists of a fuel reactor, which is operated in a fast fluidization regime using steam (at 800 °C) as fluidization agent, and an air reactor, which is operated with air (inlet temperature at 300 °C) and works in a turbulent fluidization regime. Differently from the other configurations here described, in this reactor the driving force required for solid circulation is supplied by the gas velocity in the fuel reactor. The coal is fed to the reactor using two lock hopper units. The solid that is leaving the fuel reactor is separated in the cyclone and two loop seals are used to prevent gas mixing from the reactors. The fly ash and the unconverted coal particles are separated in the first cyclone and the remaining part

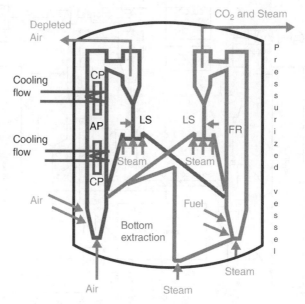

Figure 5.9   *Schematic of a 120 kW$_{th}$ unit at SINTEF as proposed in Ref. [50]. (Source: Reproduced from Ref. [50], with permission from Elsevier)*

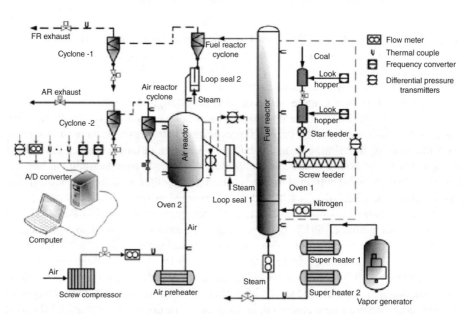

Figure 5.10   *A schematic diagram of 50 kW$_{th}$ facility in Nanjing [52]. (Source: Reproduced from Ref. [52], with permission from Elsevier)*

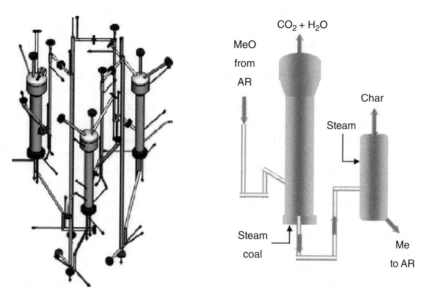

Figure 5.11   *10 kW$_{th}$ unit at IFP-lyon as proposed in Refs. [53] and [54]. (Source: Reproduced from Ref. [54], with permission from Elsevier)*

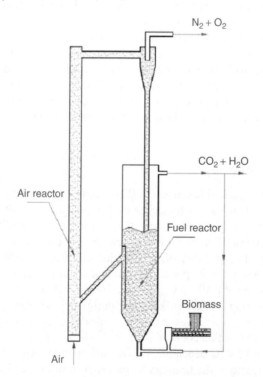

Figure 5.12   *10 kW$_{th}$ unit at Nanjing as proposed in Ref. [55]. (Source: Reproduced with permission from Ref. [55]. Copyright © 2009, American Chemical Society)*

(around 0.2%) is separated in the second cyclone of the fuel reactor. The tests have shown an improvement in the carbon conversion and thus the CO_2 capture efficiency increasing the reactor operating pressure; however, the cyclone efficiency drops and the amount of OC particle losses increases.

A $10\,kW_{th}$ prototype has been constructed and tested from IFP-Lyon [53]. Three interconnected fluidized bed reactors are considered (Figure 5.11): one reactor is operated as fuel reactor and two reactors are used as air reactors. The reactors are bubbling bed reactors. The control system is based on the use of pneumatic non-mechanical valves that allow the solid circulation to be independent of the gas flow rate in the reactors. In 2011, the same facility has been modified to be operated with coal by the addition of a carbon stripper. New analyses have been carried out in order to test the OC activity, the effect of the temperature in the coal conversion and the gasification reaction [54] that occurs in the fuel reactor.

A $10\,kW_{th}$ continuous reactor of interconnected fluidized beds has been discussed in Ref. [55] for CLC with biomass (Figure 5.12). The prototype is composed of a fast fluidized bed as air reactor, a cyclone and a spout-fluid bed as fuel reactor. In this case, the spout-fluid type reactor is adopted as fuel reactor in order to have a strong solid mixing between the biomass and OC particles and a long residence time. The spout-fluid reactor is designed to have two different compartments. In the first part, the reaction chamber is located where the OC and the biomass are combined to produce exhaust gas and solid species (metal oxide and unconverted fuel), while the second part contains the inner seal that is located at the top and it is used to allow solids movement to the air reactor. The fuel reactor is fluidized by using exhaust gas recirculation (Table 5.2).

A different configuration has been developed at the Ohio State University [56, 57]. The system has been proposed both for CLC and steam–iron processes for H_2 production. In this configuration, the fuel is fed to the reducer, which is located on the top of the system, and the gas moves in countercurrent to the OCs. After the reduction, the first oxidation occurs in a moving bed, which is operated with steam, and H_2 is produced. Subsequently, the OCs are moved to a riser where the oxidation is completed and recirculated in the loop (Figure 5.13).

5.3.2   Packed Bed Reactors

Dynamically operated packed bed reactors (PBR) have been recently proposed for CLC [59]. In this case, solids remain stationary in the bed and gases are alternatively switched. No solid circulation is required, and operation at pressurized conditions can be easily accomplished. However, HT valves are required, the gas conditions at the reactor outlet are not constant and the feasibility must be verified. Moreover, PBRs are operated with bigger particles (to minimize the pressure drop) that could lead to intra-particle diffusion limitations [60]. The use of PBR for CLC is based on the different reaction and front velocities along the bed during a gas–solid reaction: during the oxidation phase, the gas–solid conversion proceeds very fast along the reactor, while the heat front velocity is significantly slow so that the heat of reaction is stored in the bed, which is heated up to high temperature (i.e. up to more than 1200°C depending on OC and operating conditions) (Figure 5.14); after the oxidation reaction, the heat can be removed from the bed (heat removal phase) by blowing additional gas that is produced in this case at high temperature and can be

List of different fluidized bed reactor concepts for CLC.

Concept	Reactor configurations		Gas velocities (m/s)		(ΔX_s)	Fuel type	Temperature (°C)		Pressure	Inventory (kg)		Oxygen carriers	Ref.	Short Description of Setup
	AR	FR	AR	FR			AR	FR		AR	FR			
Two-interconnected fluidized bed	High velocity riser	Bubbling fluidized bed	0.75–1.5	0.1–0.3	–	–	–	–	Atmospheric	0.1–0.2		Fe, Ni based	[38]	A cold flow model is used to calculate the effect of gas leakage from the reactors with diameter of 0.19 m and length of 0.5 m
Two-interconnected fluidized bed	High velocity riser	Bubbling fluidized bed	4–10 u_t*	5–15 u_{mf}**	–	Methane	950	950	Atmospheric	4.8–8.3		Fe, Ni based	[37]	Experiments have been carried out in a cold and hot lab-scale prototype of 5–10 kW$_{th}$
Two-interconnected fluidized bed	High velocity riser	Bubbling fluidized bed	2.9/4.9	0.14	–	Coal	800 950	950 1000	Atmospheric	9	6	Ilmenite and Mn ore	[39]	The experiments have been carried in a 10 kW$_{th}$ unit with different coal and petcoke with a char conversion between 0.7 and 0.9.
Two-interconnected fluidized bed	High velocity riser	Bubbling fluidized bed	3	0.05	–	Methane	870	890	Atmospheric	33		Ni, Co based	[40]	The experiments have been carried in a 50 kW$_{th}$ unit with a gas conversion higher than 99% for 25 h.
Two-interconnected fluidized bed	Bubbling fluidized bed	Bubbling fluidized bed	0.10	0.50	41%	Methane	800	800	Atmospheric	21		Cu based	[26]	The experiments have been carried out in a hot lab-scale prototype of 10 kW$_{th}$.
Two-interconnected fluidized bed	Bubbling fluidized bed	Bubbling fluidized bed	0.11	0.40	80%	Coal	900 950	900 950	Atmospheric	1.6	0.4	Cu, Fe ilmenite	[41, 43]	The experimental setup has been tested with different oxygen carriers. The CLC and isG-CLC processes have been carried out in prototype lab scale (1.5 kW$_{th}$ and 0.5 kW$_{th}$)

Table 5.2 (continued)

Concept	Reactor configurations		Gas velocities (m/s)		(ΔXs)	Fuel type	Temperature (°C)		Pressure	Inventory (kg)		Oxygen carriers	Ref.	Short Description of Setup
	AR	FR	AR	FR			AR	FR		AR	FR			
Dual circulating fluidized bed	Fast fluidized bed	Fast fluidized bed	2–8	2	10–20%	Natural gas H_2	750 900	750 900	Atmospheric	10–15	16–23	Cu, Fe, Mn, Ni based	[16]	Experiments in continuous operation have been carried out at different temperatures, solid circulation rate and air to fuel ratio. System has been tested also for CLR.
Two-interconnected fluidized bed	Fast fluidized bed	Fast fluidized bed	–	1–5.5	–	Coal	970	950	1–5 bar	120		Fe based	[52]	Experiments have been carried out at pressurized conditions in a test facility of 50 kW$_{th}$. The unit has been continuously operated for 19 h with steady coal feeding. Stable operations at pressurized conditions have been realized for 13.5 h. The total coal conversion into CO_2 at 5 bar is about 84%.
Annular shape double circulating fluidized bed reactor	Two-section high velocity riser	Two-zone bubbling fluidized bed	–	0.05–0.07	–	Methane	650 850	650 850	Atmospheric	–	–	Fe–Ni mixture	[44]	The experimental facility is designed to achieve an efficient heat transfer between the fuel and the air reactor. The tests have been carried out in 1 kW$_{th}$ unit.

Three interconnected fluidized bed	Bubbling fluidized bed (two)	Bubbling fluidized bed	0.61	0.48/0.56	8.6–34%	Methane	800	800	Atmospheric	33	Ni based	[53]	Several solid circulation ratios, temperatures and air to fuel ratios have been tested in a 7–10 kW_{th} CLC prototype.
Double loop circulating fluidized bed	Fast fluidized bed	Fast fluidized bed	0.86–1.94	1.1	20%	Methane	1000	1000	Atmospheric	120	Mn based	[51]	The tests have been carried out in a cold flow setup and a design is proposed for a 150 kW_{th} high-temperature unit (around 1000°C)
Two-interconnected fluidized bed	High velocity riser	Spout-fluidized bed	1.56	1.77–0.06	–	Biomass char	740	920	Atmospheric	12	Fe based	[55, 58]	10 kW_{th} CLC prototype using biomass has been tested in continuous operation for 30h.
Two/three-interconnected fluidized bed	High velocity riser	Moving bed	2.05	1.39	Lower than 50%	Syngas coal	900	900	Atmospheric	3.76	– Fe based	[56, 57]	The system is designed for CLC and steam-iron production with two moving beds (the reducer and the first oxidizer with steam). An experimental facility of 25 kW_{th} has been operated with syngas and coal.

* u_t = Gas velocity at the gas terminal settling.
** u_{mf} = minimum fluidization velocity.

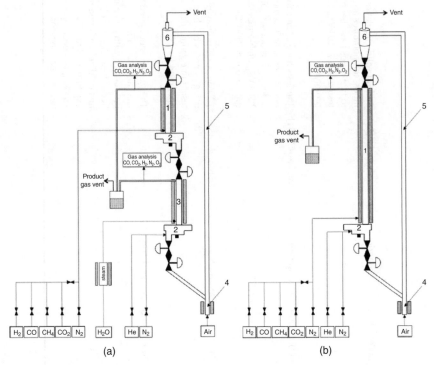

Figure 5.13   *Moving bed concept for CLC and H₂ production at ohio state university as reported in Ref. [57]. (Source: Reproduced from Ref. [57], with permission from Elsevier)*

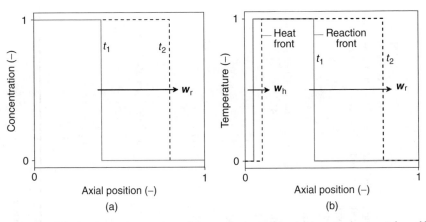

Figure 5.14   *Schematic representation of the evolution of the (dimensionless) axial profile of (a) The gaseous reactant concentration and (b) The temperature (the figure is taken from Ref. [59]). (Source: Reproduced with permission from Ref. [59]. Copyright © 2007, American Chemical Society)*

efficiently converted in a thermodynamic cycle. After the solid is completely oxidized (and heat removed), the reactor is switched to the reduction operation (reduction phase) and the solid is then reduced by converting the syngas into H_2O and CO_2. The purge cycle is also necessary to prevent that the fuel is in direct contact with air.

Differently from the interconnected fluidized bed reactors, the dynamically operated PBRs have to be operated with proper heat management strategies in order to produce hot gas that are suitable for a combined cycle. In a PBR, the maximum solid temperature, and, thus the maximum gas temperature for the thermodynamic cycle, is usually achieved during the oxidation phase (which is always strongly exothermic as shown in Figure 5.15) and the maximum solid temperature increase is calculated as follows [61]:

$$\Delta T_{MAX} = \frac{(-\Delta H_{R,ox})}{\dfrac{c_{p,s}MW_{OC,act}}{y_{OC,act}\xi} - \dfrac{c_{p,O_2}MW_{O_2}}{y_{O_2}}} \tag{5.9}$$

The maximum temperature increase in a PBR is independent of the gas flow rate and it is dependent on the heat of reaction, the heat capacity of the gas and especially the solid material weight fraction. In particular, the active weight content in the OC ($y_{OC,act}$) plays an important role in the maximum temperature increase. In the case of reduction, the maximum temperature change depends on the fuel (methane or H_2/CO species as shown in Figures 5.16 and 5.17) and the OCs used, and it can be both exothermic and endothermic. The reduction reactions between the fuel species and the OCs are mainly affected by the initial solid temperature that strongly influences the kinetics.

To efficiently integrate this technology into a power plant, an adequate heat management of the reactors is extremely important. In particular, an important boundary condition

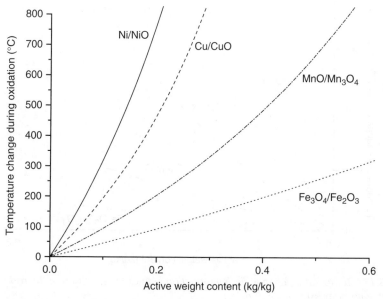

Figure 5.15 *Temperature increase during oxidation [62]. (Source: Reproduced from Ref. [62], with permission from Elsevier)*

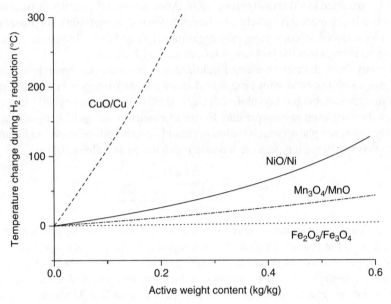

Figure 5.16   *Temperature change during reduction with H_2 [62]. (Source: Reproduced from Ref. [62], with permission from Elsevier)*

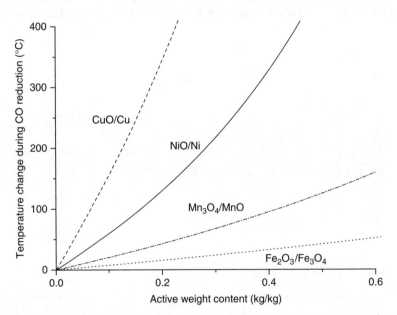

Figure 5.17   *Temperature change during reduction with CO [62]. (Source: Reproduced from Ref. [62], with permission from Elsevier)*

dictated by the power island has to be taken into account: the hot stream needs to be produced at nearly constant temperature and mass flow rate to preserve the gas turbine expander from thermal/mechanical cycling stress and from fluid dynamic instability. On the other hand, an efficient operation of the packed beds imposes that (i) solid conversion needs to be almost complete in order to increase the reactor capacity; (ii) kinetics of gas/solid reactions have to be fast enough in order to avoid fuel slip; (iii) the solid temperature cannot reach very high temperature to avoid deactivation of solid materials and reactor materials could be damaged as well. These requirements put the PBR technology for CLC in front of relevant technical challenges to assess their feasibility and competitiveness with other CCS technologies.

The high-temperature gas production can be achieved by selecting a proper OC and adjusting the active weight content. In order to have very fast kinetics for the reduction, the OC must be reactive also at low temperatures. The solid temperature profile results from the previous phase: when a gaseous stream is fed to the reactor, after the reaction front is passed, the temperature of the initial part of the bed tends to be close to the inlet gas temperature because the solid is continuously contacted with a constant temperature stream and the heat front moves to the end of the reactor. If the temperature of the inlet gas is not high enough, the OC does not react with a sufficient kinetics with the gaseous stream in the following phase and the process cannot proceed.

In the study by Hamers *et al.* [62], the heat management with Cu-based material (12.5% wt on Al_2O_3) using syngas (H_2/CO) is discussed. The solid conversion during reduction occurs properly with a complete conversion of CuO, which promptly reacts with syngas and, after the purge and the oxidation phases, it is possible to produce air at constant temperature (850 °C) as depicted in Figure 5.18. This temperature is not high enough to be used

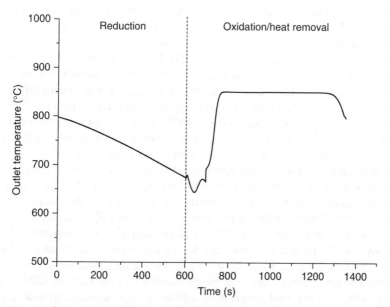

Figure 5.18 *A schematic of temperature profile at the reactor outlet for the single-stage CLC with PBR [62]. (Source: Reproduced from Ref. [62], with permission from Elsevier)*

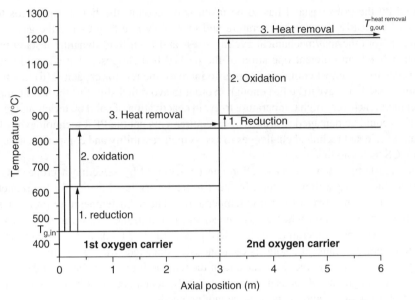

Figure 5.19   *Axial temperature profile along the PBR in the two-stage CLC concept [62]. (Source: Reproduced from Ref. [62], with permission from Elsevier)*

efficiently in a combined cycle because of the temperature limitation that concerns the low melting point of CuO. Another possibility is the two-stage CLC in which the cycle is carried out by using two reactors in series in which two different OCs are used (Figure 5.19): the first reactor is operated with CuO (12.5%wt on Al_2O_3) in the range of temperature between 450 °C and 850 °C, while the second reactor is operated with Mn_3O_4 (30%wt on Al_2O_3), which works in the range of 850–1200 °C. The minimum temperature in the second reactor is obtained by using the hot air as feed gas at 850 °C produced during the heat removal of the first bed. Due to the different heat management strategies, the high-temperature pressurized air is produced at 1200 °C during the heat removal, which is suitable for an efficient combined cycle. Another important achievement is the limited temperature increase that is required in the single reactor (respectively 400 °C in the reactor operated with CuO and 350°C in the reactor operated with Mn_3O_4) despite that the gas stream being fed at 450 °C (Figure 5.20) and the gas for the heat removal produced at 1200 °C.

A different approach for the heat management is addressed in Ref. [63] where the system is operated with ilmenite ($FeTiO_3$) as OC and syngas from coal gasification plant. The kinetics involved in the reduction reactions is not fast enough to achieve complete conversion if the minimum solid temperature is lower than 600 °C. In this case, the heat management strategy is based on a different phase sequence. The heat removal phase is carried out after the reduction instead of after the oxidation phase (Figure 5.21). In this way, the reduction reactions occur when the bed is at the maximum temperature (after the oxidation). The maximum temperature for the combined cycle is almost 1200 °C because during the reduction the solid temperature does not change significantly (about 10 °C or lower) as depicted in Figure 5.22. The heat removal cannot be carried with air because the

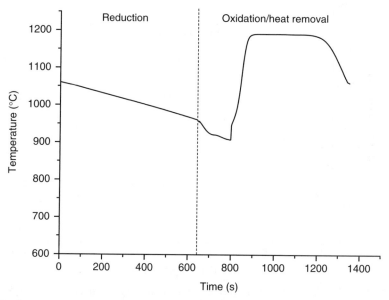

Figure 5.20   *Temperature profile at the reactor outlet for the two-stage CLC with PBR [62].*
(Source: *Reproduced from Ref. [62], with permission from Elsevier)*

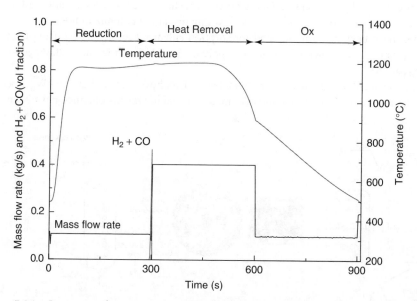

Figure 5.21   *Reactor outlet gas conditions (temperature, mass flow rate and gas concentrations) during the entire cycle Ref. [63]. (Source: Reproduced from Ref. [63], with permission from Elsevier)*

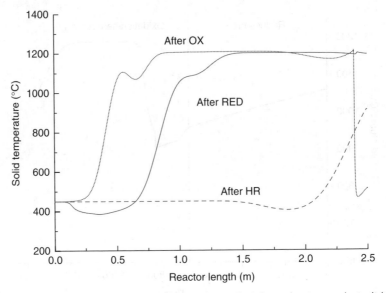

Figure 5.22 *Axial solid temperature profiles at the end of the reduction cycle (solid line), at the end of oxidation cycle (dotted line) and at the end of heat removal cycle (dashed line) [63]. (Source: Reproduced from Ref. [64], with permission from Elsevier)*

solid is completely reduced so an inert gas (i.e. N_2) has to be used. As a matter of fact, this approach is suitable for the OCs with a slow kinetics at lower temperature.

PBRs have been also investigated for sorption enhanced reforming by using Ca-based sorbent for CO_2 combined with CuO to supply the heat of reaction during the regenerative stage [64, 65].

An experimental setup has been built at the TU Eindhoven (NL) of 20–30 kW$_{th}$ operated at 10 bar and 1000 °C with ilmenite (Figure 5.23), and the results obtained have been used

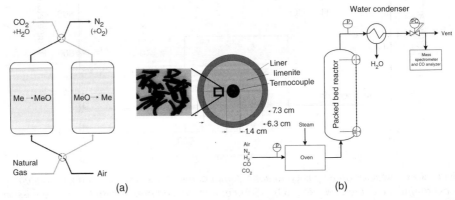

Figure 5.23 *(a) CLC combustion with packed bed reactor as discussed in Ref. [66] and (b) the experimental setup at the Technical University of Eindhoven (NL)*

to scale-up the reactor up to $500\,kW_{th}$, operated with syngas from coal gasification at the ELCOGAS Puertollano IGCC power plant.

5.3.3 Rotating Reactor

A novel reactor concept for natural gas fuelled CLC has been proposed in Ref. [67]. In this case, the solid material is trapped in the reactor, the gaseous streams are not moving while the reactor is moving like a wheel so that the OCs are converted cyclically with air or fuel. The design proposed considers the gas stream that are flowing radially outwards through the metal oxide bed. A radially directed gas flow is also preferable since the volume expansion (both for the increase in the number of moles and the increase in the temperature if the reaction is exothermic) is compensated with an increase in the volume in the radial direction. For the system proposed, some purge stages are also required between the reduction and oxidation by means of steam.

The configuration is based on using different slices of reactor for the phases that alternatively involve the solid material conversion (Figure 5.24). Due to the different mass flow rates, the natural gas feed sector is 60°, while the air feed sector is 240° over 360°.

Several tests have been carried out with a prototype laboratory-scale reactor aiming to find the best performance of the system in terms of CO_2 capture rate and methane conversion using 120 g of 1.5 mm spherical particle CuO with Al_2O_3 (10% wt of active material) [68, 69]. The operating temperature has been tested between 650 °C and 800 °C, leading to an increase in the methane conversion (up to 80%); increasing the methane flow rate (from 18 to 48 mL/min) leads to an increase in CH_4 conversion but reduces the CO_2 purity. The tests have been carried out with a reactor rotation speed ranging 1–4 rounds/min.

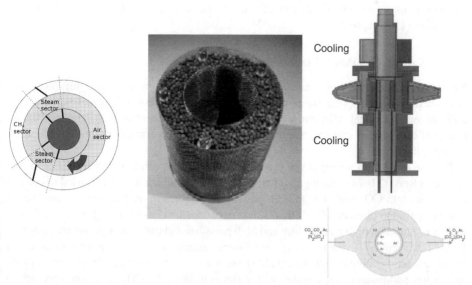

Figure 5.24 *Rotating bed reactor concept for CLC as proposed in Refs. [67, 68]. (Source: Reproduced from Refs. [67, 68], with permission from Elsevier)*

The main problem of the rotating reactor concept is the gas mixing. This is because a rotating unit must have some space between the parts moving relative to each other to avoid much friction. Due to the gas slippage, the CO_2 purity is very low (in the range of 20–65%) because the stream is diluted with unconverted methane and the N_2 from the oxidative phase.

5.4 The Integration of CLC Reactor in Power Plant

CLC has been mainly proposed for power generation and CO_2 capture. So far several studies have been carried out; most of them are based on the use of gaseous fuels (natural gas or syngas from coal gasification) using interconnected fluidized bed reactors at pressurized conditions. The simplest power plant layout for power production based on CLC process considers a combined cycle where the air reactor substitutes the combustor of the gas: the gaseous fuel enters the reactor operating with solid metal in an oxidized state giving oxygen to the fuel, which is, therefore, oxidized to CO_2 and H_2O. Reduced metal oxides then flow into a reactor operated in oxidation, where the OC is oxidized with air. The flow of solid, oxygen-rich metal oxides and vitiated air exiting the oxidation reactor enters a cyclone, where the solid oxides are separated from the gas stream and recycled to the reduction reactor.

In this section, a literature review is presented of the different power plants that have been designed and fully integrated with a CLC unit for efficient CO_2 capture. In the first part, the natural gas-fuelled plants are described and presented; in the second part, the coal-fuelled power plants operated in different configurations are also reported. In the last part, a comparison between the fluidized bed configuration and PBR is discussed. Both configurations are compared with the reference technology for coal-based power plant with and without CO_2 capture.

5.4.1 Natural Gas Power Plant with CLC

Nowadays, natural gas combined cycle (NGCC) power plant represents the most efficient thermodynamic cycle for power production. The modern unit achieves more than 60% of net electrical efficiency. NGCCs have been discussed with pre-combustion and post-combustion capture leading to around 8–10 percentage points of efficiency penalty [70].

The use of CLC in a natural gas power plant based on a combined cycle has been discussed in Refs. [71–74], leading to a first plant design in which the CLC is used to convert natural gas into CO_2 and H_2O at high pressure (8–20 bar) in the fuel reactor and the air reactor is used to produce HT air that is then expanded in a gas turbine (Figure 5.25). The CO_2 is produced at high temperature and high pressure, and two main options are proposed for the heat recovery. In the first case, the CO_2-rich stream is first expanded and then cooled to ambient temperature by generating HP steam that is combined in a multi-pressure HRSG to produce additional electric power with a steam turbine; the CO_2 is then compressed to the delivery pressure for the final storage. In the second case, the CO_2 is directly cooled down to ambient temperature with higher steam production than in the previous system and the CO_2 is sent to a final compression unit to high pressure. The systems proposed in

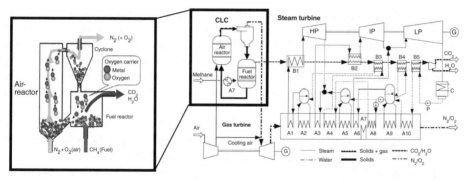

Figure 5.25 *NGCC with the CLC plant layout as proposed in Ref. [73]. (Source: Reproduced from Ref. [73], with permission from Elsevier)*

Refs. [72, 73] are based on the use of Ni- and Fe-based OCs. In Ref. [72], different GT compressor ratios are investigated in order to optimize the performance of the plant. The maximum temperature of the GT is dictated from the maximum temperature acceptable from the solid material in the air reactor. In the study by Wolf [73], a sensitivity analysis is carried out by using an air reactor temperature of 1000 °C and 1200 °C. By increasing the maximum solid temperature, the efficiency increases from around 49% up to 52.4% with a CO_2 capture rate of 100%. In the study by Consonni [72], the maximum solid temperature has been varied from 850 to 1050°C with an electrical efficiency of 43–48% and CO_2 capture rate of 100% (Figure 5.26). A supplementary natural gas post-firing has been adopted to improve the combined cycle efficiency up to 52.5% (with reduced CO_2 capture rate to 54%) by increasing the maximum turbine inlet temperature (TIT).

Similar results are also reported in the study by Chi [75], which discussed the GT performance in case of off-design operations due to a change in the ambient temperature.

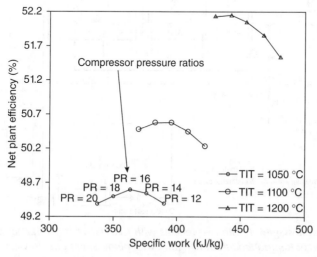

Figure 5.26 *Sensitivity analyses on TIT and compressor ratio for NGCC with CLC as discussed in Ref. [76]. (Source: Reproduced from Ref. [76], with permission from Elsevier)*

The GT-off design performance has also been investigated by Naqvi *et al.* [76]. The use of CO_2-rich stream gas turbine is also included in the analysis, which accounts for 14% of the electric gross power produced. The net electric efficiency is around 52% in the case of maximum solid temperature of 1200°C (Figure 5.26), and the electric efficiency decay due to the part-load operation is lower than that in the conventional combined cycle (higher than 94% of the nominal electrical efficiency in the case of 54% of full load) if an advanced control strategy is adopted.

The increase of maximum temperature of the gas turbine has been discussed in the study by Lozza [77], where a different configuration based on a three-stage CLC with steam-iron process is proposed. In this configuration, the maximum solid temperature is around 830 °C and H_2 is used as additional fuel to use a heavy duty natural gas–fired gas turbine, which works with TIT equal to 1335 °C. The H_2 required is produced in a steam-iron process that is carried out in a countercurrent moving bed reactor where the reduced metal (FeO) is oxidized to Fe_3O_4 and the final conversion of the OC occurs in the combustor reactor that works with air. This configuration leads to a net electrical efficiency of 51.3% CCR approaching 100%. The present configuration has the advantage of being suitable for H_2 and electricity co-production [78].

A different approach has been discussed in the studies by Ishida [79] and Naqvi [80] to improve the electrical efficiency of the thermodynamic cycle based on a multi-stage CLC (Figure 5.27). The air from the compressor is fed to an HP air reactor and the HT

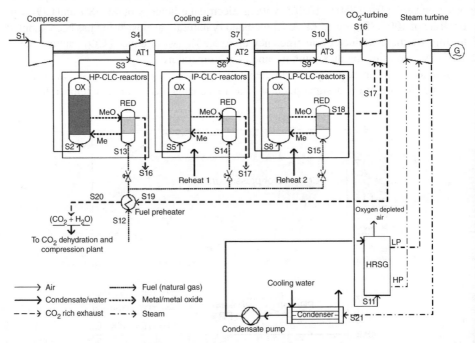

Figure 5.27 *Multi-stage GT with two reheats with CLC fed with natural gas as proposed in Ref. [80]. (Source: Reproduced from Ref. [80], with permission from Elsevier)*

O_2-depleted air is partly expanded in a GT and then sent to another CLC unit that is operated at IP pressure where the oxidant gas is heated up to 1200 °C. The N_2-enriched stream is sent to another GT cylinder, which expands the gas to a lower pressure where an additional CLC unit is operating or the HRSG is used for the efficient heat recovery. The natural gas is preheated and sent to the fuel reactors that are operated at different pressure levels. The CO_2 streams that are produced at high temperature but at different pressure levels are then expanded in a CO_2/H_2O turbine to atmospheric pressure, cooled down by preheating the fuel feeding and re-compressed up to the final pressure. The net electric efficiency of the configurations here proposed is around 51–53% depending on the number of reheats. As a matter of fact, this configuration leads to a very high efficiency due to the thermodynamic improvement that a higher amount of thermal power is transferred to the gas stream at higher temperature in comparison with the conventional gas turbine.

A different natural gas power cycle is proposed in the studies by Brandvoll [81] and Ishida [82]. The CLC is applied to a humid air turbine concept (HAT cycle) with regenerative heat exchanger. The air is compressed in a multi-stage intercooled compressor up to 20 bar, and it is sent to a saturator that is operated with water at about 200 °C, which increases the H_2O content and the temperature in the air stream. The air stream is then heated up to 500°C in a series of two different gas/gas regenerative heat exchangers and sent to the air reactor where oxidation takes place. The O_2-depleted air at 1200 °C is sent to the gas turbine, where it is expanded and subsequently cooled in the first recuperative heat exchanger and afterwards cooled to about ambient temperature by heating the water for the saturator. The natural gas is sent to the fuel reactor after being heated up to 500 °C. The CO_2 is produced at 1100/1200 °C and then expanded in a CO_2/H_2O gas turbine and gas cooling is carried out in a series of heat exchangers where the heat recovery is achieved by heating, respectively, fuel, humid air and water for the saturator. This configuration leads to an electric efficiency higher than 55% with 100% of CO_2 capture that makes the system very attractive and potentially very promising. Despite the highest net electric efficiency reported, the configuration proposed contains some components unusual for large-scale commercial power plants (i.e. highly intercooled air compression, recuperative cycle) that make the economic advantage over the conventional combined cycle configuration doubtful.

A CLC power plant operating at atmospheric conditions with steam generation can be a short-term alternative to a gas turbine with pressurized reactors as they have been discussed in the previous cases. Similarly, the HT required to perform efficiently a combined cycle may lead to some limitations in the use of different materials as OCs. A different arrangement is proposed by Naqvi [71]: the CLC reactors are operated at ambient pressure and the maximum temperature is taken equal 850 °C using cooling walls in the oxidation reactor by generating HP steam. A blower is employed for atmospheric air in order to compensate the pressure drop and air preheating is carried out in a Ljungstrom-type heat exchanger. The power unit is designed to work in an advanced steam cycle operated at 160 bar with reheat and maximum steam temperature of 560 °C. The efficiency assessed is 40.1% with a fully integrated steam cycle with CLC loop. This value is comparable with maximum efficiency of modern steam cycles fuelled with natural gas (around 41%).

5.4.2   Coal-Based Power Plant with CLC

Solid fuels such as coal represent the most important fuel source for power generation, and it is the most responsible for the CO_2 emissions in the atmosphere from power plants. Coal-fired power plants are mainly based on a USC operated with a boiler at atmospheric pressure with an electric efficiency of 45%. The most studied system for CCS from coal power plant is based on post-combustion system where the exhaust gases are washed with amine-based solvents that remove selectively CO_2. The effect on plant performance is an electrical efficiency decay of about 10 percentage points [83] mostly depending on the energetic cost of the amine regeneration. Chilled ammonia plants have been proposed as alternative to reduce the efficiency decay [84]. Another option for coal power plant is the use of the integrated gasification combined cycle (IGCC) where the fuel is decarbonized after WGS reactor unit and CO_2 separation process. The CO_2 pre-combustion capture from IGCC is expected to reach an electrical efficiency of 36% (based on coal LHV) with 90% of carbon capture rate [2].

CLC technology is particularly suitable for coal-fired power plants because the reactor loop can be used in two different configurations. In the first case, the CLC occurs with syngas that is previously produced in a coal gasification plant and the power plant is designed as the typical IGCC so that the CLC is carried out at pressurized conditions, as already discussed for NGCC. In the second case, the CLC reactors can substitute the coal boiler and the OCs react with the solid fuel as proposed for the isG-CLC and the CLOU mechanisms.

The first CLC plant layouts integrated with coal gasification unit have been proposed in Ref. [85]: the syngas from coal gasification is sent to the fuel reactor that is operated with NiO. The CO_2-rich stream produced at high pressure is sent to a turbine, expanded, cooled down in a HRSG, and after the water condensation the CO_2 is compressed for the final storage. The air is compressed and then fed to an air reactor. The O_2-depleted hot air produced in the air reactor is fed to a GT. Afterwards, heat is recovered by generating steam in a three-pressure level HRSG, which is expanded in a steam turbine (Figure 5.28a). A two-stage CLC reactor has been discussed in which the CLC unit is operated at different pressures and O_2-depleted air expansion is carried out with an intermediate reheat in the LP oxidation reactor. The electrical efficiency calculated is in the range of 34.5–36.5% depending on the arrangements (Figure 5.28b). In this work, the integration of air separation unit (ASU) that is used for the gasifier and the air compressor is also assessed with an efficiency gain of 0.5–0.7 percentage points.

An extensive thermodynamic assessment of CLC technology with coal gasification power plant has been proposed in Ref. [86] (plant layout is shown in Figure 5.29). Syngas is produced in a coal gasification process using a dry feed, oxygen-blown entrained flow gasifier, followed by a syngas cooling system and an acid gas removal unit. After the syngas is humidified and preheated, it is sent to the fuel reactor and oxidized in CO_2 and H_2O. The integrated CLC reactors work under pressurized conditions (20 bar) and two different systems are assessed for the exhaust cooling: in the first case, CO_2 is sent to a gas turbine to be expanded to ambient temperature and then, sent in a heat recovery steam generator for cooling; in the second case, the high-temperature sensible heat of the exhaust is directly recovered producing HP steam and CO_2 compression occurs starting with a gas stream at 18 bar. This analysis discusses the performance at different maximum CLC solid temperatures (1200 and 1300 °C) and different maximum pressures at which

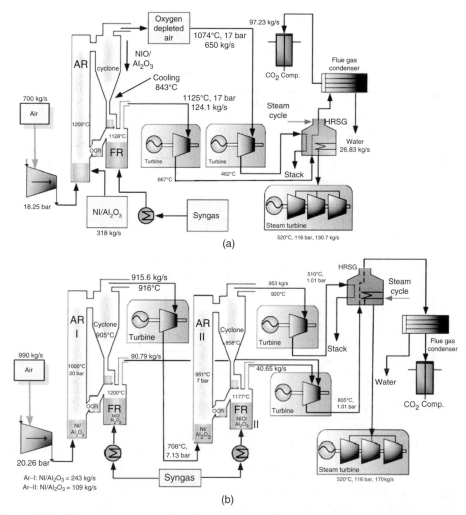

Figure 5.28 *Single-stage (a) and double-stage (b) CLC fed with syngas from coal gasification. (Source: Reproduced from Ref. [85], with permission from Elsevier)*

steam is produced for the steam cycle. The electric efficiency calculated is in the range of 39.3–40.6% (based on fuel LHV), which is about 2–3 percentage points higher than the efficiency calculated with the IGCC with pre-combustion capture using physical solvents (i.e. Selexol® process). Compared with IGCC with pre-combustion capture by physical absorption, the steam turbine power output is almost equal (or greater) to gas turbine power output so the steam cycle performance affects more the overall electric efficiency than the other IGCCs considered in the paper.

By increasing the maximum temperature at the air reactor and, thus the TIT of the gas turbine, the electric efficiency increases (about +1 percentage points of the net electric efficiency every 100 °C); the lower gain in electric efficiency compared to the natural gas power

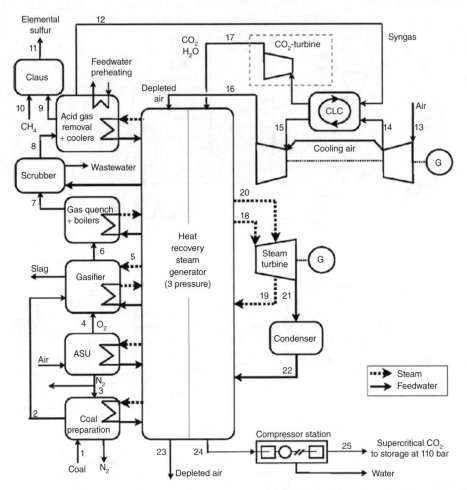

Figure 5.29   *Schematic of IGCC with CLC as proposed in Ref. [86].   (Source: Reproduced from Ref. [86], with permission from Elsevier)*

plant with CLC is the lower effect associated to the plant performance on the maximum TIT and the higher dependence on the steam cycle performance (Table 5.3).

Chemical looping process has been integrated in the IGCC to produce H_2-rich stream using the steam-iron process that is converted in a modern gas turbine heavy duty working with state-of-the-art turbine Siemens SGT5-8000H [87] as depicted in Figure 5.30. The coal is converted in two different gasifiers (Shell type and British Gas/Lurgi), and the sulfur compounds from the coal syngas are removed in a high-temperature gas desulfurization operated with zinc ferrite ($ZnFe_2O_4$). The heat recovery is based on a three-pressure level HRSG with reheat, and the net electrical efficiency of the system is in the range of 39.7–44.8% (based on LHV of coal) due to the use of the HGCU unit; an advanced gas turbine and the steam cycle have been optimized according to the heat sources of the system. The results are presented in Figure 5.31.

Table 5.3 *List of main assumptions and performance of CLC used with syngas from coal gasification plant from Ref. [86].*

	CLC-1	CLC-2	CLC-3	CLC-4	CLC-5
Air reactor temperature	1200°C	1300°C	1300°C	1200°C	1300°C
Fuel reactor temperature	1051°C	1121°C	1121°C	1051°C	1121°C
CO_2 turbine	no	no	no	yes	yes
Temperature of N_2 at HRSG inlet	455°C	498°C	498°C	455°C	498°C
Temperature of CO_2 at HRSG inlet	1051°C	1121°C	1121°C	572°C	620°C
Live steam pressure	127 bar	127 bar	280 bar	130 bar	128 bar
Live and reheat steam temperatures	600/435°C	600/478°C	600/478°C	450/429°C	500/451°C

	IGCC-1	IGCC-2	CLC-1	CLC-2	CLC-3	CLC-4	CLC-5
coal	2250.1	2250.1	2250.1	2250.1	2250.1	2250.1	2250.1
syngas to gas turbine	1784.7	1784.7	1805.9	1805.9	1805.9	1805.9	1805.9
gas turbine system power output	625.5	675.9	503.3	502.3	502.3	658.5	665.8
steam turbine power output	366.5	345.9	485.4	512.6	507.4	387.0	409.4
total power consumption	207.1	207.6	141.4	141.5	146.0	196.2	196.7
air separation unit	117.9	118.8	100.0	100.0	100.0	100.0	100.0
CO_2 compression and drying	36.3	36.3	23.8	23.9	23.9	79.9	80.1
AGR, SRU and TGT	36.5	36.5	0.4	0.4	0.4	0.4	0.4
steam cycle pumps	6.2	5.8	6.6	6.7	11.1	5.2	5.5
gasification island and coal preparation	7.3	7.3	7.1	7.1	7.1	7.1	7.1
other	3.0	3.0	3.5	3.5	3.5	3.5	3.5
net power production	785.0	814.1	847.3	873.4	863.7	849.3	878.6
cycle efficiency (HHV)	34.9%	36.2%	37.7%	38.8%	38.4%	37.7%	39.0%
specific CO_2 emissions [kg/MWh$_{el}$]	128.43	123.82	7.43	7.21	7.29	7.30	7.05

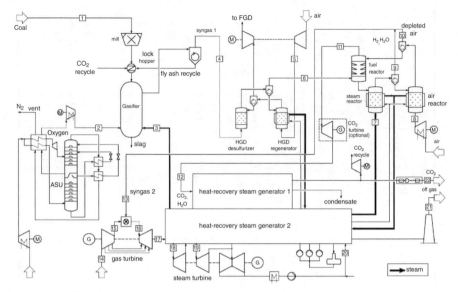

Figure 5.30   *Plant layout of IGCC integrated with CLC and steam-iron process as discussed in Ref. [87].   (Source: Reproduced from Ref. [87], with permission from Elsevier)*

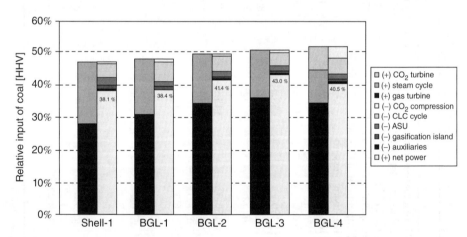

Figure 5.31   *Plant performance of IGCC integrated with CLC and steam-iron process (Results From Ref. [87]). (Source: Reproduced from Ref. [87], with permission from Elsevier)*

An alternative use of chemical looping has been discussed by Cormos [88]. The plant configuration is based on a power and H_2 co-generation from coal gasification power plant: the syngas produced is fed to a fuel reactor and converted into H_2O and CO_2 using magnetite (Fe_3O_4) as OC. The resulting Fe/FeO is then oxidized into Fe_3O_4 in a steam reactor where H_2-rich gas is produced. In this way, the original fuel is converted into H_2 that can be suitable for power generation or for other uses. The analysis is based on the main key parameters that affect the plant performance such as the effect of the gasification technology, the ASU integration with air compressor, the ratio of H_2 to power and the steam

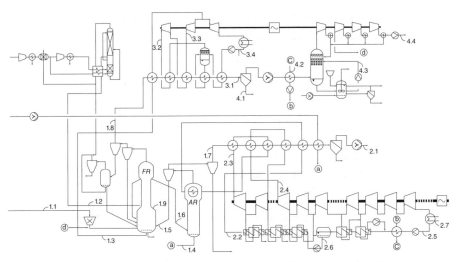

Figure 5.32 *Plant layout of direct solid CLC presented in Ref. [90]. (Source: Reproduced from Ref. [90], with permission from Elsevier)*

integration with the chemical looping unit. The electrical efficiency of the plant when it is operated only for power production is 38.8% with a carbon capture rate (CCR) of 99.5%.

The use of three-reactor CLC with the integration of a steam-iron process has also been investigated in a configuration with solid fuel feeding directly to the fuel reactor in Ref. [89]. The coal conversion occurs directly in the fuel reactor at pressurized conditions and the H_2 that is produced in the steam-iron reactor is sent to a combined cycle. The OC used is Fe_2O_3, which is partly oxidized from Fe/FeO to Fe_3O_4 in the steam reactor and is subsequently completely oxidized to Fe_2O_3 in the air reactor before to be used to convert coal. The present system achieves 41.1% of net electric efficiency with 99.3% of CCR. In order to improve the coal conversion, some O_2 is produced in an external cryogenic air separation unit and fed to the fuel reactor.

The possibility to convert directly coal with OC allows avoiding the syngas production unit as reported in Ref. [90]. In this case, the coal conversion is associated with the CLOU mechanism. The CLC reactor system is based on interconnected circulating fluidized beds at atmospheric pressure (Figure 5.32). Two different steam cycles are used for the heat recovery: steam is produced at 280 bar and 580 °C and reheat is carried out at 56 bar and 600 °C using the sensible heat of the stream exiting the air reactor, while a sub-critical steam cycle is used in order to assess an efficient fuel reactor exhaust heat recovery with steam produced at 180 bar and maximum steam temperature of 550 °C. The analysis also includes some unconverted carbon leaving the FR, which has to be recirculated in the FR to avoid energy loss for unconverted fuel. SH steam at 2 bar is used as fluidization agent in the fuel reactor. The sulfur removal occurs downstream the fuel reactor and the sub-critical steam cycle in a wet FGD producing gypsum and then CO_2 is separated and compressed up to 110 bar. The overall electric efficiency is equal to 41.6% with a CCR of 100%. (Main results are shown in Table 5.4.)

The integration of CLC with fossil fuel power plant has also been investigated by Spallina [91] using dynamically operated PBRs operated with syngas from coal gasification.

Table 5.4 *Performance of the direct solid CLC with coal as discussed in Ref. [90].*

Parameter		Value
Coal flow rate	t/h	81
Thermal power	MW_{th}	583
CO_2 flow rate	t/h	188
CO_2 capture ratio	%	100
Gross electrical power	MW_e	283.4
Gross electrical efficiency	$\%_{LHV}$	48.4
Electrical power – supercritical cycle	MW_e	251.7
Electrical power – sub-critical cycle	MW_e	31.7
Water flow rate – supercritical cycle	t/h	592
Water flow rate – sub-critical cycle	t/h	88
Auxiliaries consumption	MW_e	21.9
CO_2 compression	MW_e	14.0
O_2 production	MW_e	3.6
Net electrical power	MW_e	243.9
Net electrical efficiency	$\%_{LHV}$	41.6

Reproduced from Ref. [90], with permission from Elsevier.

Two configurations are addressed: in the first case, a gas–steam combined cycle is used where the N_2-rich stream produced during the heat removal phase at 1200 °C and 16 bar is sent to a gas turbine and then to HRSG (Figure 5.33). Due to the need of a high N_2 mass flow rate, the system is based on a semi-closed cycle in which the N_2, that is the most relevant mass flow of the power island, is partly released to the stack and partly recirculated back to the compressor inlet. The second plant is based on an advanced USC steam cycle where the syngas from coal gasification is converted in an atmospheric PBR operated at 800 °C (Figure 5.34). The oxidation phase is carried out by using preheated air in

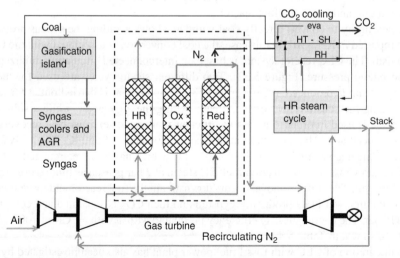

Figure 5.33 *IG-CLC schematic plant with pressurized PBRs presented in Ref. [91]. (Source: Reproduced from Ref. [91], with permission from Elsevier)*

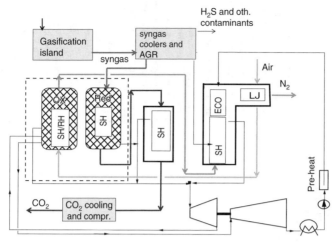

Figure 5.34 *USC schematic plant fed with syngas from coal gasification integrated with atmospheric PBRs presented in Ref. [91]. (Source: Reproduced from Ref. [91], with permission from Elsevier)*

a Ljungstrom-type heat exchanger and the O_2-depleted air is cooled down producing HP steam (320 bar) and converted in a steam cycle with RH (max steam temperature equals to 600 °C). The dynamically operated PBRs operated at pressurized conditions are adiabatic, while, in the atmospheric case, the reactors are cooled in order to reduce the footprint (in terms of heat input to the reactor cross section) of the CLC system. The footprint becomes very high if an additional phase has to be considered for the heat removal at atmospheric pressure and 800 °C. Both the configurations achieve a net electric efficiency in the range

Table 5.5 *Main performance of the studied plants as presented in Ref. [91].*

Power balance, MW_e	IG-CLC	IG-CLC-USC
Gas turbine cycle, MW_e	174.6	–
Steam cycle, MW_e	249.0	398.3
Gasification (ASU + LHs + Aux.*), MW_e	−38.5	−41.6
AGR + CO_2 compression, MW_e	−17.7	−37.5
Packed bed reactors aux., MWe	−3.1	−5.8
Syngas expander, MW_e	–	22.1
N_2 to GT compression, MW_e	–	–
Other aux., MW_e	−5.9	−4.6
Net power, MW_e	358.4	331.0
Thermal input, MW_{LHV}	896.5	896.5
Net efficiency, $\%_{LHV}$	39.98	36.92
CO_2 emission, kg/MWh_e	18.13	19.63
CO_2 avoided, %	97.52	97.32
SPECCA, MJ_{LHV}/kg_{CO2}	1.92	2.97
Max. heat input to CLC plan area**, MW_{LHV_coal}/m^2	5.7	2.0

*Coal milling, ash handling, recycling syngas blower, and so on.
**Assuming superficial gas velocity equal to 1 m/s
Reproduced from Ref. [91], with permission from Elsevier.

Table 5.6 List of published works on fossil fuel power plants integrated with CLC.

Configuration	Fuel type	Plant size	CLC pressure	Reactor temperatures		Oxygen carriers	Plant performance		Reactor type	Turbomachine adiabatic/isentropic efficiencies	Ref.	Short plant description
		MW_{LHV}		AR	FR	Red/ox:inert	net el. efficiency	CCR		GT: Compressor; Expander ST: HP/IP/LP cylinders.		
Multi-stage GT with HRSG no CO_2 capture	CH_4	802	12 atm	1200°C	1100°C	FeO/Fe_2O_3	50.2% LHV	–	i-FBRs	GT: 86%; 88.5% ST: 83.1/90.2/88.5%	[79]	The exhausts from the air reactor are expanded in the first GT and it is reheated at 1100 °C and expanded again to ambient pressure. The CO_2-rich stream is produced at high temperature and expanded in a CO_2/H_2O gas turbine. CO_2 capture is not considered.
GT cycle with IC-compressor and air saturation, no CO_2 capture	CH_4	802	19 atm	1200°C	1100°C	Ni/NiO	55.1% LHV	–	i-FBRs	GT:86%;88.5%	[82]	A three-stage intercooled compressor is used; the air preheating is carried out firstly in an air saturator and in a recuperative heat exchanger where the exhaust gases leaving the gas turbine. The CO_2-rich stream is produced at high temperature and expanded in a CO_2/H_2O gas turbine. CO_2 capture is not considered.
GT cycle with IC-compressor and air saturation	NG	50	20 bar	1200°C	900°C	Ni-NiO:YSZ	55.9% LHV	100%	i-FBRs	–	[81]	A three-stage intercooled compressor is used; the air preheating is carried out firstly in an air saturator and in a recuperative heat exchanger where the exhaust gases leaving the gas turbine. The CO_2-rich stream is produced at high temperature and expanded in a CO_2/H_2O gas turbine.

Combined cycle	NG	714	8–20 bar	1050 1200°C	–	NiO/Ni:Al$_2$O$_4$	46.3–48.5% LHV	100%	i-FBRs	GT:90%; 90% ST: 90/92/88 %	[71]	A combined cycle is considered with GT and a two pressure levels (60/5 bar) without re-heat HRSG. The analysis of the GT model is carried out focusing on the effect of blades cooling system.
Steam cycle	NG	714	1 atm	850°C	850°C	NiO/Ni:Al$_2$O$_4$	40.1% LHV	100%	i-FBRs	ST: 92%/92%	[71]	A steam cycle with steam produced at two pressure levels (160/30 bar) and maximum steam temperature of 560°C is investigated
Combined cycle	CH$_4$	800	9 bar (Fe$_2$O$_3$) 13 bar (NiO)	1000°C (Fe$_2$O$_3$) 1200°C (NiO)	–	Fe$_2$O$_3$/Fe$_3$O$_4$: Al$_2$O$_3$ NiO/Ni:Al$_2$O$_4$	48.6–52.4% LHV	–	i-FBRs	GT:85%; 90% ST: 91%	[73]	A combined cycle is considered with GT and three-pressure level HRSG (110/20/1.7 bar) with RH (max T steam 570°C). The CO$_2$-rich stream is cooled down to ambient temperature and compressed to 150 bar
Combined cycle	NG	467	8–20 bar	1050°C	986.7°C (Fe$_2$O$_3$) 707.5°C (NiO)	Fe$_2$O$_3$/Fe$_3$O$_4$ Ni/NiO	47–48% LHV	100%	i-FBRs	GT:90%; 93%	[72]	A combined cycle is considered with GT and three-pressure level HRSG (110/20/3 bar) with RH. The CO$_2$-rich stream is cooled down to ambient temperature and compressed to 85 bar. The TIT of the gas turbine is also increased up to 1200°C by combusting additional fuel with a net electrical efficiency of 52.3% (and decreased CCR to 54%). The compressor ratio has been optimized accordingly to the TIT of the GT.

(continued overleaf)

Table 5.6 (continued)

Configuration	Fuel type	Plant size MW$_{LHV}$	CLC pressure	Reactor temperatures AR	FR	Oxygen carriers Red/ox:inert	Plant performance net el. efficiency	CCR	Reactor type	Turbomachine adiabatic/isentropic efficiencies GT: Compressor; Expander ST: HP/IP/LP cylinders.	Ref.	Short plant description
Combined Cycle	NG	750	22 bar	832°C	831°C	FeO/Fe$_3$O$_4$/ Fe$_2$O$_3$; MgAl$_2$O$_4$	50.4–51.3% LHV	100%	Three i-FBRs	GT:92.4%; ST:92.6/ 94.6%	[77]	A combined cycle is considered with GT and three-pressure level HRSG (166/36/4 bar) with RH. The CO$_2$-rich stream is cooled down to ambient temperature and compressed up to 150 bar. The TIT of the gas turbine is increased up to typical value of the modern GT (1335°C) by combusting H$_2$-rich syngas produced in a steam-iron moving bed reactor.
Multi-stage GT with HRSG	NG	683.6	10–24 bar	1200°C	980°C	NiO/Ni:Al$_2$O$_4$	51–53% LHV	100%	i-FBRs	GT:91.5%; 86.2% ST: 90/92/88%	[76, 80]	The GT expansion is carried out with intermediate multi-stages of stream re-heating in additional CLC loop unit at lower pressure. The CO$_2$-rich stream is expanded and after cooling to ambient temperature, it is compressed up to 110 bar.
Combined cycle	CH$_4$	–	18 bar	1200°C	967°C	NiO/Ni:Al$_2$O$_4$	50.6% LHV	–	i-FBRs	GT:88%; 90% ST: 92%	[75]	A combined cycle is considered with GT and two-pressure level HRSG (60/35 bar) without RH. The CO$_2$-rich stream is expanded and then cooled down to ambient temperature and compressed to 85 bar. The analysis is concerning the off-design performance with different ambient temperatures.

Type	Fuel		Pressure			Oxygen carrier	Efficiency	Capture	Reactor	GT/ST	Ref	Notes
IGCC	Coal	1086	18–20 bar	1100°C 900°C	1125°C 1200°C	NiO/Ni:Al$_2$O$_3$	34.5–35.1% LHV	91–95%	i-FBRs	–	[85]	Coal syngas is used in the fuel reactor. Two different systems are implemented with CLC loop units at two different pressure levels. The system is also considered with ASU for coal gasification, which is integrated with the air compressor.
IGCC + H$_2$ production	Coal Sawdust	1148.6	30 bar	–	750°C 900°C	Fe$_3$O$_4$/Fe–FeO	38.8% LHV	–	not defined	GT based on M701G2 (HD) ST: 85%	[88]	The IGCC is coupled with a CL loop with steam-iron reactor, which is operated to oxidize the metal oxide, and fuel reactor that is used in reduction. The HRSG is based on a three-pressure level (118/34/3 bar) with RH. Different gasification technologies are discussed and the plant is designed in order to provide H$_2$ and electricity co-production.
IGCC	Coal	2250	20 bar	1200°C 1300°C	1050°C 1120°C	NiO/Ni:Al$_2$O$_3$	37.7–39% HHV	99.50%	i-FBRs	GT:92%; 91% ST: 90%	[86]	Coal syngas is used in the fuel reactor; A sensitivity analysis is carried out on the maximum temperature from the air reactor and the CO$_2$ cooling process. The system is fully integrated with three pressure levels (127/40/7 bar) with RH.

(continued overleaf)

Table 5.6 (continued)

Configuration	Fuel type	Plant size MW_{LHV}	CLC pressure	Reactor temperatures		Oxygen carriers Red/ ox:inert	Plant performance		Reactor type	Turbomachine adiabatic/isentropic efficiencies GT: Compressor; Expander ST: HP/IP/LP cylinders.	Ref.	Short plant description
				AR	FR		net el. efficiency	CCR				
Direct solid CLC + USC	Coal	583	1 atm	1000°C	985°C	Mn_3O_4/MnO: $MgAl_2O_4$	41.6% LHV	100%	i-FBRs	ST: 92/94/88%	[90]	The pulverized coal is directly converted in the fuel reactor and part of O_2 is supplied by ASU to complete the fuel conversion. The heat recovery from the air reactor is carried out in a USC (280 bar /580°C/600°C); the heat recovery from fuel reactor is carried out in a sub-critical steam cycle (180 bar/550°C/550°C)
IGCC	Coal	2080	30 bar	900°C	856.6°C	FeO/Fe_3O_4/ Fe_2O_3; $MgAl_2O_4$	39.7–44.8% LHV	100%	Three i-FBRs	GT:90.5%; 89.5/90.5% ST: 90/92/87%	[87]	Two coal gasifiers are studied (Shell and British Gas/Lurgi) and the produced syngas is fed to the fuel reactor. The TIT of the gas turbine is increased up to a typical value of the modern GT (1335°C) using H_2-rich syngas produced in a steam-iron bed reactor. After the GT, there is a three-pressure level HRSG with RH. Acid gas removal is carried out in a HT gas desulfurization unit operated at about 600°C. The CO_2-rich stream is cooled down (or expanded in a dedicated CO_2/H_2O expander) to ambient temperature and compressed up to 110 bar.

Process	Feedstock		Pressure		Temperature	Oxygen carrier	Efficiency		Reactor		Ref.	Notes
IGCC + H$_2$ production	Coal	1025.6	30 bar	–	750–900°C	Fe$_2$O$_3$/Fe$_3$O$_4$/ Fe–FeO	41.08% LHV	99.34%	not defined	–	[89]	The coal feedstock is directly converted in a fuel reactor and the OC is oxidized in two stages: in the first stage H$_2$ is produced by steam reduction, while in the second stage the hematite is formed by reacting with air. A combined cycle is used to convert the H$_2$-rich gas from the steam-iron reactor.
IGCC	Coal	859	16.5 bar	1200°C	1200°C	FeTiO$_3$	40% LHV	97.50%	PBRs	GT:92.5%; 92.1/93.1% ST: 88/94/92%	[91]	Coal syngas in produced in a Shell gasifier and used in a reactor operated in reduction. A semi-closed combined cycle is adopted to supply the proper amount of N$_2$ to HR operation. A three-pressure level (144/36/4 bar) HRSG with RH is used. The pressurized CO$_2$-rich stream is cooled down to ambient temperature and compressed up to 110 bar.
Coal gasification + USC	Coal	859	1.3 bar	800°C	800°C	FeTiO$_3$	36.9% LHV	97.50%	PBRs	ST: 91.5/92.7/ 91.5%	[91]	Coal syngas in produced in a Shell gasifier and used in a reactor operated in reduction after being expanded to 1.3 bar. A USC is adopted operating at 320 bar with maximum steam temperature of 600/620°C. PBRs are cooled in order to reduce the reactor footprint.

of 37–40% (Table 5.5), which represents an improvement compared to the performance of a pre-combustion capture IGCC (Table 5.6).

5.4.3   Comparison between CLC in packed beds and circulated fluidized beds

The influence of the reactor configuration on the performance in an IG-CLC power plant has been evaluated by Hamers *et al.* [92]. In this plant, syngas is produced in an entrained flow, oxygen-blown, dry feed Shell-type gasifier, which is operated at high temperature. CO_2 is used as feeding gas instead of N_2 to avoid syngas dilution. The syngas is cooled down producing HP superheated steam (400 °C, 144 bar), and afterwards, the cold syngas is desulfurized by using Selexol® process. Subsequently, the syngas is saturated, preheated and fed to either the circulating fluidized bed system or the packed bed CLC system, producing hot air for power generation. In both cases, hot air is produced at 1200 °C and 20 bar, which is subsequently sent to the gas turbine and a three-pressure level heat recovery steam generator (generating steam at 144, 36 and 4 bar). The simplified plant layout is presented in Figure 5.35.

The main difference between both reactor configurations is the number of operation steps. The circulating fluidized bed system contains an air and a fuel reactor (Figure 5.36a), which are connected with cyclones to separate the solids from the gas. The packed bed configuration contains at least four reactors for the following operation steps: reduction, oxidation, heat removal and purge (Figure 5.36b). Starting from an oxidized OC, it is reduced with syngas and CO_2 and H_2O is produced. Then, the reactor is purged with N_2, which is produced by the ASU. Subsequently, the carrier is oxidized with air. The oxidation and the heat removal are two different operation steps: first, the heat is produced inside the reactor and then it can be removed from the reactor. The N_2 produced during oxidation is recycled for the heat removal, and, therefore, an additional blower is required to compress the air for oxidation. At the end, the reactor is purged again with N_2, so that the bed can be reduced again.

In the circulating fluidized bed configuration, much heat is transferred from the air to the fuel reactor and vice versa, because of the circulation of the OC particles. Hence, the temperature difference between both reactors is small. So, the CO_2/H_2O is produced at

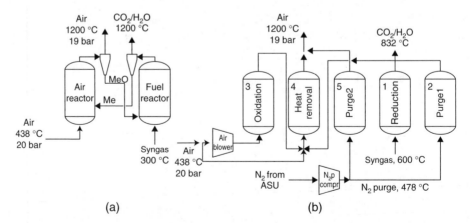

(a)   (b)

Figure 5.35   *Schematic of integrated gasification combined cycle with CLC with FBRs and PBRs*

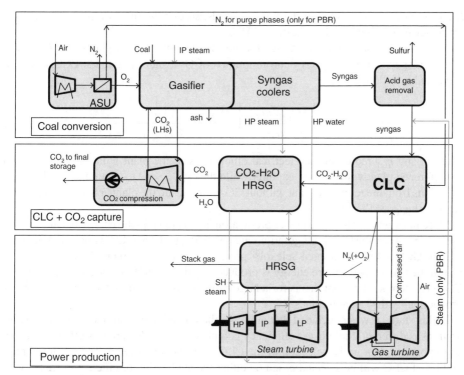

Figure 5.36   *An overview of the circulating fluidized bed (a) and the packed bed configuration (b)*

around 1200 °C. The temperature at which CO_2 and H_2O is produced in the packed bed depends on the temperature at the end of the reactor during the reduction process. This is not a constant temperature, but the temperature can be averaged if several reactors are operated in reduction mode; in this case, the average CO_2/H_2O temperature is 832 °C. High-pressure steam is produced during cooling down of the CO_2/H_2O-stream. Because of the higher CO_2/H_2O temperature, the steam can be produced at higher temperatures in the fluidized bed (565 °C) than in the packed bed (480 °C).

The temperature during reduction has also a lot of influence on the extent of carbon deposition and the conversion of the fuel in case NiO is used as OC. In the packed bed case, the beds are at 450–600 °C during reduction, and, therefore, the selectivity is not an issue, but additional steam is required to prevent carbon deposition. Because of the latter, the process efficiency is decreased. In this case, a process efficiency of 40.9% of LHV can be achieved. The fuel reactor in the circulating fluidized bed system is at 1200 °C. Hence, carbon deposition is not an issue, but here 1–2% of the fuel leaves the reactor unconverted, because of the thermodynamics. In this case, some additional oxygen is fed to the outlet stream to convert the remaining fuel, leading to a process efficiency of 41.6% of LHV.

For both reactor configurations, ideal cases were designed as well. In that case, the amount of steam produced in the saturator is sufficient to prevent carbon deposition in the PBRs (steam to carbon ratio of 0.37). This can be achieved with an OC that has slow kinetics for carbon deposition, and then a process efficiency of 42.1% of LHV can be obtained.

If a different OC is selected in the fluidized beds, so that the fuel is fully converted, the efficiency can be improved to 41.9% of LHV.

An important coefficient that is used for the comparison of power plants with CO_2 capture is the specific primary energy consumption for CO_2 avoided (named SPECCA), which is defined as follows:

$$\text{SPECCA} \left[\frac{MJ_{\text{LHV}} kg_{OC}}{kg_{CO_2}} \right] = 3600 \cdot \frac{\left(\frac{1}{\eta} - \frac{1}{\eta_{\text{ref}}} \right)}{(e_{CO_2,\text{ref}} - e_{CO_2})} \tag{5.10}$$

where η and η_{ref} represent the electrical efficiencies of the plant with CO_2 capture and the reference plant without CO_2 capture, while the $e_{CO2,\text{ref}}$ and e_{CO2} are the specific CO_2 emissions from the reference plant without CO_2 capture and the plant considered with CO_2 capture (kg_{CO_2}/MWh$_e$).

More information about the power production is illustrated in the energy balance shown in Table 5.7. Here the CLC configurations are compared with an IGCC without CO_2 capture and an IGCC with conventional Selexol® pre-combustion CO_2 capture. It is demonstrated that in the packed beds, more energy is produced by the gas turbine, because of the lower CO_2/H_2O-temperature, but this is for a large extent compensated by the higher electricity production by the steam turbines.

In this study, it was concluded that although the operation of both CLC reactor configurations is different, the process efficiency is quite close. So, the process efficiency is not an important factor in the selection of the CLC reactor configuration. The reactor configuration of CLC will be selected based on either the application of efficient cyclones with circulated fluidized beds and the possibility to operate continuously at high pressure or high temperature valves in the packed bed configuration.

5.5   Conclusions

This chapter discussed the CLC technology as possible solution for a carbon-free power production from fossil fuels. A lot of research has been carried out in the last decades to investigate the main topics that are related to the development of an efficient process. The OCs have been studied by several authors, and they have been here briefly summarized.

An extensive discussion and comparison is reported in a recent review about CLC [9]: Ni-based materials have been amply investigated for CLC and CLR; however, the very low sulfur tolerance and activity for the carbon deposition (together with high cost and toxicity) make this material suitable only to be used with clean syngas (from natural gas or coal after deep desulfurization); Fe-based materials (including natural ores such as ilmenite) have been studied in different processes such as the CLC with gaseous fuel, isG-CLC, it is also suitable in the steam-iron process for pure H_2 production and it shows generally low reactivity but is cheap and environmental friendly; Cu-based materials can be operated both with gaseous and solid fuels with high fuel conversion; however, its melting point makes it interesting only in future applications that can be operated at lower temperatures (such as direct solid CLC or two-stage CLC in PBR); Mn-based materials are cheap and

Table 5.7 *Energy balances of the different cases considered for the packed bed and the fluidized bed [92, 93].*

Power	IGCC NC N/A [93]	IGCC-Sel Selexol® [93]	PB with NiO [92]	FzB with NiO [92]	Ideal PB, no steam added [92]	Ideal FzB, full gas conversion [92]
Turbine inlet temperature, °C	1305	1261	1173	1172	1173	1173
Gas turbine compressor pressure ratio, −	18.15	18.15	20	20	20	20
Air ratio, $air_{stoichiometric}/air_{consumed, total}$	3.2	3.2	4.5	4.0	4.6	4.1
Maximum steam temperature, °C	565	532	480	565	480	565
Cold gas efficiency, %	81.6	73.2	80.6	80.6	80.6	80.6
$Air_{cooling\ in\ gas\ turbine}/air_{consumed}$, −	0.252	0.235	0.072	0.072	0.071	0.072
Heat input LHV, MW_{LHV}	812.5	898.8	853.9	853.9	853.9	853.9
Gas turbine, MW_e	261.6	263.9	225.1	192.1	232.2	197.9
Steam turbine, MW_e	179.5	161.2	180.5	224	189.2	220.3
Gross power output, MW_e	441.1	425.1	405.5	416.1	421.4	418.2
Steam cycle pumps, MW_e	−2.9	−3.5	−3.8	−4.0	−3.5	−4.0
Syngas blower, MW_e			−0.8	−0.8	−0.8	−0.8
N_2 compressor, MW_e	−34.2	−29.8				
ASU, MW_e	−29.6	−32.7	−33.9	−35.1	−33.9	−33.9
Lock-hopper CO_2 compressor, MW_e			−3.1	−3.1	−3.1	−3.1
Acid gas removal, MW_e	−0.4	−14.7	−0.4	−0.4	−0.4	−0.4
CO_2 compressor, MW_e		−19.7	−11.0	−11.0	−11.0	−11.0
N_2 intercooled compressor gasifier, MW_e			−1.3	−1.3	−1.3	−1.3
Water pumps, MW_e	−5.5	−6.3	−3.6	−3.7	−3.4	−3.6
Other auxiliaries, MW_e	−3.2	−3.6	−3.4	−3.4	−3.4	−3.4
Net power generated, MW_e	367.3	317.3	350.6	353.3	360.6	356.7
LHV efficiency, %	45.2	35.3	41.05	41.4	42.2	41.8
CO_2 capture efficiency, %		93.0	97.1	97.1	97.1	97.1
CO_2 purity, %		98.2	96.7	96.6	96.7	96.7
CO_2 emission, $kg_{CO_2\ emitted}/MWh_e$	769.8	101.4	24.6	24.5	24.0	24.3
CO_2 avoided, %	0	84.7	96.8	96.8	96.9	96.9
SPECCA, MJ_{LHV}/kg_{CO_2}		3.34	0.98	0.89	0.75	0.88

suitable for CLOU process, but they have low oxygen capacity that has to be compensated with high solid circulation; the other materials that have been studied are at the early stage of development and no general conclusions can be drawn.

The research has been devoted mostly to the study of CLC in interconnected fluidized bed reactors. Different fluidization regimes have been studied to achieve complete fuel conversion and to optimize the solid circulation between the fuel and air reactors. Several experimental facilities from $0.5\,kW_{th}$ up to $3\,MW_{th}$ have been built in recent years and the research is now at the demonstration stage. The summary of experienced time in units operated continuously for more than $4000\,h$ have been achieved using different fuel types (gaseous and solid fuels), and, therefore, a lot of experience has been accumulated both in the reactor design and management. The next stage of development is full-scale commercialization. However, the system is suitable only at atmospheric pressure (experiments at pressurized conditions have been carried out only for few hours), and most of the new research is devoted to achieve high performance operating directly with solid fuels to make CLC competitive with the new generation of fossil fuel power plants with coal with CO_2 capture. CLC has been also recently focused on PBRs, which allow the system to be easily operated at pressurized conditions. The reactor concept is now at the early stage of demonstration, and new researches are required to study the material properties, the effect of dynamic operations on the system performance and the feasibility of the technology. Despite the operation with high temperature/high pressure does not represent an issue for PBRs, the use of HT gas switching system that are far from nowadays has to be deeply investigated in the coming years. A new reactor concept based on the rotating system has also been studied recently, but some problems have been presented especially due to the gas leakage from the different sectors dedicated to the different phases.

The research on CLC integration in fossil fuel power plants has been addressed to the use of combined cycle both from natural gas and syngas from coal gasification. Power plants with direct solid conversion have been recently proposed. Summarizing, the following potentialities can be highlighted for CLC technologies:

- The electric efficiencies of natural gas power plants are in the range of $50–51\%_{LHV}$, which is 1–2 percentage points higher than competitive technologies with virtually zero CO_2 emissions. In the case of IGCC, between 2 and 3 percentage points of gain in the electrical efficiency (the electrical efficiency is in the range of $38–41\%_{LHV}$) are expected in comparison to coal power plants with CO_2 capture. Finally, the use of direct coal oxidation CLC reactor increases the net electric efficiency up to 5 percentage points (with very low penalty efficiency associated to the CO_2 capture technology).
- Fuel oxidation and CO_2 separation are carried out in a single unit composed of two (or three in one case) adiabatic reactors, and no other exotic processes are required while using solvent-based processes. Air separation unit, which is very expensive from an energy point of view, is required only if there is not complete fuel conversion in the CLC loop.
- The use of an indirect process for fuel combustion and the low temperatures allow to have zero NO_x emissions at least for the layouts proposed without any additional fuel post-firing.
- The application of the current gas turbine technology appears to be possible with no relevant re-design of state-of-the-art machines.

On the other hand, the following limits can be noted:

- Natural gas–fired power plants with CLC are competitive with other CO_2 capture plants only in the presence of pressurized operations that still to be demonstrated and pose relevant technical challenges in interconnected fluidized beds.
- In the case of coal-fired power plants, the use of direct solid CLC must be proved and several drawbacks exist: the coal conversion is still low and the additional O_2 must be provided; the material used has to be sulfur tolerant and easily to be separated from the ashes and unconverted char.
- Other configurations (packed bed and rotating fixed bed) proposed for the reactor system may be considered for pressurized operations. However, other critical technical issues arise with these designs (e.g. heat management, HT switching valve system, carbon deposition, CO_2 leakage), which require necessary deep investigations on the experimental side and on the reactor modelling.
- Plants with supplementary firing allow obtaining higher efficiencies but lead to increased CO_2 (when burning a carbon containing fuel) and NO_x emissions.
- In plant layouts without supplementary firing, TIT is limited by the resistance of the OC at high temperature, which limit the maximum temperature of the combined cycle and thus the maximum efficiency. Therefore, CLC processes will not take advantage of future advancements in the gas turbine material technology.
- The use of multi-stage gas turbine with reheat(s) may represent a possible alternative to improve the thermodynamic cycle performance, but state-of-the-art gas turbines need to be re-designed.

Nomenclature

AR	Air Reactor
ASU	Air Separation Unit
CC	Combined Cycle
CCR	Carbon Capture Rate
CCS	Carbon Capture and Storage
CLC	Chemical-Looping Combustion
CLOU	Chemical Looping with Oxygen Uncoupling
CLR	Chemical-Looping Reforming
FGD	Flue Gas Desulfurization
FR	Fuel Reactor
HAT	Humid Air Turbine
GT	Gas Turbine
HGCU	Hot Gas Clean Up
HHV	Higher Heating Value
HP/IP/LP	High/Intermediate/Low Pressure
HRSC	Heat Recovery Steam Cycle
HRSG	Heat Recovery Steam Generator
HT/IT/LT	High/Intermediate/Low Temperature
IGCC	Integrated Gasification Combined Cycle

IG-CLC   Integrated Gasification Chemical-Looping Combustion plant
isG-CLC   *in situ* Gasification Chemical Looping Combustion
LHV   Lower Heating Value
MM   Molecular Weight
NGCC   Natural Gas Combined Cycle
PBR   Packed Bed Reactor
SPECCA   Specific Primary Energy Consumption for CO_2 Avoided
SH/RH   Superheating/Reheating
ST   Steam Turbine
TS-CLC   Two Stage Chemical Looping Combustion
USC   Ultra – Super Critical

Subscripts

th   Thermal
e   electrical

References

1. IEA (2008) World Energy Outlook.
2. EBTF (2011) European best practice guidelines for assessment of CO_2 capture technologies.
3. Buhre, B.J.P., Elliott, L.K., Sheng, C.D. *et al.* (2005) Oxy-fuel combustion technology for coal-fired power generation. *Progress in Energy and Combustion Science*, **31**, 283–307.
4. Cuadrat, a., Abad, a., García-Labiano, F. *et al.* (2012) Relevance of the coal rank on the performance of the in situ gasification chemical-looping combustion. *Chemical Engineering Journal*, **195–196**, 91–102.
5. Cao, Y. and Pan, W. (2006) Investigation of chemical looping combustion by solid fuels. 1. Process analysis. *Energy & Fuels*, **20**, 1836–1844.
6. Mattisson, T., Lyngfelt, A. and Leion, H. (2009) Chemical-looping with oxygen uncoupling for combustion of solid fuels. *International Journal of Greenhouse Gas Control*, **3**, 11–19.
7. Leion, H., Mattisson, T. and Lyngfelt, A. (2009) Using chemical-looping with oxygen uncoupling (CLOU) for combustion of six different solid fuels. *Energy Procedia*, **1**, 447–453.
8. Adánez-Rubio, I., Abad, a., Gayán, P. *et al.* (2009) Performance of CLOU process in the combustion of different types of coal with CO_2 capture. *International Journal of Greenhouse Gas Control*, **12**, 430–440.
9. Adanez, J., Abad, A., Garcia-Labiano, F. *et al.* (2012) Progress in chemical-looping combustion and reforming technologies. *Progress in Energy and Combustion Science*, **38**, 215–282.
10. Hossain, M.M. and de Lasa, H.I. (2008) Chemical-looping combustion (CLC) for inherent separations—a review. *Chemical Engineering Science*, **63**, 4433–4451.

11. Lyngfelt, A., Mattisson, T., Epple, B. and Ströhle, J. (2011) Part 5 Chemical Looping for CO$_2$ Separation, in *Effic. Carbon Capture Coal Power Plants* (eds D. Stolten and V. Scherer), Wiley-VCH, Weinheim, pp. 475–524.

12. Jerndal, E., Mattisson, T. and Lyngfelt, A. (2006) Thermal analysis of chemical-looping combustion. *Chemical Engineering Research and Design*, **84**, 795–806.

13. Kuusik, R., Trikkel, A., Lyngfelt, A. and Mattisson, T. (2009) High temperature behavior of NiO-based oxygen carriers for Chemical Looping Combustion. *Energy Procedia*, **1**, 3885–3892.

14. Ishida, M. and Jin, H. (1996) A novel chemical-looping combustor without NOx formation. *Industrial & Engineering Chemistry Research*, **35**, 2469–2472.

15. Kolbitsch, P., Pröll, T., Bolhar-Nordenkampf, J. and Hofbauer, H. (2009) Characterization of chemical looping pilot plant performance via experimental determination of solids conversion. *Energy & Fuels*, **23**, 1450–145.

16. Kolbitsch, P., Bolhàr-Nordenkampf, J., Pröll, T. and Hofbauer, H. (2010) Operating experience with chemical looping combustion in a 120 kW dual circulating fluidized bed (DCFB) unit. *International Journal of Greenhouse Gas Control*, **4**, 180–185.

17. Gayán, P., de Diego, L.F., García-Labiano, F. *et al.* (2008) Effect of support on reactivity and selectivity of Ni-based oxygen carriers for chemical-looping combustion. *Fuel*, **87**, 2641–2650.

18. Linderholm, C., Mattisson, T. and Lyngfelt, A. (2009) Long-term integrity testing of spray-dried particles in a 10-kW chemical-looping combustor using natural gas as fuel. *Fuel*, **88**, 2083–2096.

19. Fan, L. (2010) *ChemicalLooping Systems for Fossil Energy*, John WIley & Son Inc..

20. Leion, H., Lyngfelt, A., Johansson, M. *et al.* (2008) The use of ilmenite as an oxygen carrier in chemical-looping combustion. *Chemical Engineering Research and Design*, **86**, 1017–1026.

21. Abad, A., Adánez, J., Cuadrat, A. *et al.* (2011) Kinetics of redox reactions of ilmenite for chemical-looping combustion. *Chemical Engineering Science*, **66**, 689–702.

22. Corbella, B.M. and Palacios, J.M. (2007) Titania-supported iron oxide as oxygen carrier for chemical-looping combustion of methane. *Fuel*, **86**, 113–122.

23. Adánez, J., Cuadrat, A., Abad, A. *et al.* (2010) Ilmenite activation during consecutive redox cycles in chemical-looping combustion. *Energy & Fuels*, **24**, 1402–1413.

24. Pröll, T., Mayer, K., Bolhàr-Nordenkampf, J. *et al.* (2009) Natural minerals as oxygen carriers for chemical looping combustion in a dual circulating fluidized bed system. *Energy Procedia*, **1**, 27–34.

25. Berguerand, N. and Lyngfelt, A. (2008) The use of petroleum coke as fuel in a 10kWth chemical-looping combustor. *International Journal of Greenhouse Gas Control*, **2**, 169–179.

26. De Diego, L.F., García-Labiano, F., Gayán, P. *et al.* (2007) Operation of a 10 kWth chemical-looping combustor during 200 h with a CuO–Al$_2$O$_3$ oxygen carrier. *Fuel*, **86**, 1036–1045.

27. Abad, a., Mattisson, T., Lyngfelt, a. and Ryden, M. (2006) Chemical-looping combustion in a 300W continuously operating reactor system using a manganese-based oxygen carrier. *Fuel*, **85**, 1174–1185.

28. Leion, H., Mattisson, T. and Lyngfelt, A. (2009) Use of ores and industrial products as oxygen carriers in chemical-looping combustion. *Energy & Fuels*, **23**, 2307–2315.

29. Leion, H., Larring, Y., Bakken, E. *et al.* (2009) Use of CaMn 0.875 Ti 0.125 O 3 as oxygen carrier in chemical-looping with oxygen incoupling. *Energy & Fuels*, **23**, 5276–5283.

30. Wang, B., Yan, R., Lee, D.H. *et al.* (2008) Thermodynamic Investigation of Carbon Deposition and Sulfur Evolution in Chemical Looping Combustion with Syngas. *Energy & Fuels*, **22**, 1012–1020.

31. García-Labiano, F., de Diego, L.F., Gayán, P. *et al.* (2009) Effect of fuel gas composition in chemical-looping combustion with Ni-based oxygen carriers. 1. fate of sulfur. *Industrial & Engineering Chemistry Research*, **48**, 2499–2508.

32. Forero, C.R., Gayán, P., García-Labiano, F. *et al.* (2010) Effect of gas composition in Chemical-Looping Combustion with copper-based oxygen carriers: Fate of sulphur. *International Journal of Greenhouse Gas Control*, **4**, 762–770.

33. Lyngfelt, A. (2011) Oxygen carriers for chemical looping combustion - 4 000 h of operational experience. *Oil & Gas Science and Technology*, **66**, 161–172.

34. Lyngfelt, A. (2014) Chemical-looping combustion of solid fuels – Status of development. *Applied Energy*, **113**, 1869–1873.

35. Lyngfelt, A., Leckner, B. and Mattisson, T. (2001) A uidized-bed combustion process with inherent CO 2 separation; application of chemical-looping combustion. *Chemical Engineering Science*, **56**, 3101–3113.

36. Kronberger, B. and Lyngfelt, A. (2005) Design and fluid dynamic analysis of a bench-scale combustion system with CO_2 separation-chemical-looping combustion. *Industrial & Engineering Chemistry Research*, **44**, 546–556.

37. Kronberger, B., Johansson, E., Löffler, G. *et al.* (2004) A two-compartment fluidized bed reactor for CO_2 capture by chemical-looping combustion. *Chemical Engineering & Technology*, **27**, 1318–1326.

38. Johansson, E., Lyngfelt, a., Mattisson, T. and Johnsson, F. (2003) Gas leakage measurements in a cold model of an interconnected fluidized bed for chemical-looping combustion. *Powder Technology*, **134**, 210–217.

39. Linderholm, C., Lyngfelt, A., Cuadrat, A. and Jerndal, E. (2012) Chemical-looping combustion of solid fuels – Operation in a 10kW unit with two fuels, above-bed and in-bed fuel feed and two oxygen carriers, manganese ore and ilmenite. *Fuel*, **102**, 808–822.

40. Ryu, H.-J., Gin, G.-T., Bae, D.-H. and Yi, C.K. (2004) Demonstration of inherent CO_2 separation and NOx emission in a 50 kW chemical looping combustion: continuous reduction and oxidation experiment. 7th Int Conf Greenh. Gas Control Technol., pp. 1907–1910.

41. Abad, A., Adánez-Rubio, I., Gayán, P. *et al.* (2012) Demonstration of chemical-looping with oxygen uncoupling (CLOU) process in a 1.5kWth continuously operating unit using a Cu-based oxygen-carrier. *International Journal of Greenhouse Gas Control*, **6**, 189–200.

42. Cuadrat, A., Abad, A., García-Labiano, F. *et al.* (2011) The use of ilmenite as oxygen-carrier in a 500Wth Chemical-Looping Coal Combustion unit. *International Journal of Greenhouse Gas Control*, **5**, 1630–1642.

43. Mendiara, T., Izquierdo, M.T., Abad, A., Diego, L.F., De García-labiano F., Gayán, P., *et al.* (2012) Performance of a Fe-based residue using different coals in a 500 W th CLC unit. Proc. 2nd Int. Conf. Chem. Looping, Darmstadt, pp. 26–28.

44. Son, S.R. and Kim, S.D. (2006) Chemical-looping combustion with NiO and Fe2O3 in a thermobalance and circulating fluidized bed reactor with double loops. *Industrial & Engineering Chemistry Research*, **45** (8), 2689–2696.

45. Kolbitsch, P., Pröll, T., Bolhar-Nordenkampf, J. and Hofbauer, H. (2009) Design of a chemical looping combustor using a dual circulating fluidized bed (DCFB) reactor system. *Chemical Engineering & Technology*, **32**, 398–403.

46. Sit, S., Reed, A., Hohenwarter, U., Horn, V., Marx, K. and Pröll, T. (2012) 10 MW CLC field pilot. Proc. 2nd Int. Conf. Chem. Looping, Darmstadt, pp. 26–28.

47. Andrus, H., Chiu, J., Thibeault, P., Brautsch, A. (2009) Alstom's Calcium Oxide Chemical Looping Combustion Coal Power Technology Development. 34th Int. Tech. Conf. Clean Coal Fuel Syst., Florida USA.

48. Andrus, H., Chui, J., Thibeault, P., Edberg, C., Turek, D., Kennedy, J., *et al.* (2012) Alstom's Limeston Based (LCL™) Chemical Looping Process. Proc. 2nd Int. Conf. Chem. Looping, Darmstadt.

49. Orth, M., Ströhle, J., Epple, B. (2012) Design and Operation of a 1 MW th Chemical Looping Plant. Proc. 2nd Int. Conf. Chem. Looping, Darmstadt, pp. 26–28.

50. Bischi, A., Langørgen, Ø., Morin, J.-X. *et al.* (2011) Performance analysis of the cold flow model of a second generation chemical looping combustion reactor system. *Energy Procedia*, **4**, 449–456.

51. Bischi, A., Langørgen, Ø., Saanum, I. *et al.* (2011) Design study of a 150kWth double loop circulating fluidized bed reactor system for chemical looping combustion with focus on industrial applicability and pressurization. *International Journal of Greenhouse Gas Control*, **5**, 467–474.

52. Xiao, R., Chen, L., Saha, C. *et al.* (2012) Pressurized chemical-looping combustion of coal using an iron ore as oxygen carrier in a pilot-scale unit. *International Journal of Greenhouse Gas Control*, **10**, 363–373.

53. Rifflart, S., Hoteit, A., Yazdanpanah, M.M. *et al.* (2011) Construction and operation of a 10 kW CLC unit with circulation configuration enabling independent solid flow control. *Energy Procedia*, **4**, 333–340.

54. Sozinho, T., Pelletant, W., Stainton, H., Guillou, F., Gauthier, T. (2012) Main results of the 10 kW th pilot plant operation. Proc. 2nd Int. Conf. Chem. Looping, Darmstadt, pp. 26–28.

55. Shen, L., Wu, J., Xiao, J. *et al.* (2009) Chemical-looping combustion of biomass in a 10 kWth reactor with iron oxide as an oxygen carrier. *Energy & Fuels*, **23**, 2498–2505.

56. Sridhar, D., Tong, A., Kim, H. *et al.* (2012) Syngas chemical looping process: design and construction of a 25 kWth subpilot unit. *Energy & Fuels*, **26**, 2292–2302.

57. Tong, A., Sridhar, D., Sun, Z. *et al.* (2013) Continuous high purity hydrogen generation from a syngas chemical looping 25kWth sub-pilot unit with 100% carbon capture. *Fuel*, **103**, 495–505.

58. Song, T., Shen, L., Zhang, H., Gu, H., Zhang, S., Xiao and J. (2012) Chemical looping combustion of two bituminous coal / char with natural hematite as oxygen carrier in 1 kW th reactor. Proc. 2nd Int. Conf. Chem. Looping, Darmstadt, pp. 26–28.
59. Noorman, S., Annaland, M.V.S. and Kuipers, H. (2007) Packed Bed Reactor Technology for Chemical-Looping Combustion. *Industrial & Engineering Chemistry Research*, **46**, 4212–4220.
60. Noorman, S., Gallucci, F., van Sint, A.M. and Kuipers, J.a.M. (2011) A theoretical investigation of CLC in packed beds. Part 2: Reactor model. *Chemical Engineering Journal*, **167**, 369–376.
61. Noorman, S., Gallucci, F., van Sint, A.M. and Kuipers, J.a.M. (2011) Experimental investigation of chemical-looping combustion in packed beds: A parametric study. *Industrial & Engineering Chemistry Research*, **50**, 1968–1980.
62. Hamers, H.P., Gallucci, F., Cobden, P.D. *et al.* (2013) A novel reactor configuration for packed bed chemical-looping combustion of syngas. *International Journal of Greenhouse Gas Control*, **16**, 1–12.
63. Spallina, V., Gallucci, F., Romano, M.C. *et al.* (2013) Investigation of heat management for CLC of syngas in packed bed reactors. *Chemical Engineering Journal*, **225**, 174–191.
64. Fernandez, J.R., Abanades, J.C. and Murillo, R. (2013) Modeling of Cu oxidation in an adiabatic fixed-bed reactor with N2 recycling. *Applied Energy*, **113**, 1–7.
65. Fernández, J.R., Abanades, J.C., Murillo, R. and Grasa, G. (2012) Conceptual design of a hydrogen production process from natural gas with CO_2 capture using a Ca–Cu chemical loop. *International Journal of Greenhouse Gas Control*, **6**, 126–141.
66. Noorman, S. (2009) *Packed Bed Reactor Technology for Chemical - looping Combustion*, University of Ywente.
67. Dahl, I.M., Bakken, E., Larring, Y. *et al.* (2009) On the development of novel reactor concepts for chemical looping combustion. *Energy Procedia*, **1**, 1513–1519.
68. Håkonsen, S.F., Grande, C.a. and Blom, R. (2013) Rotating bed reactor for CLC: Bed characteristics dependencies on internal gas mixing. *Applied Energy*, **5**, 9619–26.
69. Håkonsen, S.F. and Blom, R. (2011) Chemical looping combustion in a rotating bed reactor--finding optimal process conditions for prototype reactor. *Environmental Science & Technology*, **45**, 9619–9626.
70. Manzolini, G., Macchi, E., Binotti, M. and Gazzani, M. (2011) Integration of SEWGS for carbon capture in Natural Gas Combined Cycle. Part B: Reference case comparison. *International Journal of Greenhouse Gas Control*, **5**, 214–225.
71. Naqvi, R., Bolland, O., Brandvoll, Ø., Helle, K. (2004) Chemical looping combustion-analysis of natural gas power cycles with inherent CO_2 capture. Proc ASME Turbo … , pp. 1–9.
72. Consonni, S., Lozza, G., Pelliccia, G. *et al.* (2006) Chemical-looping combustion for combined cycles with CO[sub 2] capture. *Journal of Engineering for Gas Turbines and Power*, **128**, 525.
73. Wolf, J., Anheden, M. and Yan, J. (2005) Comparison of nickel- and iron-based oxygen carriers in chemical looping combustion for CO capture in power generation. *Fuel*, **84**, 993–1006.

74. Kvamsdal, H.M., Jordal, K. and Bolland, O. (2007) A quantitative comparison of gas turbine cycles with CO_2 capture. *Energy*, **32**, 10–24.

75. Chi, J., Wang, B., Zhang, S. and Xiao, Y. (2010) Off-design performance of a chemical looping combustion (CLC) combined cycle: effects of ambient temperature. *Journal of Thermal Science*, **19**, 87–96.

76. Naqvi, R., Wolf, J. and Bolland, O. (2007) Part-load analysis of a chemical looping combustion (CLC) combined cycle with CO_2 capture. *Energy*, **32**, 360–370.

77. Lozza, G., Chiesa, P., Romano, M.C., Savoldelli, P, (2006) Three Reactors Chemical Looping Combustion For High Efficiency Electricity Generation With CO_2 Capture From Natural Gas. Proc ASME Turbo Expo 2006, GT2006-90345, pp. 1–11.

78. Chiesa, P., Lozza, G., a, M. *et al.* (2008) Three-reactors chemical looping process for hydrogen production. *International Journal of Hydrogen Energy*, **33**, 2233–2245.

79. Ishida, M., Zheng, D. and Akehata, T. (1987) Evaluation of a chemical-looping-combustion power-generation system by graphic exergy analysis. *Energy*, **12**, 147–154.

80. Naqvi, R. and Bolland, O. (2007) Multi-stage chemical looping combustion (CLC) for combined cycles with CO_2 capture. *International Journal of Greenhouse Gas Control*, **1**, 19–30.

81. Brandvoll, Ø. and Bolland, O. (2002) *Inherent CO_2 capture using Chemical Looping Combustion in a Natural Gas Fired Power Cycle*, Asme Turbo Expo 2002, Land, Sea Air, Amsterdam, pp. 1–7.

82. Ishida, M. and Jin, H. (1994) A new advanced power-generation system using chemical-looping combustion. *Energy*, **19**, 415–422.

83. Goto, K., Yogo, K. and Higashii, T. (2013) A review of efficiency penalty in a coal-fired power plant with post-combustion CO_2 capture. *Applied Energy*, **111**, 710–720.

84. Valenti, G., Bonalumi, D. and Macchi, E. (2012) A parametric investigation of the Chilled Ammonia Process from energy and economic perspectives. *Fuel*, **101**, 74–83.

85. Rezvani, S., Huang, Y., McIlveen-Wright, D. *et al.* (2009) Comparative assessment of coal fired IGCC systems with CO_2 capture using physical absorption, membrane reactors and chemical looping. *Fuel*, **88**, 2463–2472.

86. Erlach, B., Schmidt, M. and Tsatsaronis, G. (2011) Comparison of carbon capture IGCC with pre-combustion decarbonisation and with chemical-looping combustion. *Energy*, **36**, 3804–3815.

87. Sorgenfrei, M. and Tsatsaronis, G. (2013) Design and evaluation of an IGCC power plant using iron-based syngas chemical-looping (SCL) combustion. *Applied Energy*, **113**, 1958–1964.

88. Cormos, C. (2010) Evaluation of iron based chemical looping for hydrogen and electricity co-production by gasification process with carbon capture and storage. *International Journal of Hydrogen Energy*, **35**, 2278–2289.

89. Cormos, C.-C., Cormos, A.-M. and Petrescu, L. (2013) Assessment of chemical looping-based conceptual designs for high efficient hydrogen and power co-generation applied to gasification processes. *Chemical Engineering Research and Design*, **92**, 1–11.

90. Authier, O. and Le Moullec, Y. (2013) Coal Chemical-Looping Combustion for Electricity Generation: Investigation for a 250 MWe Power Plant. *Energy Procedia*, **37**, 588–597.
91. Spallina, V., Romano, M.C., Chiesa, P. and Lozza, G. (2013) Integration of Coal Gasification and Packed Bed CLC process for High Efficiency and Near-zero Emission Power Generation. *Energy Procedia*, **37**, 662–670.
92. Hamers, H.P., Romano, M.C., Spallina, V. *et al.* (2014) Comparison on process efficiency for CLC of syngas operated in packed bed and fluidized bed reactors. *International Journal of Greenhouse Gas Control*, **28**, 65–78.
93. Spallina, V., Romano, M.C., Chiesa, P. *et al.* (2014) Integration of coal gasification and packed bed CLC for high efficiency and near-zero emission power generation. *International Journal of Greenhouse Gas Control*, **27**, 28–41.

6

Sorption-Enhanced Fuel Conversion

G. Manzolini,[1] D. Jansen[2] and A. D. Wright[3]

[1]Politecnico di Milano, Department of Energy, Milano (MI), Italy
[2]Utrecht University, Energy & Resources, Copernicus Institute of Sustainable
Development, Faculty of Geosciences, Utrecht, The Netherlands
[3]Air Products PLC, Energy Technology, Surrey, UK

6.1 Introduction

This chapter deals with the advancements in enhanced fuel conversion by sorbents. Particular attention will be devoted to sorbent application in power plants for H_2 production and/or CO_2 capture using both natural gas and coal as feedstock.

Sorbents are materials that under specific conditions of temperature and pressure can adsorb a particular molecule separating it from a gaseous stream. The adsorption consists of a chemical (covalent bonding) or physical (van der Waals) bond between the sorbent and the considered molecule. Compared to solvents, sorbents have lower load capacity as well as regeneration energy demand. The bond between the molecule and the sorbent must be broken during the so-called regeneration process. The regeneration is usually performed by varying the thermodynamic conditions of the sorbent compared to the adsorption step in terms of temperature, pressure and gas composition. In order to make the process continuous, two or more reactors in parallel must be adopted: one or more reactors dedicated to the adsorption process, with the remaining to the regeneration of the sorbent.

The sorbent can also be used in a chemical reaction, with the adsorbed molecule as one of the reaction products. In sorption-enhanced reaction processes, equilibrium-constrained reactions are driven to the product side, that is, to completion by removing one of the products by using a solid sorbent. The advantage of sorption-enhanced reaction is that the conversion and the separation are combined in just one reactor. Apart from potential equipment savings, reduced energy use for the purification of the main product is the driving

Process Intensification for Sustainable Energy Conversion, First Edition.
Edited by Fausto Gallucci and Martin van Sint Annaland.
© 2015 John Wiley & Sons, Ltd. Published 2015 by John Wiley & Sons, Ltd.

force for the further development of this technology. The concept of sorption enhanced fossil fuel conversion has been developed in the last three decades mainly for the purpose of production of pure hydrogen on industrial scale and for pre-combustion CO_2 capture in natural gas or coal gas fuelled combined cycles for power generation.

Among several enhanced fuel conversion processes, the most investigated are the Sorption-Enhanced Reforming (SER) process and Sorption-Enhanced Water Gas Shift (SEWGS). The first combines the separation of one product, typically CO_2, within the reforming process, while the second applies it to the water-gas shift (WGS) reaction.

This chapter is organised as follows: firstly, enhanced processes concepts are presented (Section 6.2), then the sorbent properties and performances (Section 6.3), thirdly enhanced reactor configurations (Section 6.4) are discussed and finally, examples of performances and economic benefits of sorption-enhanced processes for CO_2 capture are shown (Section 6.5).

6.2   Development in Sorption-Enhanced Processes

Modern-day industrial-scale facilities for hydrogen or syngas production use multiple reactions to convert carbonaceous feedstock (e.g. natural gas) into the required composition. These reactions are typically equilibrium limited and this drives the operating temperatures and pressures that are used.

For gaseous (e.g. methane) or liquid feedstocks (e.g. methanol), the first reaction performed is a reforming step in which the feed stream is heated to 700 °C–1000 °C, mixed with steam and reacted over a catalyst to form a syngas mixture of H_2O, CO, H_2 and CO_2. With solid fuels (e.g. coal), a similar process known as gasification is used, but this is carried out at higher temperatures and without the need of a catalyst.

As an example, in a steam-methane reformer (SMR), the following two dominating reactions take place:

$$CH_4 + H_2O \rightleftharpoons CO + 3H_2 \quad \Delta H_f^\circ = 206.2 \, kJ/mol \tag{6.1}$$

$$CO + H_2O \rightleftharpoons CO_2 + H_2 \quad \Delta H_f^\circ = -41.2 \, kJ/mol \tag{6.2}$$

The overall conversion of CH_4 to H_2 is favoured by a low operating pressure and high temperature. However, for large-scale process, the economics for equipment sizing require high pressures and this elevates the temperature needed to achieve sufficient conversion. Alternatively, by Le Chatelier's principle, removing CO_2 from the gas via the second reaction would help shift the equilibrium of all the reactions towards the production of hydrogen.

When hydrogen production for fuel is required, the second reaction (known as water-gas shift or WGS) can also be carried out again downstream of the reforming/gasification step to further convert residual H_2O and CO into CO_2 and H_2. Enhancement of the conversion can again be achieved by removing the CO_2 from the gas in situ to the reactor.

For both the reforming and WGS processes, various methods have been proposed and tested that utilise a sorbent material for removing the CO_2 from the gas phase as the reactions occurs. These have been given various names in the literature but are generically known as Sorption Enhanced Reaction Processes (SERP) and more specifically for fuel conversion as SER and SEWGS. While there are many other types of sorption-enhanced reactions that have been considered in the literature (e.g. ammonia by Nikačević *et al.* [1] and CO production by Carvill *et al.* [2]), the focus here is on SER and SEWGS for the removal of CO_2 from syngas to produce H_2 as a clean fuel for either a gas turbine or fuel cell. In respect to gas turbine applications, the use of a SERP is particularly advantaged in that there is no need for an ultra-pure H_2 product gas and the fuel can therefore be prepared without the need to inefficiently cool and condense out water [3].

6.2.1   Enhanced Steam Methane Reformer

The SMR reaction process (1) is strongly endothermic and industrially operated at high temperatures (850–1000 °C) in order to obtain high methane conversions. In industrial hydrogen plants, SMR is followed by two water-gas shift reactors (2) (one at high temperature, 350–450°C, and one at low temperature, 210–250°C), and finally a PSA to purify the hydrogen. Therefore, a commercial plant for hydrogen production from natural gas requires four different steps: three reactors for fuel conversion and one adsorber for hydrogen separation.

In the sorption-enhanced steam methane reforming process (SER), a steam reforming catalyst is combined with a CO_2 sorbent and the overall equilibrium is shifted to the product side. SER can be operated at lower temperatures than ordinary SMR making the two shift reactors possibly redundant. Sorption-enhanced reforming was investigated by Air Products in the 1990s as an alternative technology for the industrial production of pure hydrogen using hydrotalcite-based sorbents [4]. With this sorbent, the process was operated successfully at temperatures between 450 and 550 °C. A major obstacle to industrial implementation of the SER concept for hydrogen production was the relatively low purity of the hydrogen, caused by incomplete conversion of methane. In the SMR process (Eq. (6.1)), the CO_2 removal with sorbents has a relative small effect on the equilibrium methane conversion. Also, elevated pressures lessen the enhancement effect of the sorbent at the temperatures investigated. Another disadvantage was the high steam demand for regeneration of the hydrotalcite-based sorbent.

SER has been investigated as a CO_2 capture technology in power plants, and in particular, the pre-combustion decarbonization of natural gas combined cycles (NGCC) has been looked into by the Energy Centre of the Netherlands (ECN) [5, 6]. Part of the produced hydrogen was used as fuel in the gas turbine combined cycle. The other part was used as fuel for the burner producing the heat for the steam reforming. The steam for the regeneration is extracted from the heat recovery steam cycle of the combined cycle. In this pre-combustion CO_2 capture concept, the hydrogen purity was much less important than for the industrial production of pure hydrogen: Gas turbine combustor usually works with diffusive flame requiring a low H_2 purity ranging from 50% to 70%. However, the regeneration of the sorbent, which reduces the efficiency of power plant, is an important parameter in the development of SERP process (Figure 6.1).

6.2.2   SEWGS

The use of a sorption process for the removal of CO_2 during the water-gas shift reaction is not a recent idea and Gluud *et al.* [7], for example, describe the combination of a catalyst

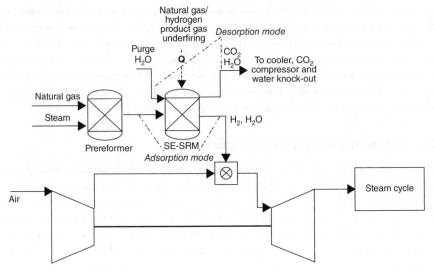

Figure 6.1 *Natural gas combined cycle with pre-combustion CO_2 capture on the basis of the SERP concept [5, 6]. (Source: Reproduced from Ref. [5], with permission from Elsevier and reproduced with permission from Ref. [6], Copyright © 2006, american chemical society)*

with a metal oxide sorbent such as calcium or magnesium oxide. The idea did not take off commercially at the time, but interest has grown strongly over the last two decades due to concerns over climate change and the drive to improve the efficiency of decarbonised fossil fuel power plants. For example, Air Products started to look at improving the sorbents and process cycles in the 1990s and this work spawned into feasibility work for the CO_2 capture project (CCP), laboratory-scale testing for the EU framework 6 (FP6) project CACHET [8] before branching out into high-performance materials and new applications as part of the EU FP7 project CAESAR [9].

The reaction is an exothermic equilibrium reaction. The equilibrium lies at the product side at lower temperatures (around 200 °C), but shifts to the reactant side at increasing temperatures. To reach high CO conversion, the WGS reaction is normally carried out in two stages: a high-temperature shift and a low-temperature shift, thus benefiting from high reaction rates at high temperature and a more favourable equilibrium at low temperature. Typical operating temperatures range from 200 °C to 500 °C with various catalysts. The minimum molar H_2O/CO ratio is around 2.

The SEWGS process combines CO_2 adsorption with the water-gas-shift (WGS) reaction. When syngas (containing CO, H_2, CO_2, H_2O and inerts) is fed at high pressure and temperature (30 bar, 400 °C), CO_2 is removed by the sorbent. Hence, the WGS equilibrium is shifted to the right-hand side, thereby completely converting the CO and maximizing the production of H_2. This effectively removes CO and CO_2 from the feed gas, producing a high-pressure, hydrogen-rich product stream until the adsorbent is saturated and CO_2 appears in the product stream. At a predetermined level of CO_2 breakthrough, the sorbent is regenerated.

The SEWGS process is particularly attractive for pre-combustion decarbonisation applications in IGCC and blast furnace gas. Here the desire is to convert as much CO in the fuel gas to H_2 as possible and then separate the CO_2 from the H_2. The H_2 can then be fed to a gas turbine to generate power, while the CO_2 is sequestered. Conventional approaches to this require multiple cooling steps. The first cools the hot gas from the high-temperature WGS reactor to around 200 °C so that a second low-temperature WGS reaction can be carried out to achieve the required conversion of CO. Further cooling is then necessary to enable the capture of CO_2 by absorption with a physical solvent.

A more elegant and straightforward approach is offered by incorporating the SEWGS process. Partially shifted fuel gas from the high-temperature WGS reactor is fed to a SEWGS unit, and a hot, high-pressure, hydrogen-rich product stream is produced. This H_2-rich product can be fed directly to a gas turbine at around 400 °C. This removes the inefficiency of cooling/heating of the H_2 that is an inherent part of the conventional process and in particular it eliminates the condensation of steam, which must be replaced in the gas turbine feed with another diluent, for example, nitrogen.

The advantages of combining the water-gas shift reaction with separation of CO_2 are as follows:

- Conversion of CO to H_2 in the equilibrium-limited shift reaction is increased through the removal of a product of reaction.
- Separation of CO_2 at high temperature and utilization of the hot hydrogen product minimises heat exchange equipment.
- Hydrogen exits the reactor at high temperature with surplus process steam, which increases the overall efficiency and reduces NO_x emissions in the gas turbine.

The ideal application for the SEWGS process should fully exploit the following SEWGS characteristics:

- High-temperature hydrogen (typically 400 °C)
- High-pressure hydrogen (typically 30 bar)
- Medium-purity hydrogen (typically 90–95 mol%)
- High-temperature CO_2 (typically 400 °C)
- CO_2 at near-ambient pressure
- High CO conversion.

In a combined cycle power production unit with CO_2 capture by pre-combustion, the process will be enhanced by the production of high-temperature hydrogen, making the SEWGS process potentially advantaged for this application. As the requirement for CO_2 removal (and hence hydrogen purity) is not very stringent, processes such as SEWGS that do not produce high separation factors would not be penalised by the requirements for further polishing steps. So SEWGS fits with power production with CO_2 capture.

Non-power applications have been assessed on the basis of the typical characteristics of the SEWGS technology mentioned earlier. The applications that are being assessed are (i) distributed refinery hydrogen, (ii) refinery process hydrogen, (iii) refinery fuel gas, (iv) ammonia production, (v) coal to liquid chemicals and liquid fuels. The assessment made clear that these non-power applications cannot fully exploit the SEWGS characteristics, and therefore, these non-power applications are less obvious with likely lower economic benefits.

6.3   Sorbent Development

As briefly introduced in the previous sections, the sorbent is the most important component in an enhanced reactor since it determines the reactor design and the performance of the process. The sorbent material must have the following characteristics:

1. Able to adsorb and desorb CO_2 over several hundred thousands of cycles from gases with high partial pressures of both steam and CO_2;
2. Show both good mechanical and chemical stability that is a lifetime of at least 5 years;
3. High cyclic capacity for CO_2 and high selectivity over hydrogen and other (trace) components in the gas;
4. Low adsorption of steam;
5. Kinetics of adsorption and desorption sufficiently fast
6. Low steam use during desorption (purge and rinse);
7. Low specific costs.

In particular, the high cyclic capacity of the sorbent is one of the key characteristics. The higher the capacity, the more CO_2 can be adsorbed per m^3 reactor volume and the less steam for sorbent regeneration is needed reducing the primary energy consumption per ton of CO_2 captured. Lifetime and specific sorbent cost are the other two key parameters predominantly affecting the operational cost of the CO_2 capture process.

In this section, the advancement in sorbent development for SERP and SEWGS is presented.

6.3.1   Sorbent for Sorption-Enhanced Reforming

In SERP, CO_2 is adsorbed over a solid sorbent while SMR and WGS reactions occur.

Sorbents that are considered in the literature for sorption-enhanced reforming process are potassium promoted hydrotalcite (K-HTC), lithium orthosilicate ($LiSiO_4$), lithium zirconate ($LiZrO_3$), sodium zirconate (Na_2ZrO_3) and calcium oxide (CaO). The affinity of a sorbent to a molecule can be expressed by the equilibrium partial pressure at different temperatures. For example, the equilibrium partial pressure of carbon dioxide for different sorbents is shown in Figure 6.2.

Lithium- and sodium-based sorbents have the highest capacities and good stability, in particular, at lower temperatures [10]. They also show a high heat of reaction with CO_2, allowing for adiabatic reformers. On the other hand, lower reforming temperatures are needed to take advantage of their sorption capabilities and higher S/C required to obtain good methane conversions. Because of these reasons and the slow kinetics, Li- and Na-based sorbents do not seem to be suitable for enhanced reactors applied to power plants with satisfactory cost and CO_2 capture (> 80%).

Another type of sorbent is based on hydrotalcite, which shows good cyclic stability, even if lower than competitive sorbents, and fast sorption kinetics. On the other hand, CO_2 sorption on hydrotalcite has a low heat of reaction, making it necessary to sustain the endothermic reforming reaction by an external source of heat or the application of an autothermal system. In addition, these materials are suitable for operations at 400–450 °C where methane conversion is rather low. For these reasons, hydrotalcites are more promising for the sorption-enhanced water-gas shift concept, which is described in the following section.

A promising sorbent for SER applications is calcium oxide (as pure CaO or dolomite), which can react with CO_2 generating $CaCO_3$ according to the following carbonation reaction (6.3):

$$CaO_{(s)} + CO_2 \leftrightarrow CaCO_{3(s)} \quad \Delta H_r^\circ = -179.2\,kJ/mol \qquad (6.3)$$

Equation (6.4), reported in Ref. [11], is an example equation expressing the equilibrium CO_2 partial pressure with temperature.

$$p_{CO_2,eq} \; [Pa] = 4.137 * 10^{12} * \exp(-20,474/T) \qquad (6.4)$$

The overall calcium-based SER reaction, which results from the single carbonation (Eq. (6.3)), reforming and water-gas shift reactions is reported in Eq. (6.5):

$$CH_4 + 2H_2O + CaO_{(s)} \leftrightarrow 4H_2 + CaCO_{3(s)} \quad \Delta H_r^\circ = -14.5\,kJ/mol \qquad (6.5)$$

As the enthalpy balance of the overall reaction is almost neutral, the carbonation reaction, in addition to CO_2 removal from the gaseous phase, provides the heat required for the steam reforming reaction, allowing for the use of adiabatic reactors, or with very limited heat duties.

An example of the hydrogen yield[1] and carbon capture ratio at chemical equilibrium considering pure methane as the primary fuel at different SER operating conditions is shown in Figure 6.2 [14]. In the curves in the graph, at given pressure and S/C, a temperature range can be identified where H_2 yield and CCR experience limited variations. In this range, effects of temperature on SMR, WGS and carbonation reactions counterbalance.

At higher temperatures, the equilibrium CO_2 partial pressure of the carbonation reaction increases (6.4) and a higher CO_2 fraction will hence be present in the gaseous phase, leading to lower $CaCO_3$ formation and decreasing carbon capture ratio. Despite the lower CO_2 sorption, CH_4 conversion increases, provided that exothermic SMR reaction is favoured, while H_2 yield varies depending on the SMR and WGS equilibria. By further increasing temperature, a point is reached where the CO_2 pressure in the gaseous phase is below that predicted by Eq. (6.4) and no further CO_2 can be adsorbed by generating $CaCO_3$. At such temperatures, no sorption occurs and reactions behave like in conventional steam reformers. By reducing the temperature below the flat zone section, a point is reached where steam in the gaseous phase reacts with calcium oxide producing solid $Ca(OH)_2$.

Sorbent regeneration in SERP is usually performed via calcination reaction (the reverse of reaction (6.5)). This is achieved by reducing the CO_2 partial pressure in the gaseous phase below the CO_2–$CaCO_3$ equilibrium either by reducing the actual CO_2 partial pressure (pressure swing) or by increasing the temperature and hence the equilibrium pressure (temperature swing). In any case, a large amount of thermal power must be supplied to the endothermic calcination reaction.

The main limit of natural CaO and dolomite as the sorbent is the loss of cyclic capacity due to sintering after repeated carbonation and calcination cycles [15]. For this reason, research activity has been focused on the development of improved synthetic Ca-based sorbents as well as reactivation techniques. As the latter, it was demonstrated that CaO can

---

[1] Defined as the moles of hydrogen generated per mole of methane, whose maximum theoretical value is 4 according to reaction (6.5).

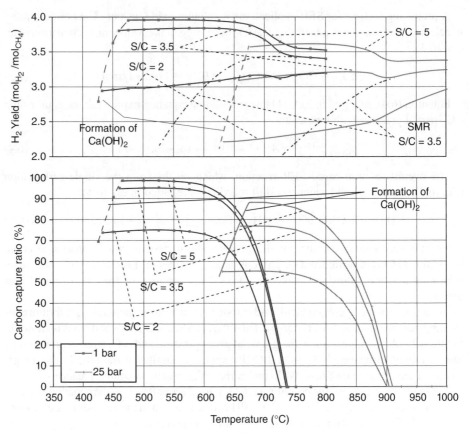

Figure 6.2 *Equilibrium partial pressure of carbon dioxide for different CO$_2$ Acceptors (Ref. [14]). (Source: Reproduced from Ref. [14], with permission from Elsevier)*

be reactivated with steam at conditions favourable to Ca(OH)$_2$ formation [16–18], which leads to particle break-up and an increased porosity. Synthetic sorbents, despite their higher cost, may also be considered due to their superior performance in cyclic operations (Figure 6.3). For advancements on material issues, the reader can refer to recent reviews [19, 20].

6.3.2   Sorbent for Enhanced Water-Gas Shift

For the last 10–12 years, the development of sorption-enhanced fuel conversion has mainly focused on the further improvement of sorption-enhanced water-gas shift using hydrotalcite-based sorbents. Starting with the sorption-enhanced steam reforming work of Air Products, hydrotalcite-based materials were identified as potentially attractive sorbents for a pressure swing-based sorption-enhanced water-gas shift reactor system.

Hydrotalcite-based materials (Figure 6.4) impregnated with K$_2$CO$_3$ have shown favourable properties for adsorbing CO$_2$ at 400 °C in the presence of steam [4–6, 21–25]. Potassium-promoted hydrotalcite CO$_2$ sorbents are based on the hydrotalcite structure

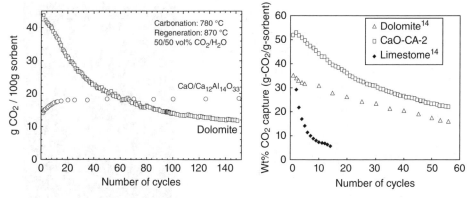

Figure 6.3 *Multicyclic capacity of natural and synthetic sorbents [12, 13]. (Source: Reproduced with permission from Refs. [12, 13]. Copyright © 2006 and 2011, american chemical society)*

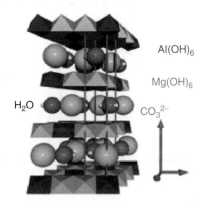

Figure 6.4 *Chemical structure of hydrotalcite material*

depicted earlier in Figure 6.5. The impregnation with potassium carbonate boosts the CO_2 adsorption capacity by a factor of seven [6]. Promoted HTC sorbents are particularly well suited for application in the SEWGS process. The main interaction of CO_2 with such materials at 400 °C is due to an aluminium potassium-carbonate species, subject to modification by the presence of magnesium [25].

In the European FP6 project CACHET, the SEWGS technology was demonstrated using potassium-promoted, hydrotalcite-based materials under realistic conditions in a dedicated single-column experimental rig comprised of a of 2 m length reactor vessel (Figure 6.5). In this rig, batches of calcined pellets of K_2CO_3-promoted hydrotalcite material, designated as K-Mg70 (HTC material with 70 weight percent Mg), were extensively tested for several thousand cycles [27]. Figure 6.5 shows the scheme of the single-column SEWGS test rig.

The K-Mg70 sorbent showed reversible take up of CO_2 at temperatures near 400 °C with breakthrough capacities of 1.3–1.4 mmol/g measured under realistic conditions (see Figure 6.6). Total capacities for this material could exceed 8 mmol/g (see Figure 6.7), if

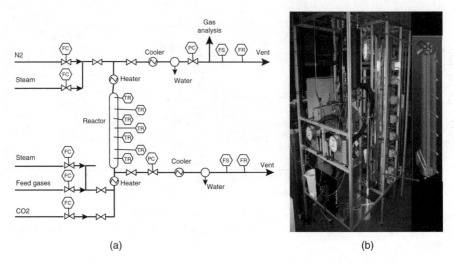

Figure 6.5   *Flow scheme (a) and picture of the test rig (b) at ECN*

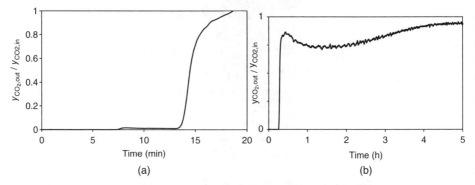

Figure 6.6   *Breakthrough curve for CO_2 at a pressure of 28 bar and a temperature of 400°C (a). Slow uptake of CO_2 continues for hours after breakthrough. Total pressure 28 bar, temperature 400°C (b). [26]*

feed partial pressures of CO_2 and H_2O were sufficiently high. However, this high capacity was associated with formation of $MgCO_3$ and the kinetics of this chemisorption is too slow to exploit in a pressure swing adsorption process such as SEWGS [27].

The chemical stability of the K-Mg70 sorbent material was shown for more than 1500 cycles of adsorption and desorption at the envisaged industrial operating conditions. In Figure 6.7, the CO_2 level in the product gas is plotted for the first 1400 cycles. The low slip of CO_2 in the product gas confirmed that carbon capture levels of well above 95% can be realised with the SEWGS technology. In additional experiments, the chemical stability was demonstrated for more than 4000 cycles.

Although the K-Mg70 HTC sorbent showed good chemical stability and relatively high CO_2 uptake, the high steam consumption for regeneration of this sorbent, resulting in a S/C ratio of more than 4, was one of the key factors for the relatively high efficiency penalty of

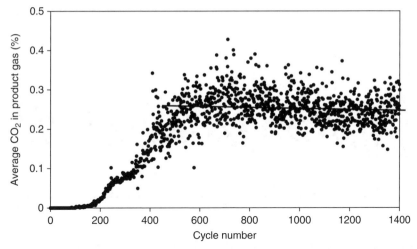

Figure 6.7   *Cycle-averaged CO₂ level of the top product over 1400 cycles, showing cyclic stability after 250 cycles. Feed at 400°C, 28 bar, 20% CO₂. Refs. [28, 29]. (Source: Reproduced from Ref. [28], with permission from Elsevier and Reproduced with permission from Ref. [29], Copyright © 2009, american chemical society)*

almost 13% points [27]. The S/C high ratio also caused the high capture costs [27] observed for the SEWGS process using the K-Mg70 sorbent. In order to make the SEWGS process economically competitive, at least a 50% reduction of the CO_2 capture cost was essential. To achieve this breakthrough, a considerable increase in sorbent capacity per unit volume with a substantial reduction in sorbent cost was required.

Another critical issue observed after unloading the single column test rig at the end of the CACHET project was that the tested K-Mg70 HTC sorbent pellets were fractured and powdered (i.e. showed mechanical degradation [27]). Since friable materials may lead to valve sealing failures, blocked filters, and increased particulates to the gas turbine, the mechanical stability of the pellets is critical for the SEWGS process. Interestingly, the degradation did not affect the CO_2 sorption capacity that remained high during all the tests.

Another issue arising from the long-term testing of the K-Mg70 (4000 cycles performed) under realistic process conditions, that is, elevated partial pressure of CO_2 and H_2O, was that a steadily increasing CO_2 slip in the hydrogen product gas appeared with increasing cycle number. Both the above-mentioned mechanical degradation and the increasing CO_2 slip were found to be related to the formation of considerable amounts of $MgCO_3$ [30]. The formation of $MgCO_3$ in this K-Mg70 HTC sorbent was reversible but much slower than the CO_2 adsorption and desorption. As a result of the carbonate formation, the capture rate decreased over time. Complete regeneration of the $MgCO_3$ was possible but at the cost of a high steam use resulting in a high-efficiency penalty [27].

In the FP7 CAESAR project [9], sorbent development focused on addressing these issues with the hydrotalcite-based sorbent. The objectives here were to increase the cyclic capacity, to reduce the amount of steam needed for regeneration, to increase the mechanical strength of the sorbent pellets and finally to reduce the specific sorbent cost. The key issue was to develop a HTC-based sorbent material that did not form notable amounts of $MgCO_3$

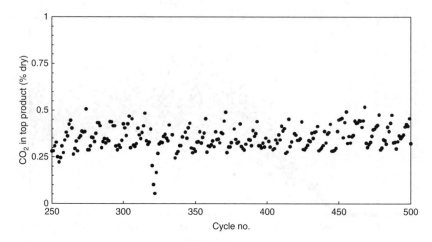

Figure 6.8   *Cyclic stability of the carbon recovery using a steam rinse cycle. [ref?]*

under relevant process conditions. The new sorbent chosen for testing was a potassium carbonate-promoted hydrotalcite-based material with only 30 weight per cent Mg, known as K-Mg30 [26].

Figure 6.8 shows with K-Mg30 the cycle-averaged CO_2 fraction in the product gas as a function of cycle number for cycles 250 to 500. The carbon slip was so low that CO_2 recoveries of 99% were obtained, similar to CO_2 recoveries with the K-Mg70 sorbent. The tests showed that the sorbent material was mechanically stable over at least 1200 adsorption–desorption cycles without loss of activity. Due to its higher density, the cyclic working capacity of the sorbent on volume basis was 27% higher compared to K-Mg70. Equivalent CO_2 removal of K-Mg30 during cyclic testing was at a lower steam consumption than K-Mg70 and thus the performance was better.

The bulk formation of $MgCO_3$ previously observed in K-Mg70, which led to mechanical degradation of the sorbent tablets, was not observed for the K-Mg30 sorbent. The new K-Mg30 sorbent pellets also had 65% higher crush strength than K-Mg70 on an as received basis. After the cyclic tests, crush tests revealed that the pellets had lost 25 to 50% of their initial crush strength but had remained intact, in comparison to the K-Mg70 sorbent pellets, which were fractured and powdered throughout the entire reactor upon repeated cycling (see Figure 6.10). Hence, it was concluded that the new sorbent pellets remain mechanically stable during operation of more than 1200 cycles, as shown in Figure 6.9. The mechanical stability is attributed predominantly to the absence of $MgCO_3$ formation.

Furthermore, due to the absence of the slow CO_2 uptake process by $MgCO_3$ formation, cyclic steady state was reached rapidly, which is important for a PSA type of process. For the new sorbent, cyclic steady state was reached within a few cycles (see Figure 6.8), whereas for the reference sorbent, it took several hundred cycles to reach cyclic steady state.

As a result of all the improvements, the K-Mg30 sorbent had substantially lower steam consumption for the regeneration resulting in a steam/CO_2 ratio of 1.6 at the test conditions used. Consequently, the efficiency penalty was reduced almost 50% from 13% points (K-Mg70) to about 7.6% points (K-Mg30) for a 95% CO_2 capture ratio in an NGCC [31].

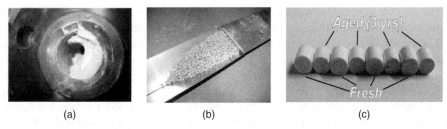

(a)   (b)   (c)

Figure 6.9 *Sorbent unloaded from the reactor: (a) fractured K-Mg70 HTC sorbent after more than 1200 Cycles; (b) mechanically stable K-Mg30 sorbent after more than 2000 cycles and (c) the sorbent after three years testing in the SC Unit*

In conventional shift applications, the catalyst operates under reducing conditions. For application in a SEWGS process, the catalyst should be able to withstand the oxidising conditions during the cycle, such as when the sorbent is regenerated by steam. It should also remain active at the SEWGS operating conditions where only a limited amount of steam in the feed is present. Commercially available catalysts on supports have been benchmarked at different temperatures and in the presence of actual CO_2 sorbents [32]. However, during breakthrough experiments with promoted hydrotalcite under realistic conditions, it was observed that even in the absence of a catalyst the carbon monoxide in the feed gas was completely shifted to carbon dioxide (see Figure 6.10). Testing showed the stability of the sorbent working capacity as well as shift activity during 5000 cycles, at a minimal S/C ratio (2 mole/mole). Hence, it was demonstrated that the SEWGS process does not require a shift catalyst, which brings substantial economic and technical benefits for this technology. Additional tests have shown that the water-gas shift activity of HTC sorbents is also sufficient to leave out the separate water-gas shift catalyst in a SEWGS reactor if the syngas contains H_2S [34].

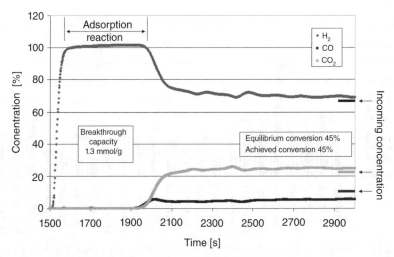

Figure 6.10 *ALKASORB breakthrough tests without catalyst showing sufficient WGS activity before CO_2 breakthrough*

The effect of contaminants in the syngas on the CO_2 adsorption capacity of the HTC sorbents has also been investigated. A test with H_2S containing syngas showed clearly that the HTC sorbents capture H_2S along with the CO_2 without significant loss of capacity. In separate tests, full COS hydrolysis and adsorption of the H_2S followed by a simultaneous breakthrough of H_2S and CO_2 have been observed. Methane slips through completely, whereas NH_3 is partially captured. Finally, HCH is partially converted into NH_3 and partially captured. These test results clearly showed that K-Mg30 sorbent is a robust sorbent and capable of simultaneous decarbonisation and desulphurization of sour syngas originating from the gasification of coal. Accordingly, it produces a hot pressurised H_2-rich product stream with low contents of CO_2, COS and H_2S and next to a CO_2-rich stream with H_2S [34, 35].

In addition to the efforts to improve the performance of potassium-promoted hydrotalcite sorbents, new alumina-based sorbents have been screened via high throughput testing on their applicability in the SEWGS process [33].

In total, 432 new sorbent formulations were prepared, partly characterised and more than 300 sorbents evaluated under realistic conditions in a three-cycle adsorption–desorption test. For the evaluation, a comparison with the K-Mg70 HTC sorbents has been made (see Figure 6.11) and four sorbent leads were selected for up-scaling and testing for sorption performance and particle stability under realistic SEWGS conditions. However, none of these four sorbents performed sufficiently well compared with K-Mg30 to justify scale up of the sorbent.

6.4   Process Descriptions

For sorption processes, the difficult part is typically not the adsorption step where one or more components are removed from a gas stream but instead how to regenerate the sorbent economically and recover the adsorbed component(s). For SERP, different techniques have

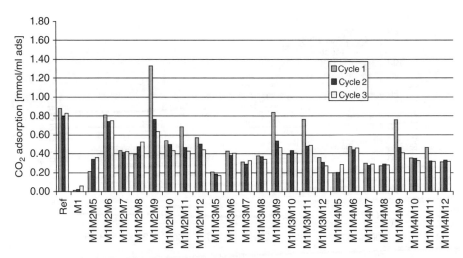

Figure 6.11   *Example of a high-throughput test result. CO_2 adsorption capacity of new sorbents is benchmarked against the reference K-Mg70 HTC sorbent Ref. [33]. (Source: Reproduced from Ref. [33], with permission from Elsevier)*

been proposed to enable this to happen and can be roughly separated into two groups. In the first of these, the sorbent is moved back and forth between an adsorber vessel and a regeneration vessel in a similar manner to absorption-based processes (like MEA removal of CO_2), but using a solid carrier rather than a liquid. The other group uses stationary packed beds of sorbent and the feed and regeneration gases are switched between the vessels in a semi-batchwise manner.

6.4.1   Fluidised Beds

Figure 6.12 shows the most commonly proposed fluidised bed design where two fluidised beds are used with a first as the adsorber/reactor for carrying out the SERP process and a second fluidised bed for regenerating the spent sorbent [36].

The fluidised bed approach is advantageous in that it is a continuous process producing product streams at a steady flow rate and of constant composition. For simplicity, the regeneration step is typically carried out at the same pressure as the syngas and therefore the resulting CO_2 is obtained at elevated pressure. This is an important consideration if the CO_2 is ultimately to be compressed and captured. However, there are a number of challenges with the fluidised bed approach that need to be considered in its design. The first is that the loading of CO_2 on the sorbent is dependent on the partial pressure of CO_2 in the outlet syngas. As the required CO_2 capture rate increases, then the allowable CO_2 partial pressure in the outlet syngas decreases and with it the maximum loading on the sorbent material. This can be compensated by increasing the recycle rate of material between the adsorber/reactor and regenerator, but at the expense of additional operational costs. Alternatively, by adding additional complexity, multiple fluidised beds can be used in series to mimic a truly counter-current process with the gas passing through the vessels in reverse order to the sorbent.

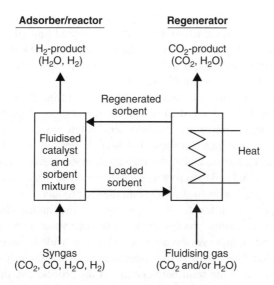

Figure 6.12   *Schematic of a dual fluidised bed for SEWGS*

Another consideration that needs to be taken into account, particularly for power generation with CO_2 capture, is where the regeneration heat is sourced from. If this is a natural gas burner, then a secondary source of CO_2 emissions is created and if from steam or electric heating then the overall efficiency of the plant will be reduced.

There is also a challenge in that the material needs to be physically robust and not degrade over time. The rate of abrasion of the sorbent must be taken into account in any economic analysis to evaluate the cost of replenishing the material over time. The same also goes for the catalyst particles and whether they are stable over the range of operating conditions between the adsorber/reactor and regenerator. Liu *et al.* [37] note that there is a potential problem if there is a need to separate out the sorbent particles from the catalyst particles before sending them to the regenerator. However, their work along with [38] show that at least for SEWGS the sorbent itself may have sufficient catalytic activity without the need for a separate specialist catalyst. An alternative idea by Ref. [39] for SEWGS is to simply use a series of fluidised beds with WGS reactors in-between, but this is a less efficient approach.

When low-pressure reactors are considered, the technical feasibility of fluidised beds has been demonstrated by commercial application (fluid catalytic cracking process) and by research projects focused on other technologies under development (post-combustion Ca-looping, CLC, sorption-enhanced gasification of biomass) in a number of pilot installations. The limit of this layout is related to the low pressure of the H_2 delivered, which would require relevant energy penalties associated with hydrogen cooling and compression, in case of a high-pressure final utilization (e.g. combustion in a gas turbine).

The use of pressurised interconnected fluidised bed is an alternative investigated in some works. However, stable operation of interconnected pressurised fluidised beds has not been demonstrated yet and requires technology breakthrough.

6.4.2   Fixed Beds

In fixed bed designs, the sorbent and catalyst particles are mixed together and formed into a stationary packed bed. The reaction gas mixture is passed through the vessel allowing the sorbent to remove CO_2 as the catalyst further reacts with CO and H_2O to form H_2. At some point, however, the sorbent will become saturated with CO_2 and start to contaminate the H_2-product gas as shown in Figure 6.11.

Unlike the fluidised bed, the sorbent cannot be removed from the vessel and so the process conditions need to be changed to regenerate the sorbent. Therefore, the fixed bed process is not continuous but is effectively a semi-batch process requiring at least two vessels. Figure 6.13 shows how one vessel is used to perform the SEWGS process on the syngas stream and a second undergoes regeneration to remove the adsorbed CO_2. The regeneration of the second bed therefore needs to be timed so that it is ready to be put back online once the first bed begins to slip CO_2 into the H_2-product.

The fixed bed approach suffers from complexity over the fluidised bed system as valve switching is required to change process streams from one vessel to another. It can also be higher capital because multiple vessels are required and each additional one adds to the total cost in a linear fashion. There is also a time dependency to the compositions and flows from fixed bed processes due to the semi-batch nature of the process and this may require the addition of mixing tanks to be added to reduce the fluctuations for downstream equipment.

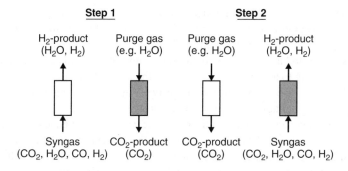

Figure 6.13   *Schematic of a parallel fixed bed process for SEWGS*

There are, however, advantages of fixed bed processes over fluidised beds. Firstly, the cyclic capacity of the sorbent can be much higher with the equilibrium loading of CO_2 depending more on the partial pressure of CO_2 at the inlet rather than the outlet. The required mechanical stability of the material is also less challenging as the requirement on abrasion resistance is much lower. These two factors are critically important and these are the reasons why in general the majority of (ad)sorption-based processes in the field are fixed bed rather than moving bed.

There are multiple ways in which fixed bed processes can be put together in order to optimise between operating and capital costs. The designs can be quite different, and the best one can only be determined from a full-cost analysis over the entire plant.

Lee *et al.* [40] have proposed a five-step scheme shown in Figure 6.14, which uses the addition of heat (i.e. temperature swing adsorption (TSA)) to drive CO_2 from the sorbent during regeneration. The cycle operates by first passing syngas over a packed bed of sorbent and catalyst to produce a desired high-purity H_2-product gas. In the experiments performed, a pure mixture of CO and H_2O was used as the feed, although the concept could be extended to other SERP processes.

When CO_2 and CO breakthrough into the H_2-product stream, the bed is taken offline and rinsed co-currently with CO_2. This means that a CO_2-rich stream is fed into the vessel

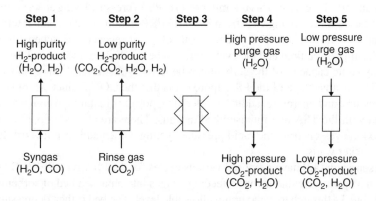

Figure 6.14   *Temperature swing regeneration for a SERP*

from the feed end of the process (i.e. using part of the CO_2-product gas). The reason for doing this is to flush out as much of the residual H_2 and CO inside the vessel as possible. The product gas from this step is rich in H_2, but contains a substantial quantity of CO_2. It may be possible to find a use for this hydrogen stream, or alternatively it can be mixed in with the syngas going to another bed to minimise the loss of CO_2. In principle, this type of process could also be used to produce high-purity H_2 during the syngas feed step followed by a lower purity H_2 suitable for a gas turbine during the rinse step.

Following the rinse step, in step 3 of the cycle, the vessel is isolated and heated externally to cause the CO_2 to desorb from the sorbent and the pressure to increase. High-pressure steam is then introduced counter-currently to the feed direction in order to push out the CO_2 where it is then cooled to condense out water. This resulting CO_2 stream is stated as having a purity of up to 99% and part of this can be used for the CO_2 rinse step on another bed.

After the vessel has been sufficiently purged at high pressure to remove CO_2, the pressure is dropped back to the syngas pressure as additional steam is supplied to remove the remaining CO_2 from the bed.

The benefit of the TSA approach is that it produces a CO_2 stream at high pressure that reduces the cost for subsequent compression and capture. However, there are a number of difficulties with this technology. For example, the batch heating step can take several minutes to complete, which places a limit on the minimum cycle time that can be used. With a relatively long cycle time, the dynamic capacity of the sorbent must be very high to keep the vessels small. For the materials looked at by Ref. [41], near-complete regeneration of the sorbent was required, which resulted in regeneration steam requirements around 6–10 times the amount of carbon in the feed gas. The vessels required for batch heating are also expensive as poor radial heat transfer within a packed bed means that a shell-and-tube arrangement (with the sorbent and catalyst in the tubes) is required to maximise heat transfer surface area. There is also the same problem with the TSA as for the fluidised bed arrangements, in that high-temperature heat must be supplied externally to remove the CO_2 from the sorbent and this must be factored into the economics.

Overall, the thermal regeneration approach in its current state of development may be suitable for small-scale systems but is probably not applicable for large-scale power generation and H_2 generation processes.

An alternative approach to using temperature for regenerating the sorbent is to use a reduction in partial pressure. This is similar to a H_2 PSA (pressure swing adsorption), which can also be used for removing CO_2 from H_2, but SERPs require a much higher temperature and include a shift reaction. The benefits of this type of regeneration scheme are that no external heating is required and the cycle time can be much shorter than for TSA cycles so that the dynamic capacity of the sorbent can be lower, yet still result in reasonably sized vessels. The disadvantage of the PSA approach is that the CO_2-product steam is obtained at low pressure, and, therefore, additional electrical power is required to compress the CO_2 compared with the TSA and fluidised bed designs. There are also additional mechanical stresses on the vessels from the rapid pressure cycling, which adds capital cost due to the need for thicker walls.

Anderson *et al.* [42] have proposed various cycles for the PSA-type SEWGS cycle that centres on a basic cycle that involves feeding syngas into a packed bed of sorbent/catalyst until CO_2 breaks through to a maximum allowable level. The bed is then depressurised and a vacuum pulled to remove most of the CO_2 from the sorbent. Once regeneration has been

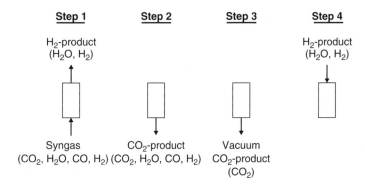

Figure 6.15   *Temperature swing regeneration for a SERP*

completed, the vessel is then repressurised up to the syngas pressure and put back on feed. Figure 6.15 shows all these steps for one vessel, but multiple vessels operating in parallel would be required in practice to achieve a constant H_2-product and CO_2-product flow.

Unfortunately, Reynolds *et al.* [43] found a couple of problems with this simple design. The first is that residual CO and H_2 in the vessel at the end of the feed step ends up in the CO_2-rich product stream. This not only means that further clean-up of the CO_2 is required to separate out the H_2 before it can be used for sequestration, but it also results in a loss of valuable hot H_2-product gas. The second problem with this design is that it is difficult to remove CO_2 from the sorbent without dropping the pressure sub-atmospherically. This has both cost implications and challenges from a safety perspective in exposing H_2-containing vessels to vacuum where air (and hence O_2) can potentially leak into the system and cause an explosive atmosphere to form. While it is possible to design the process to monitor and manage such a scenario safely from an inherently safe design perspective, it is best avoided if at all possible.

In order to resolve the problem of H_2-rich gas ending up in the CO_2-product stream, there are two possible approaches that have been proposed. The first described by Ref. [44] is to use equalisation steps as shown in Figure 6.16 so that as the pressure of one vessel is reduced and the H_2-rich gas flows out, that instead of adding it to the CO_2-product gas it is instead fed into another vessel being repressurised. This allows the H_2-rich gas to be recovered and reduces the quantity that ends up in the CO_2-product gas. The downside to this method is that the cycle must be tailored to pair up two beds to perform the equalisation step. This often cannot be done without adding another vessel into the cycle, which therefore increases complexity and capital cost. It may also not be possible to achieve the required CO_2 purity with just one equalisation step and multiple equalisation steps may be required, each adding an additional vessel to the process.

A second approach that can be used in conjunction with or instead of the equalisation step(s) is to use a rinse gas after the syngas feed step. Similar to the method used for fixed-bed TSA regeneration scheme, Reynolds *et al.* [43] propose taking part of the CO_2 product gas, recompressing it to the syngas pressure and adding it counter-currently after the feed step (as shown in Figure 6.17). This has the advantages of both increasing the amount of H_2 product produced and the CO_2 purity. The downside though is additional

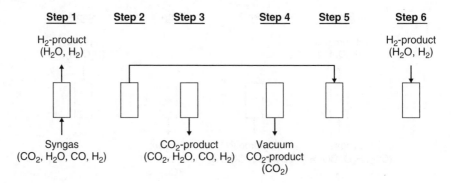

Figure 6.16   *SEWGS with an equalisation step*

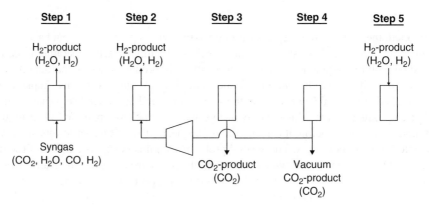

Figure 6.17   *SEWGS with a CO₂ rinse step*

capital and operating cost associated with recompressing part of the CO_2-product stream before it is fed back into the PSA process.

Along with the cost of obtaining the CO_2 for the rinse gas, there is also a limitation on how much can be used for rinsing before it begins to breakthrough to the other end of the bed and contaminates the H_2-product gas. Ying *et al.* [45] therefore proposed the use of a steam rinse in which steam at syngas pressure is passed counter-currently (with respect to the syngas feed direction) through the bed to push out the H_2-rich gas and this is fed into the syngas feed to another vessel. Van Selow *et al.* [27] provide experimental results of this in action and Wright *et al.* [29] showed through modelling that the overall plant efficiency was improved in using a steam rinse rather than a CO_2 rinse. Cobden *et al.* [46] also propose the option of providing the steam rinse in a co-current direction instead of counter-current.

The advantage of using steam is that it is far less of a concern if it contaminates the H_2-product stream. It is also easily separable from the CO_2 in the CO_2-product

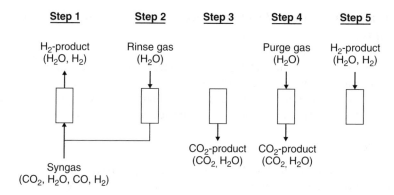

Figure 6.18   *SEWGS with counter-current rinse and purge steps using steam*

stream by simple cooling and condensation, which would be required in any case before compression.

As an alternative to using vacuum for regenerating the sorbent of CO_2, Sircar and Golden [47] propose the use of a steam purge step as shown in Figure 6.18. At a minimum, this reduces the partial pressure of the CO_2 and helps it to desorb while Couling *et al.* [48] note that for some materials the steam can actively displace the CO_2 from the surface of the sorbent. This means that the same work can be done in regenerating the sorbent material but at a higher pressure than when no purge gas is used. This saves having to install and use a vacuum pump on the SERP process and eliminates the safety implications of running sub-atmospherically. The penalty, instead, is the cost of the steam needed for the purge gas.

6.4.3   Design Optimisation of Fixed Bed Processes

The optimisation of fixed bed processes is complicated by the large number of adjustable parameters and cycle designs possible. The cycle time, operating pressures, operating temperatures, rinse gas flow, purge gas flow and vessel size are just some of the variables that need to be considered. Jang *et al.* [49] have even looked at the optimal distribution of catalyst throughout the sorbent bed for SEWGS.

Reijers *et al.* [50] completed a case study where a number of these parameters were varied for a SEWGS process to determine the resulting performance. Starting with a base case of 90% carbon capture with a 98% pure CO_2 product gas, it was shown that both CO_2-product recovery and purity can be increased or decreased fairly easily depending on the plant requirements.

With fixed bed designs, there are some constraints on the system such as making sure the particles do not fluidise, but these are not sufficient to explicitly define the process design. It is only through the integration of the SERP process into the full fuel production and conversion process that the optimal cycle arrangement and operating conditions can be obtained.

6.5   Sorption-Enhanced Reaction Processes in Power Plant for CO_2 Capture

This section gives an overview of the possible integration of SERP into a power plant with CO_2 capture. However, as there is no facility worldwide that implements all the components, the examples are calculated from work undertaken by several research groups with the aim of demonstrating the potential of this technology.

6.5.1   SER

The SER process can be employed to produce a hydrogen-based fuel for low-emission power generation. As already discussed, the adoption of a SER reactor greatly simplifies

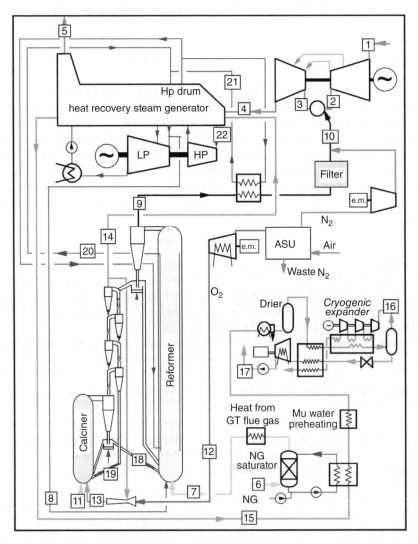

Figure 6.19   *SER-based combined cycle assessed by Ref. [14]. (Source: Reproduced from Ref. [14], with permission from Elsevier)*

the H_2 production section with respect to conventional systems, by condensing methane reforming, water-gas shift and CO_2 separation into one single adiabatic reactor operating at moderate temperatures. However, sorbent regeneration (i.e. decomposition of $CaCO_3$ into CaO and CO_2) is the critical step for the process. It can be obtained by (i) increasing the temperature, (ii) reducing the total pressure with a reactor depressurization or (iii) only reducing the CO_2 partial pressure by means of steam purging. At any rate, the calcination step requires a significant amount of thermal power that can be provided either by means of heat exchangers or by direct combustion in the calciner. If direct combustion is the selected option, it must be oxy-fuel kind, so that the CO_2 released from calcination is not diluted with nitrogen.

Solieman *et al.* and Romano *et al.* [14, 51] applied the SER process into a combined cycle-based power plant. The first case considered a packed bed system, while in the second case, pressurised interconnected fluidised beds with regeneration by oxy-fuel combustion were adopted (see Figure 6.19). This configuration implies the solids pre-heating by the hot CO_2-rich stream in a multistage suspension heat exchanger and oxygen dilution with gas stream recycling to prevent high local temperatures in the calciner. After cooling and filtering to remove solid particles, the hydrogen-rich fuel produced in the SER process is burned in a state-of-the-art gas turbine. The main results of the simulation are reported in Tables 6.1 and 6.2 .

A net plant efficiency of 50.2% was obtained, 8.4% points less than the reference NGCC, with about 88% CO_2 capture, resulting in a specific emission of 14% of the NGCC plant. The efficiency is close to the one estimated for the 'base case' of the competing technology based on an Autothermal Reformer (ATR) and CO_2 capture with MDEA, and 2% points higher than a similar ATR+MDEA case calculated with more conservative assumptions, closer to the standards of the current technology. Adding to this, a slightly lower carbon capture ratio was obtained for the SER-based plant leading to almost 40% higher CO_2 emissions.

Table 6.1 *Power balance of the plants considered by Ref. [14]*

	NGCC	SE-SMR	ATR+MDEA 'base'	ATR+MDEA 'simplified'
Electric power, MW				
Gas turbine	273.4	316.1	287.7	289.0
Gas turbine auxiliaries	−0.96	−1.11	−1.01	−1.01
Steam turbine	150.7	148.7	157.2	217.7
Steam cycle pumps	−1.98	−2.86	−3.17	−3.68
Auxiliaries for heat rejection	−2.31	−2.02	−2.89	−3.97
ASU	–	−14.75	–	–
O_2 compressor / air boost compressor	–	−8.99	−7.09	−10.54
N_2 compressor	–	−10.83	–	–
MDEA process auxiliaries	–	–	−3.58	−4.31
CO_2 compression	–	−5.10	−15.14	−18.00
Net power, MW$_e$	418.8	419.2	412.0	465.1
Net efficiency, %	58.59	50.19	50.65	48.18
Cold gas efficiency, %	–	83.09	88.87	79.78
Carbon capture ratio, %	–	87.96	91.56	91.71
Specific emission, g_{CO_2}/kWh	350.2	49.2	34.2	35.3

Reproduced from Ref. [14], with permission from Elsevier.

Table 6.2 Temperature, pressure, flow rate and composition of the main plant streams shown in Figure 6.20, as reported by Ref. [14]

Point	G, kg/s	T, °C	p, bar	M, kmol/s	CH_4	C_{2+}	CO	CO_2	H_2	H_2O	O_2	N_2	Ar	He	CaO	$CaCO_3$
												Molar composition, %				
1	571.3	15.0	1.01	19.80				0.03		1.04	20.73	77.28	0.92			
2	455.1	419.8	18.36	15.77				0.03		1.04	20.73	77.28	0.92			
3	519.1	1416	17.81	19.94				0.54		25.18	9.05	64.50	0.73			
4	635.3	617.0	1.04	23.97				0.46		21.12	11.01	66.65	0.76			
6	13.38	15.0	30.00	0.71	82.88	10.38		1.10				5.49		0.15		
9	46.27	700.0	25.00	4.82	1.88		0.11	0.15	53.16	43.87		0.81		0.02		
11	5.54	15.0	30.00	0.29	82.88	10.38		1.10				5.49		0.15		
13	55.08	579.1	26.04	1.65				35.92		21.54	40.00	0.61	1.91	0.02		
14	55.91	775.0	25.00	1.63				59.87		35.90	2.00	1.00	1.20	0.03		
17	43.15	38.5	150.0	0.99				96.12			2.01	0.69	1.17	0.01		
18	247.4	1200		4.41											100	
19	276.5	911.4		4.41											85.00	15.00

Reproduced from Ref. [14], with permission from Elsevier.

A SER based on low-pressure interconnected bubbling fluidised beds applied to power plants with CO_2 capture is proposed by Ref. [54]. In this case, the H_2-rich gas produced is converted in a high-temperature SOFC and regeneration is carried out by recovering the waste heat from the fuel cell through an internal heat transfer loop.

Unfortunately, at the moment, there is not any work available in literature on the economic assessment of SER; therefore, the evaluation and comparison can only be performed in terms of efficiencies and carbon capture ratio.

6.5.2   SEWGS case

The large number of papers and experimental activities on sorbent development for SEWGS are followed only by a limited number of integration assessments. The only studies available are about SEWGS integration in power plants with CO_2 capture [8, 29, 55–57]. All these cases are based on the same concept of SEWGS with PSA as developed in CACHET project [8], first, and CAESAR project later [9]. In particular, the application of SEWGS to three different kinds of power plants was considered: NGCC, integrated coal gasifier cycle and blast furnace plants. In general, SEWGS reduces the energy penalties[2] for CO_2 capture in any considered plant compared to reference commercially available technologies. Differences among the different applications arise when the cost of CO_2 avoided[3] is adopted as a term of comparison. In NGCC plants in spite of the SPECCA[4] reducing by 25%, the cost of CO_2 avoided is slightly higher than the reference capture case by amine scrubbing [57]. This is because in an NGCC plant the adoption of SEWGS requires a reforming section upstream, which significantly penalises plant cost, balancing the efficiency gain.

The most significant advantages of SEWGS are in coal-based plants such as IGCCs. IGCCs are power plants where the coal instead of being burnt in air at ambient pressure is gasified (burnt in oxygen atmosphere with sub-stoichiometric oxygen quantity) at high pressure. The syngas is produced at high temperature (above 1000 °C), and main components are CO, H_2 and H_2O and CO_2 depending on the gasifier technology adopted [59]. Thanks to the presence of H_2 and CO, the syngas can be used to fuel a gas turbine of a combined cycle. However, because of the sulphur contents in the coal, a low-temperature sulphur abatement system or acid gas removal (AGR) sulphur removal system is required. Commercial AGR (e.g. Rectisol®, Selexol™ or Sulfinol) work at or below ambient temperature, hence a cooling section after the gasifier is required with consequent exergy losses. The same commercial system with small modifications can be also used to capture the CO_2. On the contrary, the SEWGS system can separate the CO_2 at high temperature and a recent sorbent developed at ECN laboratories showed its capability of co-capturing H_2S and CO_2. Simultaneous CO_2 and H_2S capture allows (i) exergy loss reduction and (ii)

---

[2] Energy penalties are measured by the SPECCA parameter, which represents the additional primary energy required to avoid one kg of CO_2 [58].

[3] The cost of CO_2 avoided is the additional cost of electricity to avoid the emission of one unit mass of CO_2.

[4] SPECCA is the Specific Energy consumption for CO_2 avoided which evaluates the amount of the energy spent to avoid one unit mass of CO_2 emission [58].

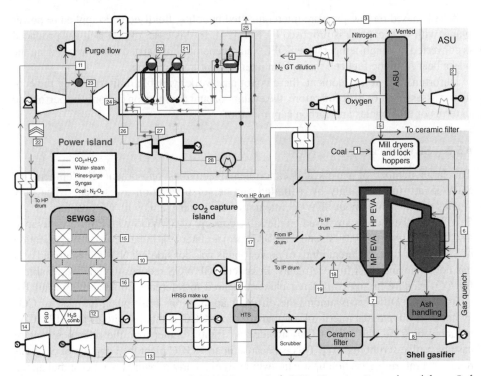

Figure 6.20 *Layout of investigated SEWGS case Ref. [55]. (Source: Reproduced from Ref. [55], with permission from Elsevier)*

layout simplification, by avoiding syngas cooling/preheating sections. The application of SEWGS with the ECN sorbent applied to a Shell gasifier is shown in Figure 6.20.

After the gasification section, steam is added to the syngas for the water-gas shift reactor (WGSR). The adiabatic WGSR upstream of the SEWGS is adopted in order to limit the temperature rise inside the SEWGS reactors. After WGS, the syngas is cooled down to 400 °C for the SEWGS unit, producing high-pressure steam that is sent to the heat recovery steam generator (HRSG). Further cooling of the SEWGS vessels could be considered, but this is not considered economically attractive. After CO_2 and sulphur separation in the SEWGS unit, the hydrogen-rich syngas produced at about 450 °C can feed the gas turbine. However, because of a fuel temperature limit assumed equal to 350 °C, a cooling step between SEWGS and the combustor is introduced. During the desorption process, CO_2 and sulphur are released at the same pressure and a further separation step must be adopted. H_2S catalytic combustion is indicated as the most suitable solution for CO_2-H_2S separation because it is commercially ready [60, 61]. The rinse and purge streams for the SEWGS unit are steam at a temperature of 400°C and at a pressure of 25 bar and 1.1 bar, respectively. In both cases, the steam is bled from the steam turbine and needs to be heated to achieve the target temperature (Table 6.3).

Table 6.3 *Mass flow rate, temperature, pressure and composition of the main fluxes reported in Figure 6.20 Ref. [55]*

Point	G, kg/s	T, °C	p, bar	Composition, %mol								
				CH_4	CO	CO_2	H_2	H_2O	Ar	N_2	O_2	H_2S
1	37.60	15.0	–	Dry coal 2%, coal composition in Ref. [53]								
2	69.72	15.0	1.01	Air as in Ref. [53]								
3	69.72	30.0	5.76									
4	31.37	260.6	27.23	–	–	–	–	–	–	100	–	–
5	10.46	80.0	88.00	–	–	–	–	–	–	100	–	–
6	33.00	180.0	48.00	–	–	–	–	–	3.1	1.91	95	–
7	131.9	298.0	40.97	–	55.9	2.9	25.9	5.36	0.8	8.89	–	0.2
8	55.56	214.1	44.44	–	51.7	2.7	23.9	12.5	0.9	8.22	–	0.2
9	148.2	500.0	40.16	–	5.39	23.8	35.3	30.5	0.4	4.42	–	0.1
10	148.2	426.1	24.11	–	5.39	23.8	35.3	30.5	0.4	4.42	–	0.1
11	77.42	350.0	23.16	–	0.42	1.4	50.3	23.1	0.5	24.2	–	–
12	124.2	445.4	1.11	–	–	53.9	0.51	45.3	–	0.06	–	0.2
13	123.9	57.4	0.49	–	–	53.9	0.51	45.32	–	0.06	–	0.2
14	95.60	25.0	110.0	–	0.01	98.8	0.94	–	0.02	0.11	0.08	–
15	11.39	400.0	23.77	–	–	–	–	100	–	–	–	–
16	10.66	400.0	1.11	–	–	–	–	100	–	–	–	–
17	60.26	400.1	53.56	–	–	–	–	100	–	–	–	–
18	11.86	300.0	54.00	–	–	–	–	100	–	–	–	–
19	3.37	300.0	54.00	–	–	–	–	100	–	–	–	–
20	99.38	341.7	144.0	–	–	–	–	100	–	–	–	–
21	17.26	300.0	54.00	–	–	–	–	100	–	–	–	–
22	657.3	15.0	1.01	Air, as in Ref. [53]								
23	525.1	COT: 1440.3	17.60	–	–	0.6	–	22.8	0.9	67.24	8.47	–
		TIT: 1360.0		–	–	–	–	–	–	–	–	–
		TIT_{ISO}:1257.6		–	–	–	–	–	–	–	–	–
24	665.0	605.0	1.04	–	–	0.5	–	18.7	0.9	69.14	10.8	–
25	665.0	100.0	1.01	–	–	0.5	–	18.7	0.9	69.14	10.8	–
26	133.9	550.4	144.8	–	–	–	–	100	–	–	–	–
27	106.0	559.1	44.23	–	–	–	–	100	–	–	–	–
28	88.61	32.2	0.05	–	–	–	–	100	–	–	–	–

For SEWGS integration into gas turbine-based plants, no significant improvements are required to the power section. The only difference compared to a commercial NGCC is the volumetric heating value of the H_2-rich stream produced by the SEWGS that is lower than natural gas (7.5 MJ_{LHV}/Nm^3 vs 35 MJ_{LHV}/Nm^3). It must be outlined that this volumetric heating value is close to the one produced by a gasifier, and there are several gas turbines with syngas running around the world fed.

The performance impact of SEWGS vessel size and sorbent capacity are shown in Figure 6.21 and Table 6.4.

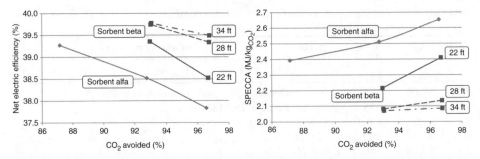

Figure 6.21 *Net electric efficiency and SPECCA for sorbent alfa and different vessel length with sorbent beta (Ref. [52]). (Source: Reproduced from Ref. [52], with permission from Elsevier)*

Table 6.4 *Performances of sorbent beta for different vessel volumes and CCR [62]*

CCR [%]	95			98		
Vessel length [ft/m]	22/6.7	28/8.5	34/10.4	22/6.7	28/8.5	34/10.4
Net power output, [MW]	**401.5**	**404.2**	**403.3**	**393.3**	**400.7**	**403.2**
Thermal power input$_{LHV}$, [MW]	**1020.3**	**1017.0**	**1013.9**	**1021.1**	**1018.8**	**1021.1**
Net electric efficiency$_{LHV}$, [%]	**39.35**	**39.75**	**39.78**	**38.52**	**39.33**	**39.49**
Net electric efficiency$_{HHV}$, [%]	37.76	38.14	38.17	36.96	37.74	37.89
CO$_2$ avoided, [%]	**93.0**	**93.0**	**93.1**	**96.7**	**96.7**	**96.8**
SEWGS costs	35.9	39.7	43.5	35.9	39.7	43.5
Total equipment costs [M€]	440.0	443.7	448.8	442.5	444.0	448.4
Total plant costs [M€]	1043.6	1052.3	1064.3	1049.6	1053.1	1063.4
Specific investment costs [€/kW]	2599.2	2603.2	2639.2	2668.3	2627.9	2637.6
Investment cost, [€/MWh]	43.03	43.09	43.68	44.17	43.50	43.66
Fixed O&M costs, [€/MWh]	8.27	8.26	8.34	8.47	8.33	8.33
Consumables, [€/MWh]	3.54	3.75	4.00	3.63	3.79	3.99
Fuel costs, [€/MWh]	27.44	27.17	27.15	28.04	27.46	27.35
COE, [€/MWh]	**82.28**	**82.27**	**83.17**	**84.30**	**83.08**	**83.33**
Cost of CO$_2$ avoided [€/t$_{CO2}$]	**23.5**	**23.4**	**24.7**	**25.4**	**23.7**	**24.0**

Bold is used to outline the most significant lines in the table.

Although the K-Mg30 sorbent showed good overall performance, the techno-economic assessments showed that sorbent cyclic capacity must be increased by at least another 50% in order to make the SEWGS process substantially (25%) cheaper than the more conventional capture technologies. In the CAESAR project [9], sorbent development finally resulted in a further improvement with respect to the sorbent capacity. This Beta type K-Mg30 sorbent that was developed by ECN features a 100% higher capacity at 6 bar CO$_2$ total pressure compared to the state-of-the-art Alfa type K-Mg30.

The higher capacity of Sorbent Beta reduces the efficiency penalties related to CO$_2$ capture by about 20%. In regard to vessel size, a taller vessel increases the efficiency because the greater quantity of sorbent used lowers the amount of steam required for regenerating the sorbent and consequently the efficiency penalties. However, this has penalties on SEWGS costs as shown in Table 6.4. The best configurations achieve a cost of CO_2 avoided below 24 €/t_{CO2}, which is significantly lower than the 36 €/t_{CO2} calculated for the reference CO_2 capture plant with Selexol [53, 62]. These results outline the potential of SEWGS when applied to coal-based plants.

SEWGS application in integrated steelworks, which are large stationary plants suitable for application of carbon capture processes, has also been considered. The main characteristic of integrated steel works is the high specific CO_2 emissions (1338 g_{CO2}/kWh$_{el}$) as a consequence of the large amount of CO and CO_2 in the syngas produced in the blast furnace. This characteristic strongly penalises conventional CO_2 separation systems based on amine scrubbing because of the large regeneration energy required, which brings out the potentiality of SEWGS as discussed in Ref. [52].

6.6   Conclusions

This chapter described the development and application of sorption-enhanced reaction processes to power plants with low CO_2 emissions. In particular, the application of two specific sorption-enhanced processes was discussed: SER and SEWGS. For both cases, an overview of the most promising sorbents and process applications was discussed together with the significant processes made over the last two decades in sorbent development and fuel conversion. For example, in the CAESAR project, the strength and the capacity of the SEWGS sorbent were significantly improved together with the capability of capturing H_2S together with CO_2, which makes it really attractive for sour conditions. The calculated cost of CO_2 for SEWGS in IGCC was about 24 €/t_{CO2}, which is 33% lower than the reference technology. Unfortunately, IGCC still suffers from a higher cost of electricity than competitive coal-based power plants limiting its adoption. Only a serious CO_2 emissions reduction policy can push the growth of IGCCs and, consequently, SEWGS.

In regard to SER development, some steps forward have been made, but there are still challenges in terms of stability of the sorbent and process design to be overcome.

Despite all these efforts and advancements, there is still much work that needs to be done to turn these processes into commercial reality. While obtaining a satisfactory material is clearly a mandatory step in moving the SERPs to larger scale, a difficult hurdle that also needs to be addressed is in the mechanical design of the vessels and valves. The combination of non-steady operation at high operating temperature and pressures along with gaseous components that have the capability to quickly damage and corrode most of the commonly used stainless steels is a huge challenge. The actual costs and performance capabilities associated with large-scale mechanical equipment still need to be fully evaluated for these applications. To give an idea of the SEWGS equipment required in the proposed IGCC application covered in Section 6.5.2, it consisted of 45 vessels.

Therefore, the future of SERP technologies requires the construction of pilot plants so that both sorbents and mechanical equipment can be tested and developed.

Nomenclature

AGR	Acid Gas Removal
ATR	Autothermal Reformer
CCR	Carbon Capture Ratio [%]
CLC	Chemical Looping Combustion
HRSG	Heat Recovery Steam Generator
HTL	Hydrotalcite
IGCC	Integrated Gasifier Combined Cycle
NGCC	Natural Gas Combined Cycle
PSA	Pressure Swing Adsorption
S/C	Steam to Carbon Ratio
SER	Sorption-Enhanced Reforming
SERP	Sorption-Enhanced Reaction Processes
SEWGS	Sorption-Enhanced Water-Gas Shift
SMR	Steam Methane Reformer
SOFC	Solid Oxide Fuel Cell
SPECCA	Specific Primary Energy Consumption for CO_2 avoided [MJ/kg_{CO2}]
TSA	Temperature Swing Adsorption
WGS	Water-Gas Shift
WGSR	Water-Gas Shift Reactor

References

1. Nikačević, N., Jovanović, M. and Petkovska, M. (2011) Enhanced ammonia synthesis in multifunctional reactor with in situ adsorption. *Chemical Engineering Research and Design*, **89** (4), 398–404.
2. Carvill, B.T., Hufton, J.R., Anand, M. and Sircar, S. (1996) Sorption-enhanced reaction process. *AIChE Journal*, **42** (10), 2765–2772.
3. Allam, R., Chiang, R., Hufton, F. *et al.* (2005) Development of the sorption enhanced water gas shift process, in *Carbon Dioxide Capture for Storage in Deep Geologic Formations*, vol. **1** (eds D. Thomas and S. Benson), Elsevier Ltd.
4. Hufton, J.R., Allam, R.J., Chiang, R., Middleton, P., Weist, E.L. and White, V. (2004) Development of a Process for CO_2 Capture from Gas Turbines Using a Sorption Enhanced Water Gas Shift Reactor System. Proc. 7th Int. Conf. Greenhouse Gas Technologies. Vancouver (Canada).
5. Cobden, P.D., Van Beurden, P., Reijers, H.T.J. *et al.* (2007) Sorption enhanced hydrogen production for pre-combustion CO_2 capture: Thermodynamic analysis and experimental results. *International Journal of Greenhouse Gas Control*, **1**, 170–179.
6. Reijers, H.T.J., Valster-Schiermeier, S.E.A., Cobden, P.D. and Van den Brink, R.W. (2006) Hydrotalcite as CO_2 sorbent for sorption-enhanced steam reforming of methane. *Industrial & Engineering Chemistry Research*, **45**, 2522–2530.
7. Gluud, W. and Ritter, H. (1931) *Berichte der Gesellschaft fuer Kohlentechnik*, **3**.
8. Beavis, R. (2011) The EU FP6 CACHET project - Final results. *Energy Procedia*, **4**, 1074–1081. doi: 10.1016/j.egypro.2011.01.157

9. Jansen, D. (2012) EU FP7 Project CAESAR: SEWGS technology is now ready for scale-up! Proc. 11th Int. Conf. Greenhouse Gas Technologies. Kyoto, Japan.

10. Ochoa-Fernandez, E., Haugen, G., Zhao, T. *et al.* (2007) Process design simulation of H_2 production by sorption enhanced steam methane reforming: Evaluation of potential CO_2 acceptors. *Green Chemistry*, **9**, 654.

11. García-Labiano, F., Abad, A., De Diego, L. *et al.* (2002) Calcination of calcium-based sorbents at pressure in a broad range of CO_2 concentrations. *Chemical Engineering Science*, **57**, 2381–2393.

12. Li, Z.S., Cai, N.S. and Huang, Y. (2006) Effect of preparation temperature on cyclic CO_2 capture and multiple carbonation-calcination cycles for a new Ca-based CO_2 sorbent. *Industrial & Engineering Chemistry Research*, **45**, 1911.

13. Mastin, J., Aranda, A. and Meyer, J. (2011) New synthesis method for CaO-based synthetic sorbents with enhanced properties for high-temperature CO_2-capture. *Energy Procedia*, **4**, 1184–1191.

14. Romano, M.C., Cassotti, E.N., Chiesa, P. *et al.* (2011) Application of the sorption enhanced-steam methane reforming process in combined cycle-based power plants. *Energy Procedia*, **4**, 1125–1132.

15. Sun, P., Lim, C. and Grace, J.R. (2008b) Cyclic CO_2 capture of limestone-derived sorbent during prolonged calcination/carbonation cycling. *AIChE Journal*, **54**, 1668.

16. Fennell, P., Davidson, J., Dennis, J. and Hayhurst, A. (2007) Regeneration of sintered limestone sorbents for the sequestration of CO_2 from combustion and other systems. *Journal of the Energy Institute*, **80** (2), 116–119.

17. Manovic, V. and Anthony, E.J. (2007) Steam reactivation of spent CaO based sorbent for multiple CO_2 capture cycles. *Environmental Science & Technology*, **41**, 1420.

18. Sun, P., Grace, J., Lim, C. and Anthony, E. (2008a) Investigation of attempts to improve cyclic CO_2 capture by sorbent hydration and modification. *Industrial & Engineering Chemistry Research*, **47** (6), 2024–2032.

19. Blamey, J., Anthony, E., Wang, J. and Fennell, P. (2010) The calcium looping cycle for large-scale CO_2 capture. *Progress in Energy and Combustion Science*, **36**, 260–279.

20. Harrison, D. (2008) Sorption-enhanced hydrogen production: A review. *Industrial & Engineering Chemistry Research*, **47**, 6486–6501.

21. Ebner, A.D., Reynolds, S.P. and Ritter, J.A. (2006) Understanding the adsorption and desorption behavior of CO_2 on a K-promoted hydrotalcite-like compound (HTlc) through non equilibrium dynamic isotherms. *Industrial & Engineering Chemistry Research*, **45**, 6387–6392.

22. Ebner, A.D., Reynolds, S.P. and Ritter, J.A. (2007) Non-equilibrium kinetic model that describes the reversible adsorption and desorption behavior of CO_2 in a K-promoted HTC. *Industrial & Engineering Chemistry Research*, **46**, 385–397.

23. Moreira, R.F.P.M., Soares, J.L., Casarin, G.L. and Rodrigues, A.E. (2006) Adsorption of CO_2 on Hydrotalcite-like Compounds in a Fixed Bed. *Separation Science and Technology*, **41**, 341–357.

24. Oliveira, E.L.G., Grande, C.A. and Rodrigues, A.E. (2008) CO_2 sorption on hydrotalcite and alkali-modified (K and Cs) hydrotalcites at high temperature. *Separation and Purification Technology*, **62**, 137.

25. Walspurger, S., Boels, L., Cobden, P.D. *et al.* (2008) The crucial role of the K+-aluminium oxide interaction in K+-promoted alumina- and hydrotalcite-based materials for CO_2 sorption at high temperatures. *ChemSusChem*, **1**, 643–650.
26. Selow, E.R., Cobden, P.D., Van den Brink, R., Wright, A. and Jansen, D. (2010) Improved sorbent for the sorption enhanced water gas shift process. Proc. 10th Int. Conf. Greenhouse Gas Technologies,. Amsterdam, The Netherlands.
27. Van Selow, E.R., Cobden, P.D., Van den Brink, R.W. *et al.* (2009a) Pilot-scale development of the sorption enhanced water gas shift process, in *Carbon Dioxide Capture for Storage in Deep Geologic Formations*, vol. **3** (ed L.I. Eide), CPL Press, Berks, pp. 157–180.
28. Van Selow, E.R., Cobden, P.D., Verbraeken, P.A. *et al.* (2009b) Carbon capture by sorption enhanced water–gas shift reaction process using hydrotalcite-based material. *Industrial & Engineering Chemistry Research*, **48** (9), 4184–4193.
29. Wright, A., White, V., Hufton, J. *et al.* (2009) Reduction in the cost of pre-combustion CO_2 capture through advancements in sorption-enhanced water-gas-shift. *Energy Procedia*, **1**, 707–714.
30. Walspurger, S., Cobden, P., Haije, W. *et al.* (2010) In situ XRD detection of reversible dawsonite formation on alkali promoted alumina: a cheap sorbent for CO_2 capture. *European Journal of Inorganic Chemistry*, **2010** (17), 2461–2464.
31. Manzolini, G., Macchi, E., Binotti, M. and Gazzani, M. (2011) Integration of SEWGS for carbon capture in Natural Gas Combined Cycle. Part B: Reference case comparison. *International Journal of Greenhouse Gas Control*, **5** (2), 214–225. doi: 10.1016/j.ijggc.2010.08.007
32. Van Dijk, H.A., Walspurger, S., Cobden, P.D. *et al.* (2009) Performance of water-gas shift catalysts under sorption-enhanced water-gas shift conditions. *Energy Procedia*, **1** (1), 639–646.
33. Bakken, E., Cobden, P., Henriksen, P., Håkonsen, S.E., Spjelkavik, A., Stange, M., Stensrød, R., Vistad, Ø., *et al.* (2010) Development of CO_2 sorbents for the SEWGS process using high throughput techniques. Proc. 10th Int. Conf. Greenhouse Gas Technologies. Amsterdam, The Netherlands.
34. Van Dijk, E., Walspurger, S.., Cobden, P. and Van den Brink, R. (2011b) Sour SEWGS application in IGCC. 6th Trondheim Conference on CO_2 Capture, Transport and Storage TCCS-6. Trondheim, June 14–16 2011.
35. Van Dijk, H.A., Walspurger, S., Cobden, P.D. and Van den Brink, R.W. (2011c) Testing of hydrotalcite based sorbents for CO_2 and H_2S capture for use in sorption enhanced water gas shift. *International Journal of Greenhouse Gas Control*, **5**, 505–511.
36. Johnsen, K., Grace, J., Elnashaie, S. *et al.* (2006) Modeling of sorption-enhanced steam reforming in a dual fluidized bubbling bed reactor. *Industrial & Engineering Chemistry Research*, **45**, 4133–4144.
37. Liu, Y., Li, Z., Xu, L. and Cai, N. (2012) Effect of sorbent type on the sorption enhanced water gas shift process in a fluidized bed reactor. *Industrial & Engineering Chemistry Research*, **51**, 11989–11997.
38. Van Dijk, E., Walspurger, S., Cobden, P. and Van Den Brink, R. (2011a) Testing of hydrotalcite based sorbents for CO_2 and H_2S capture for use in sorption enhanced water gas shift. *International Journal of Greenhouse Gas Control*, **5**, 505–511.

39. Fisher, J. II,, Siriwardane, R. and Stevens, R. Jr., (2012) Process for CO_2 capture from high-pressure and moderate-temperature gas streams. *Industrial & Engineering Chemistry Research*, **51**, 5273–5281.
40. Lee, K., Beaver, M., Caram, H. and Sircar, S. (2007) Reversible chemisorption of carbon dioxide- Simultaneous production of fuel-cell grade H_2 and compressed CO_2 from synthesis gas. *Adsorption*, **13**, 385–397.
41. Lee, K., Beaver, M., Caram, H. and Sircar, S. (2008) Effect of reaction temperature on the performance of thermal swing sorption-enhanced reaction process for simultaneous production of fuel-cell-grade H_2 and compressed CO_2 from synthesis gas. *Industrial & Engineering Chemistry Research*, **47**, 6759–6764.
42. Anderson, C., Scholes, C., Lee, C. *et al.* (2011) Novel pre-combustion capture technologies in action - Results of the CO_2CRC-HRL Mulgrave capture project. *Energy Procedia*, **4**, 1192–1198.
43. Reynolds, S., Ebner, A.D. and Ritter, J.A. (2006) Carbon dioxide capture from flue gas by pressure swing adsorption at high temperature using a K-promoted HTlc - Effects of mass transfer on the process performance. *Environmental Progress*, **25** (4), 334–342.
44. Wright, A., White, V., Hufton, J. *et al.* (2011) CAESAR-development of a SEWGS model for IGCC. *Energy Procedia*, **4**, 1147–1154.
45. Ying, D., Nataraj, S., Hufton, J. *et al.* (2008) *Simultaneous Shift-Reactive and Adsorptive Process to Produce Hydrogen*.
46. Cobden, P., Van Dijk, H. and Walspurger, S. (2013) *Regeneration of Gas Adsorbents*.
47. Sircar, S. and Golden, T.C. (2000) Purification of hydrogen by pressure swing adsorption. *Separation Science and Technology*, **35** (5), 667–687. doi: 10.1081/SS-100100183.
48. Couling, D.J., Das, U. and Green, W. (2012) Analysis of hydroxide sorbents for CO_2 capture from warm syngas. *Industrial & Engineering Chemistry Research*, **51**, 5273–5281.
49. Jang, H., Kang, W. and Lee, K. (2013) Sorption-enhanced water gas shift reaction using multi-section column for high-purity hydrogen production. *International Journal of Hydrogen Energy*, **38** (14), 6065–6071.
50. Reijers, R., Van Selow, E., Cobden, P. *et al.* (2011) SEWGS process cycle optimization. *Energy Procedia*, **4**, 1155–1161.
51. Solieman, A.A.A., Dijkstra, J.W., Haije, W.G. *et al.* (2009) Calcium oxide for CO_2 capture: Operational window and efficiency penalty in sorption-enhanced steam methane reforming. *International Journal of Greenhouse Gas Control*, **3** (4), 393–400.
52. Gazzani, M., Romano, M. and Manzolini, G. (2012a) Application of sorption enhanced water gas shift for carbon capture in integrated steelworks. Proc. 11th Int. Conf. Greenhouse Gas Technologies, Kyoto, Japan.
53. Franco, F., Anantharaman, R., Bolland, O. *et al.* (2011) *European Best Practice Guide for Assessment of CO_2 Capture Technologies*.
54. Meyer, J., Mastin, J., Bjørnebøle, T. *et al.* (2011) Techno-economical study of the Zero Emission Gas power concept. *Energy Procedia*, **4**, 1949–1956.
55. Gazzani, M., Macchi, E. and Manzolini, G. (2012b) CO_2 capture in integrated gasification combined cycle with SEWGS – Part A: Thermodynamic performances. *Fuel*. doi: 10.1016/j.fuel.2012.07.048.

56. Gazzani, M., Macchi, E. and Manzolini, G. (2012c) CO_2 capture in natural gas combined cycle with SEWGS. Part A: Thermodynamic performances. *International Journal of Greenhouse Gas Control*. doi: 10.1016/j.ijggc.2012.06.010.

57. Manzolini, G., Macchi, E. and Gazzani, M. (2012a) CO_2 capture in natural gas combined cycle with SEWGS. Part B: Economic assessment. *International Journal of Greenhouse Gas Control*, 4–11. doi: 10.1016/j.ijggc.2012.06.021.

58. Chiesa, P., Campanari, S. and Manzolini, G. (2011) CO_2 cryogenic separation from combined cycles integrated with molten carbonate fuel cells. *International Journal of Hydrogen Energy*, **36** (16), 10355–10365. doi: 10.1016/j.ijhydene.2010.09.068.

59. DOE/NETL 2014 -2011/1498. (n.d.). Cost and Performance of PC and IGCC Plants for a Range of Carbon Dioxide Capture.

60. Al, 2014 M. et. (n.d.). Selective combusting of hydrogen sulfide in carbon dioxide injection gas.

61. Enel. 2014 (n.d.). Processo per la rimozione dell'idrogeno solforato da gas non condensabili di un impianto geotermico. Roma.

62. Manzolini, G., Macchi, E. and Gazzani, M. (2012b) CO_2 capture in Integrated Gasification Combined Cycle with SEWGS – Part B: Economic assessment. *Fuel*. doi: 10.1016/j.fuel.2012.07.043.

7

Pd-Based Membranes in Hydrogen Production for Fuel cells

Rune Bredesen,[1] *Thijs A. Peters,*[1] *Tim Boeltken*[2] *and Roland Dittmeyer*[2]

[1]*SINTEF Materials and Chemistry, P.O. Box 124, Blindern, N-0314, Oslo, Norway*
[2]*Institute for Micro Process Engineering (IMVT), Karlsruhe Institute of Technology (KIT), 76344, Eggenstein-Leopoldshafen, Germany*

7.1   Introduction

Hydrogen is one of the most important chemicals used in industry today. It is mainly produced from reactions of steam with fossil feedstocks; see Figure 7.1 [1]. Hydrogen is largely used in the chemical and petroleum industries. Worldwide, most of the hydrogen produced is directly converted in downstream processes, such as production of ammonia and fertilizers, synthesis gas chemistry, or the processing of fossil fuels; see Figure 7.2 [2].

Membranes with high hydrogen permeance and selectivity have been identified as a promising enabling technology for efficiency improvement and cost reduction of hydrogen production. In particular, Pd and certain Pd-alloy hydrogen separation membranes are known to have 100% selectivity and high permeability, and, thus, allow for direct production of high-purity hydrogen for use in fuel cells. Combining these membranes with appropriate catalysts in membrane reactors to produce hydrogen from different sources has been described in numerous studies [3, 4]. The current main interest behind this research is the prospective use of hydrogen as an energy carrier. Two major driving forces have directed the research during the last two decades, that are the concerns about (i) CO_2 emissions and (ii) the development of fuel cell vehicles (FCV).

The development along the first direction has led to research that focuses on the integration of Pd-based membrane technology in pre-combustion decarbonization (PCDC) power

Process Intensification for Sustainable Energy Conversion, First Edition.
Edited by Fausto Gallucci and Martin van Sint Annaland.
© 2015 John Wiley & Sons, Ltd. Published 2015 by John Wiley & Sons, Ltd.

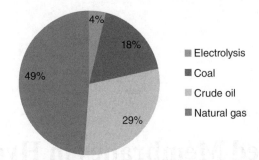

Figure 7.1   *Feedstock share of global hydrogen production, (Source: Reproduced with permission from Ref. [1]. Copyright © 2013, Pergamon)*

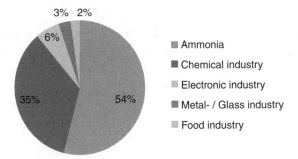

Figure 7.2   *Hydrogen consumers in the world, (Source: Reproduced with permission from Ref. [2]. Copyright © 2013, Pergamon)*

generation processes from natural gas or coal. In such applications, attaining a high H_2 selectivity is not a main issue as hydrogen gas turbines are expected to tolerate impurities; it is rather obtaining a high flux as the scale of operation makes membrane module cost considerations more important. This research has focused on membrane integration in a few main processes, that is, the steam or autothermal reforming (SR or ATR) of methane, the water-gas shift (WGS) reaction, or plain hydrogen separation after the WGS reactor. Optimization of the combination of membrane flux and stability in the presence of main gas components and impurities, such as H_2S, has been the main focus of the research. The Pd-alloy membrane technology in PCDC processes is a relatively better option compared to other CO_2 capture technologies, because of its combined low cost/high efficiency, and because it is applicable to hydrogen (co-)production as well [5, 6].

The second development direction is towards the production of high-purity hydrogen fuel as required by the current proton exchange membrane fuel cell (PEMFC) technology. This research focuses on membrane separation or membrane reactors operating in processes employing readily available primary fuels such as methane, liquid hydrocarbons and ammonia. In recent years, different options have been researched, particularly PEMFC for transportation where both on-board reforming of liquid fuels and storage of compressed hydrogen are considered. Current on-board storage of compressed hydrogen (700 bar) in light storage tanks enables the same driving range for light duty vehicles as internal combustion engines fuelled with gasoline. Thus, on-site hydrogen production and/or pipeline transportation from central production to the filling station are alternative possibilities. The source of hydrogen and production method may therefore vary depending on the local situation in a future hydrogen-fuelled transport system. Moreover, the growing climate concern has directed this research towards the use of available renewable sources such as biogas or converted biomass for hydrogen production [7]. The different sources open up a large parametric window for the research related to primary fuel type and quality, process scheme and operation conditions, catalyst, membrane and module design when assessing an optimum hydrogen production process.

In this chapter, we enter into an in-depth discussion on some of the main issues related to hydrogen production from various primary fuels by employing Pd-based membrane reactors. Following this survey, we look at process intensification by focusing on issues limiting the hydrogen production rate in membrane reactors. Here, we introduce microstructured membrane reactors that enhance heat management, reduce gas phase diffusion limitations and increase the membrane area to reactor volume ratio compared to traditional tubular reactors. Finally, we give some examples on the performance of systems comprising Pd-based membrane fuel processor technology integrated with PEMFC. But first a brief description of existing fuel cell technology, centralized and distributed hydrogen production, and an introduction to Pd-based membrane technology follow.

7.2 Characteristics of Fuel Cells and Applications

A fuel cell is an electrochemical device that converts fuel into electrical energy (and heat) continuously as long as reactants are supplied to its electrodes. The implication is that neither the electrodes nor the electrolyte is consumed by the operation of the cell [8]. Figure 7.3 illustrates the operation of a fuel cell with an electrolyte conducting hydrogen ions and electrodes that conduct electrons. Hydrogen is consumed at the anode and oxygen (from air) on the cathode side. The total reaction is formation of water from these reactants while current is produced for work in the outer electrical circuit.

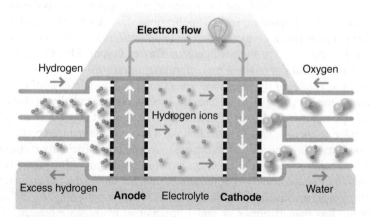

Figure 7.3 *Working principle of a hydrogen-powered fuel cell with a proton-conducting electrolyte (Ref. [9])*

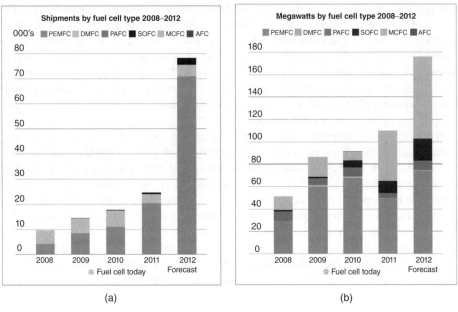

(a)   (b)

Figure 7.4 *(a) unit shipments by fuel cell type 2008–2012, (b) installed power by fuel cell type 2008–2012 (For a color version of this figure, please see color plate section.)*

Fuel cells may be divided into six main types [10]. These are PEMFC, including high-temperature variant HT PEMFC, direct methanol fuel cells (DMFC), phosphoric acid fuel cells (PAFC), molten carbonate fuel cells (MCFC), solid oxide fuel cells (SOFC) and alkaline fuel cells (AFC). By far the most commercially produced in terms of units is PEMFC (see Figure 7.4a), for small stationary and vehicle applications and in consumer electronics. A few, large installations of MCFC plants contribute to a large part of the totally installed electrical power as indicated in Figure 7.4b. SOFC is growing, particularly in the market of small combined heat and power (CHP) systems. AFC is the most mature technology, but is mainly used for special applications such as submarine and space craft.

The operation temperature, typical power and applications for various fuel cell systems can be seen in Table 7.1. The high-temperature fuel cell types, MCFC and SOFC, may conveniently use natural gas (and internal reforming) or reformate directly as fuel, that is, pure hydrogen as produced by membrane separation is not required. DMFCs use, as the name indicates, methanol as fuel, which results in an anode process that is slower than for the PEMFC operating on hydrogen as fuel. PAFCs working at about 200°C with Pt-type anodes can tolerate about 1.5% CO in the hydrogen fuel without reduced performance due to anode poisoning. This means that reformate gas, which is sufficiently converted in the WGS reaction, has low enough CO content to avoid anode poisoning. The PEMFC and AFC also require Pt-based catalysts at the anode side; but due to the lower operating temperature, these types are much more vulnerable to CO anode poisoning. As fuel, hydrogen should contain less than 10 ppm CO, which means that hydrogen produced from fossil sources by reforming of hydrocarbons and subsequent water-gas shift reactions is not sufficiently low in CO content. Required downstream hydrogen purification by adsorption,

Table 7.1   *Types, operation temperature, typical power and applications for fuel cell systems*

	PEMFC	HT PEMFC	DMFC	MCFC	PAFC	SOFC	AFC
Operating temperature	80°C	120–200°C	120–200°C	650°C	200°C	1000°C	60–90°C
Electrical efficiency	40–60%	60%	40%	45–60%	35–40%	50–65%	45–60%
Typical electrical power	<250 kW	<100 kW	<1 kW	>200 kW	>50 kW	<200 kW	>20 kW
Applications	Vehicles, small stationary	Small stationary	Portable	Stationary	Stationary	Stationary	Submarines, spacecraft

partial oxidation or methanation processes to reduce the CO level thus adds to cost and complexity. Concerning AFC that employs a KOH-solution as electrolyte, the CO_2 level at both electrodes can only be some few tens of ppm to avoid carbonate formation that reduces performance. Because of the hydrogen purity required, high-quality Pd-based membrane technology offers an attractive possibility in combination with the PEMFC and AFC. In this chapter, we primarily present Pd-membrane development that is directed towards or relevant to PEMFC applications, but not limited to, as such studies are currently by far the most abundant.

7.3   Centralized and Distributed Hydrogen Production for Energy Applications

The scale of hydrogen production and envisaged ways of distribution vary considerably and are strongly linked to cost considerations. The dominant large-scale route for the centralized production of hydrogen today is natural gas conversion. Most commonly used is the steam methane reforming (SMR) process (Eq. 7.1, Table 7.2), which is endothermic and operates typically over a Ni-based catalyst at around 800–900°C and between 15 and 30 bars pressure to produce a reformate (synthesis gas) [11].

The operation conditions (T,p) and the steam to carbon ratio (S/C) determine the synthesis gas composition. The external heat required for the SMR process can be avoided by adding oxygen to the reactants. When the heat released by the partial oxidation (Eq. 7.2) balances the heat demand of the steam reforming reaction, the process is called ATR, which requires less amount of catalyst and enables smaller and more compact reactors. Synthesis gas may also be produced by gasification (Eq. 7.6) of coal, heavy oil and biomass. To increase the hydrogen content, the synthesis gas temperature is reduced for carrying out the slightly exothermic WGS reaction (Eq. 7.5). This is usually performed in two stages: at high temperature (~350°C) over a Fe–Cr catalyst and at low temperature (~200°C) over a Cu-based catalyst. To produce high-purity hydrogen of 99.999%, a final pressure swing adsorption (PSA) process is carried out. Many other liquid hydrocarbons and oxygenates may alternatively be used for production of hydrogen at smaller scale in reforming or gasification processes, for example, propane, methanol, ethanol, glycerol or diesel, originating from either biomass or fossil sources.

Table 7.2   *Main chemical reactions in the production of hydrogen from natural gas, liquid fuels, coal and biomass*

Reaction	ΔH^0_{298} (kJ/mol)	Equation number
$CH_4 + H_2O \leftrightarrow CO + 3H_2$	206	7.1
$CH_4 + \frac{1}{2}O_2 \rightarrow CO + 2H_2$	−36	7.2
$CH_4 + 2O_2 \rightarrow 2CO_2 + 2H_2O$	−803	7.3
$CH_4 + CO_2 \leftrightarrow 2CO + 2H_2$	247	7.4
$CO + H_2O \rightarrow H_2 + CO_2$	−41	7.5
$C_xH_yO_z + nO_2 + mH_2O = CO + H_2$	–	7.6
$C_2H_5OH + 3H_2O \leftrightarrow 2CO_2 + 6H_2$	157	7.7
$CH_3OH + H_2O \leftrightarrow CO_2 + 3H_2$	49	7.8
$CH_3OH \leftrightarrow CO + 2H_2$	90	7.9
$CH_3OH + 1/2O \leftrightarrow CO_2 + 2H_2$	−184	7.10
$C_3H_8O_3 + 3H_2O \leftrightarrow 3CO_2 + 7H_2$	128	7.11
$C_3H_8O_3 \leftrightarrow 3CO + 4H_2$	251	7.12
$C + H_2O \leftrightarrow CO + H_2$	131	7.13
$2C + O_2 \leftrightarrow 2CO$	−283	7.14
$NH_3 \leftrightarrow 0.5N_2 + 3/2H_2$	55	7.15
$CO + 3H_2 \leftrightarrow CH_4 + H_2O$	−206	7.16

Even though the energy efficiency of the centralized process can be as high as 90% if the export of the high-level steam is included in the calculation [12], the separation of hydrogen from the gas mixture accounts for over 50% of the total plant costs [13]. For distribution, hydrogen produced at these large-scale reforming plants is either compressed in gaseous form or liquefied by cryogenic processes before transported by trucks to the end-user. Physically this means that trucks of 40 tons are transporting only 400 kg of hydrogen over distances of several hundred kilometres, and roughly two thirds of the energy content of hydrogen produced is evenly wasted as compression energy and the required transportation energy. On-site hydrogen generation systems are intended to overcome these drawbacks. Due to the scale, this solution is, however, more challenging with respect to the overall energy efficiency of the process [12]. Moreover, for distributed hydrogen production for PEMFC applications, hydrogen separation and purification by pressure swing adsorption is economically not attractive; hence, other purification methods such as preferential oxidation, selective methanation (Eq. 7.16) or membrane separation may be used.

Process intensification, the main topic of this Wiley Volume, is important for improving the efficiency and reducing the cost of distributed hydrogen production in modular units adapted to local capacity needs. In the following, the topic will be treated in terms of microstructured reactors [14, 15] and in relation to microstructured membrane reactors in the Section 7.6. Microstructured reactors are most commonly based on planar plate-type heat exchangers, which consist of thin metal foils with catalyst-containing channels in the sub-millimetre range for fluids. This technology is receiving much attention for chemical energy conversion due to high heat and mass transport rates that can be achieved [16].

Heat transfer is promoted by thin walls between the reaction channels and the heating or cooling channels, which leads to an improved energy efficiency [17], excellent temperature control [18] often enables higher selectivity and catalyst lifetime [19]. Another promising feature of microstructured reactors is their modularity, which facilitates the scale-up and the adaptation of the plant to the changing process needs.

The operation of integrated microstructured reformers based on metal plate-type heat exchangers has been demonstrated with good mechanical durability for high power ranges, and several types of primary fuels, such as methanol, ethanol, methane and diesel (-substitutes) have been suggested for production of hydrogen [14]. For the very low power class, reactors integrated with microelectromechanical systems (MEMS) fabricated of silicon offer great potential [20, 21]. Methanol is often considered the fuel of choice due to its low reforming temperature and high energy density. Catalyst systems for methanol steam reforming (Eq. 7.8) are mainly based on Cu/ZnO due to their high activity and low selectivity towards CO [22]. Nevertheless, CO concentrations of $>1\%$ are usually reported, which makes an additional hydrogen purification step necessary for the reformate to be used in a PEMFC [14].

Water electrolysis is an established, but also expensive, alternative method for distributed hydrogen production; however, this technology will not be discussed in this chapter. Focus will be on Pd-based membrane technology for hydrogen production. Since the technology represents a viable alternative to conventional reformers and is scalable over a membrane surface area range of 10 orders of magnitude, it may potentially play an important role in many different sectors, such as PCDC, heavy oil upgrading, ammonia production, transportation, stationary power generation, small portable products (see Figure 7.5). Note here that with "microtechnology" we refer to the integration of membranes in microsystems based on MEMS or glass processing and not the modular metallic microstructured reactors, which are described in Section 6.

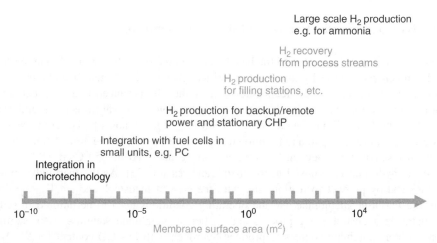

Figure 7.5 *Potential applications of Pd-membrane technology scalable over a membrane surface area range of 10 orders of magnitude*

7.4   Pd-Based Membranes

Palladium (Pd) and many Pd-alloys have high solubility and diffusivity of hydrogen and therefore show promise as membranes for medium-to-high-temperature hydrogen separation (250–550°C) [23, 24]. Compared to other types of hydrogen separation membranes [24], Pd-based membranes provide the best selectivity–flux combination, and possible applications are typically found in hydrogen production, recovery and purification [23, 25–30]; hydrogenation [31] and dehydrogenation [31, 32] processes. Current membrane technology comprises self-supported and mechanically supported composite structures. Self-supported membranes are usually tubes or foils with a thickness range of 50–100 μm with thicker membranes used at higher total transmembrane pressure [33]. To reduce the cost and increase the flux, composite membranes consisting of membranes of 2–10 μm thickness deposited on porous steel or ceramics are being developed [23]. The most common membrane compositions are pure Pd [34], alloys of Pd-Ag [34–36], Pd-Cu [37–42] and Pd-Au [43, 44], where the alloying element first of all adds to stability improvement without sacrificing permeability. Composite membranes have successfully been operated in laboratory scale and small pilot scale producing hydrogen for up to 13 h [45], and the technology appears ready for up-scaling.

Some gas components, most relevant being sulphur-containing species and CO, may reduce the flux by surface adsorption (poisoning) that competes with hydrogen adsorption [38, 46–48]. Sulphur is a particular challenge since only small concentrations already affect the membrane performance and may even lead to irreversible deterioration of the membrane [47, 49, 50]. This has led to the development of a number of new alloys with higher sulphur tolerance [50] or protective membrane coatings [51], particularly for use in PCDC IGCC processes. Still, flux targets [52] remain a challenge for new alloys in large-scale applications in the presence of sulphur impurities.

7.5   Hydrogen Production Using Pd-Based Membranes

A number of studies have verified the beneficial potential of Pd-based membrane reactors for hydrogen production by investigating critical parameters. A main benefit from the integration of membranes in steam reforming is that the produced hydrogen extracted from the reaction zone is of ultra-high purity. But hydrogen removal shifts the equilibrium of these equilibrium-limited endothermic reactions, which enables higher conversion at lower temperatures compared to traditional reactors. For example, theoretical calculations show that satisfactory methane conversion can be achieved at 550°C, if 80% of the produced hydrogen is removed in the membrane reactor [53]. As the thermodynamics, kinetics and by-product formation vary with the type of feedstock and catalyst applied, both the Pd-based membrane and the operating conditions need to be tailored in order to guarantee a cost-effective process and sufficient membrane stability. Also, starting from carbon-containing sources to produce hydrogen with low CO content for PEMFC imposes stringent requirement on the membrane quality. To illustrate this point, one may look at a reformate from methane steam reforming (Eq. 7.1), with a composition ratio of $H_2:CO \sim 5:1$ upstream a WGS-membrane reactor [54]. For simplicity, assuming that the CO conversion (Eq. 7.5) and preferential hydrogen permeation keep the $H_2:CO$ ratio constant, then setting the permeate CO concentration equal to 10 ppm, we estimate a required global selectivity $\alpha_{H_2/CO} = 1.0/(1\times10^{-5} \times 5) \sim 20{,}000$. This is a very high value for membrane systems, setting stringent demand not only on the membrane selectivity but also on the sealing system in the membrane reactor module. However, if membrane separation is done after the WGS reaction where the CO content in the feed stream may be lower than 0.6%, then only moderate selectivity will be required. This illustrates some flexibility, but also that a membrane reactor with high selectivity may replace either both or the second WGS reactor, while a low-quality module may only be used for final separation. In the following section, we come back to experimental studies on hydrogen production.

7.5.1   Hydrogen from Natural Gas and Coal

Pd-based membrane reactors have been studied in SMR [53, 55–60], ATR [61–63] and dry reforming [64–67] (Eq. 7.4) processes. The SMR is reported over Ni-based catalysts down to temperatures in the range of 450–550°C. Thinner membranes with higher hydrogen permeance and higher feed gas space velocity may render the catalyst activity insufficient in the lower temperature region as observed experimentally by Tong and Matsumura when comparing 8 μm and 11 μm thick membranes in feed gas $H_2O/CH4$ (S/C) = 3/1 and space velocity in the range ~1100–3300 h^{-1} [55]. In a later study [68], the authors found that Ru-supported on Al_2O_3 has higher activity than Ni catalysts at comparable conditions. Increasing the temperature to achieve higher catalyst activity and methane conversion and thereby higher hydrogen partial pressure as driving force for separation is difficult because of limited membrane stability. Still, as long as the membrane permeance is not limiting, the hydrogen production rate of membrane reactors surpasses that of traditional reactors at comparable and relevant conditions. Moreover, higher methane conversion at lower temperature reduces the cost and improves the efficiency of the SMR process.

Since the slightly exothermic WGS reaction occurs simultaneously with the MSR in the membrane reactor, the two reactions are efficiently thermally coupled. The hydrogen recovery also increases the conversion of the intermediate product CO (Eq. 7.5) by shifting the equilibrium. Lin *et al.* [69] found that for studies with a 20 μm thick Pd-membrane CO yields are less than 2% at 500°C, reaching 0.6% CO at low weight hourly space velocity (WHSV) of 0.3 h^{-1} at 20 bar feed pressure and S/C = 3. Thus, if the fast WGS reaction keeps the CO content less than 2% and hydrogen content about 30% in the reactor, a global $\alpha_{H_2/CO}$ membrane selectivity of about 7000 would be required to keep the CO content below 10 ppm in the hydrogen permeate. This is within the range of state-of-the-art thin composite Pd-based membranes.

The high permeability of Pd-based membranes may change the gas composition radially in the catalyst bed towards the membrane surface. Local variations in gas composition and low hydrogen content could lead to carbon deposition via endothermic decomposition of methane by $CH_4 = C + 2H_2$ or by the exothermic Boudouard reaction $CO = C + CO_2$ [59, 70]. An excess of steam reduces the thermodynamic driving force for coking, and an S/C ratio of about 3 is regarded as sufficient to avoid the problem. The excess steam has a negative effect on the feed-side hydrogen partial pressure driving the hydrogen permeation, lowering the hydrogen recovery and energy efficiency, and increasing the overall cost; hence, steam usage should be kept as low as possible.

Tokyo Gas has demonstrated the world's largest membrane reformer so far with a rated hydrogen production capacity of $40\,Nm^3/h$ ($150\,kW_{th}$) from natural gas [71], achieving a remarkable hydrogen production efficiency of 81.4%. The hydrogen production efficiency was, however, calculated on the basis of a hydrogen permeate pressure of 0.04 bar and subsequent compression of the hydrogen for the desired application, a hydrogen refuelling station, was not accounted for. The membrane unit, operating at about 550°C [72], is made by rolling Pd-alloy sheets to a thickness of about 15–20 µm. Since the measured hydrogen flux equals only 50% of the expected value on the basis of permeability, and CO adsorption effects are insignificant at this operating temperature [73], it is believed that the main cause of the flux reduction is the so-called concentration polarization due to depletion of hydrogen in the gas-phase layer next to the membrane surface [74–76]. This topic is further discussed in relation to the membrane reactor design later.

Chang *et al.* studied Pd-based membrane reactors in the ATR process at a low temperature between 350–470°C over a commercial Ni-based catalyst [61]. Methane conversion increases with O_2/CH_4 ratio in the range of 0–0.6 reaching 95%, and is always higher than that for the traditional reactor. Autothermal conditions were obtained for ratios $H_2O/CH_4 = 1.3$ and $O_2/CH_4 = 0.36$–0.37 in the temperature range studied, and somewhat surprisingly reported without coking or catalyst deactivation. In order to avoid potential hot spots in ATR reactors, Gallucci *et al.* [77] used an ATR fluidized bed membrane reactor (FBMR), which also reduces mass transfer limitations between the catalyst and the membrane surface [78]. They studied different reactor concepts by applying 10 tubular supported dead-end Pd-membranes with a thickness of 8–10 µm, outer diameter of 3.2 mm and length of 20 cm. A second type of Pd-membrane in the reactor was used to extract hydrogen, which reacted with air on the permeate side to generate heat for the ATR (see Figure 7.6). Some variations of the concept were studied, and CH_4 conversion of 75% and hydrogen yield (H_2 extracted/CH_4 fed) of 2.92 were obtained at 500°C [77].

Dry reforming (Eq. 7.9) has been studied in combination with and without oxygen addition [35, 64, 80, 81] employing Ru, Rh and Ni-based catalysts. Generally, an improved

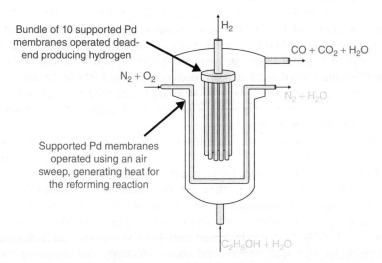

Figure 7.6   *Schematic of the fluidized bed membrane reactor for autothermal reforming of ethanol. Same concept as for methane as feed gas,* (Source: after Ref. [79])

performance of the membrane reactor over traditional packed-bed reactors is reported. For example, investigations at 550°C with co-feeding of oxygen at high CO_2/CH_4 ratio (equal to 1.9) reached methane conversion of about 70%, while coking was hardly detected [80, 81].

Gasification of coal (combination of Eqs. 7.13 and 7.14) is an effective hydrogen production method though it is more challenging than gasification of natural gas due to a wide range of possible impurities and associated by-products. Moreover, integration of Pd-based membranes requires a sulphur removal process installed upstream the membrane module due to their limited sulphur stability. Therefore, high-temperature gas clean-up processing appears attractive to fully explore and harness the Pd-based membrane technology. For the same reason, for employing biogas that typically contains CH_4 (55–70 vol%) and CO_2 (30–45 vol%) with traces of H_2S (500–4000 ppm) and NH_3 (100–800 ppm) [82], or gasification of biomass [7] as primary feedstock for hydrogen, upfront gas cleaning would be required. The importance of hydrogen produced from bio-sources is expected to increase in the coming years [83]; thus, development of more robust membranes and appropriate gas cleaning technology are critical issues.

7.5.2   Hydrogen from Ethanol

Steam reforming of (bio-)ethanol (Eq. 7.7) in Pd-based membrane reactors is reported in many studies [84–97]. Traditionally, this highly endothermic reaction operates at above 600°C for optimal hydrogen yield [85]. Noble metal–based catalysts (Pd, Pt, Rh, Ru), Ni and Co on various supports [84, 98] are used, but the reforming reactions may also involve a large number of side reactions that can lead to a significant formation of methane and CO, and coking if the process conditions are not carefully controlled [85]. The shift of the equilibrium to a higher yield at lower temperature and higher pressure by the hydrogen extraction may also change the kinetics and the driving force of the many competing side reactions [85, 87]. The need for parametric investigations to obtain optimum H_2 production conditions is reflected in recent research.

Simulation studies by Gallucci *et al.* [92] show that the ethanol steam reforming in a membrane reactor operating at 400°C can achieve ethanol conversion as high as 95.3% and hydrogen production and recovery of 90% and 94%, respectively. Generally, higher pressure and sweep gas increase the ethanol conversion and hydrogen recovery, as expected, and the counter-current feed–sweep gas mode is found superior to the co-current mode [90]. Yet in another study Gallucci *et al.* [79] have suggested the fluidized bed concept for ethanol reforming, as shown in Figure 7.6. Their simulation of ethanol conversion follows predictions by thermodynamics, but only high pressure gives significant hydrogen recovery in the membrane process.

Oxidative reforming [63, 99–102], by adding oxygen or air to the ethanol–steam mixture, reduces endothermicity and may render the reaction autothermal, as well as reduces carbon formation [103]. Use of air is cheaper and more convenient than pure oxygen, but the hydrogen partial pressure is lowered by nitrogen dilution. Santucci *et al.* [100] compared hydrogen production by employing a Pd-Ag membrane reactor filled with a Pt-based catalyst both in steam reforming and oxidative reforming of ethanol by using air. Using a self-supported 60-µm-thick membrane and molar ratios of oxygen/ethanol and steam/ethanol in the range of 0.3–0.7 and 4–13, respectively, with few exceptions, the hydrogen yield was found higher for oxidative steam reforming compared to steam reforming.

7.5.3   Hydrogen from Methanol

Steam reforming of methanol (Eq. 7.8) employing Pd-based membrane reactors is reported in many papers [91, 96, 104–106]. The reaction is much less endothermic than ethanol or methane conversion; thus, conversion is readily achieved at lower temperatures, 200–350°C over a Cu-based catalyst [107]. The lower temperature, however, reduces permeability and generates potential for flux reduction due to adsorption of methanol and/or CO [96], the latter compound formed by direct methanol decomposition (Eq. 7.9) or reverse WGS. In their assessment of fuel processors for PEMFC, Harold *et al.* [108] assessed methanol conversion and hydrogen production employing packed-bed tubular Pd-membrane reactors along three routes: methanol decomposition, steam reforming and partial oxidation (Eq. 7.10) using air. Direct methanol decomposition was found too slow to be interesting at the temperatures investigated (<260°C), while partial oxidation had potential to exceed steam reforming in hydrogen productivity (rate of hydrogen permeated per reactor volume) and utilization (ratio of hydrogen permeated/hydrogen in feed) if efficient preferential CO oxidation over hydrogen could be achieved. The reason is less CO adsorption, and furthermore, oxidation also drives the methanol conversion. Looking at the minimum reactor volume needed to sustain the hydrogen production for a $50\,kW_e$ fuel cell stack, maximum hydrogen productivity occurs at an intermediate gas hourly space velocity for the appropriate catalyst activity and membrane permeance combination. By varying the ratio of membrane diameter to catalyst bed diameter for different tubular composite membranes and Pd-membrane thicknesses, they calculated the minimum required mass of Pd for different composite membranes (see Figure 7.7).

For example, for a composite membrane of 1 mm outer diameter, feed space velocity of $24{,}000\,h^{-1}$, and 4-μm-thick Pd membrane, about 122 g Pd is needed for sustaining the $50\,kW_e$ fuel cell. And, for a 1–μm-thick Pd membrane on a 1 mm fibre, a required fuel

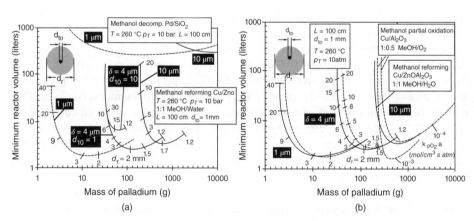

Figure 7.7   (a) a plot of minimum reactor volume versus Pd requirements compares methanol reforming on Cu/ZnO/Al₂O₃ and methanol decomposition on Pd/SiO₂. Thin Pd membranes and fibre supports lead to reduced volume and Pd needs. (b) a plot of minimum reactor volume versus Pd requirements compares methanol reforming on Cu/ZnO/Al₂O₃ with methanol partial oxidation on Cu/Al₂O, (Source: After Ref. [108])

processor volume of 2 L was estimated. Even though such fibre membranes are far from commercialization, the figures indicate the potential of the technology.

More recent experimental studies comparing traditional packed-bed [105] and fluidized bed [106] reactors with Pd-based membrane reactors in methanol steam reforming show that the latter gives better performance. CO selectivity decreases with increasing pressure due to hydrogen removal [105], which also improves methanol conversion. This is exemplified by complete methanol conversion obtained at 300°C in the membrane reactor, while under similar conditions an FBR reaches only 55% [106].

7.5.4   Hydrogen from Other Hydrocarbon Sources

Steam reforming of glycerol (Eq. 7.11), a by-product of bio-diesel production, has been studied in membrane reactors for hydrogen production [109–112]. Iulianelli *et al.*, using commercial Ru-Al$_2$O$_3$ catalyst at 400°C and H$_2$O:glycerol feed ratio 6:1 at 5 bar inlet pressure, found that 60% conversion and hydrogen recovery were achievable at WHSV = 0.1 h^{-1} for their 50-μm-thick Pd/Ag tubular membrane reactor [111]. The membrane reactor was always found superior to the traditional one, but carbon deposition was found despite the high steam content. Other examples of hydrocarbon and oxygenate reforming in different Pd-based membrane reactor concepts showing advantageous performance in hydrogen production are reported for acetic acid [113, 114], propane [115], heptane [116, 117], naphtha [118], light distillates [119] and dodecane [120].

7.5.5   Hydrogen from Ammonia

Hydrogen production by ammonia decomposition (Eq. 7.15) is an interesting alternative to hydrocarbon reforming. The potential problem of PEMFC catalyst poisoning by CO can be avoided and no steam is required in the reaction, which reduces energy efficiency. Ammonia is also produced in large quantities worldwide, and safe handling of this toxic product is established. The hydrogen content is 17.6% on mass basis compared to methanol of 12.5%. Decomposition occurs readily at 700–900°C over catalysts of iron oxide, Mo, Ru and Ni [121]. García-García *et al.* [122] used a 40 μm stainless steel–supported tubular Pd-membrane in combination with a Ru-based catalyst and found that 100% ammonia decomposition could be achieved at 635 K compared to a thermodynamically limited conventional reactor without hydrogen extraction reaching 65%. It has also to be noted that Pd-based membrane reactors have been studied in hydrogen removal from purge gas in ammonia plants [123, 124], coal gasification streams [125] and purification of urea wastewater [126].

7.6   Process Intensification by Microstructured Membrane Reactors

As described above, the principle of an extractor-type membrane reactor has been extensively demonstrated, most often with tubular membranes of 5–10 mm diameter and catalysts having a particle size of around 1 mm in small laboratory reactors where the catalyst is placed as a packed bed inside or around the membrane tube. For Pd-alloy membranes on porous supports used in methane steam reforming, more elaborate design concepts for larger units have been worked out and tested in pilot operations, for example, by Tokyo Gas, Shell or MRT [127, 128]. In this section, some main aspects for designing reliable membrane module systems providing high conversion, high hydrogen recovery, high space-time yield and high energy efficiency are discussed.

A good match of the hydrogen production rate and the hydrogen flux is required to shift reaction equilibria, which requires a sufficiently large membrane area per catalyst volume [31]. Tubes with large diameter offer only a low membrane area per catalyst volume, for example, $400 \, \text{m}^2/\text{m}^3$ for a 1 cm tube having the catalyst inside. For high reaction rates (high temperature, highly active catalyst), this might not be sufficient. Moreover, the transport of hydrogen from the catalyst surface, where it is produced to the membrane surface, where it is removed may come at a significant resistance depending on the design of the system. In general, the tighter the catalyst and membrane are connected, the lower the mass transfer resistance for hydrogen is. On the other hand, direct contact between the sensitive membrane and the catalyst should preferentially be avoided to prevent mechanical damage and undesired chemical interactions. While it could be done, in principle, by placing the catalyst in contact with the membrane support, this may lead to a substantial decrease in the hydrogen partial pressure across the support as the hydrogen undergoes counter-diffusion towards the other constituents of the gas mixture in the pores to reach the membrane surface, and there is no possibility to enhance the transport by convective flow. The result is a reduced driving force for hydrogen permeation. This effect is known as concentration polarization and may be observed even without or outside the membrane support if a gap exists between the catalyst bed and the membrane (or support) surface and no specific measure is in place to promote transverse mixing. It may also be observed when the support is placed on the permeate side whenever a sweep gas is used. Concentration polarization becomes more important as the membrane flux increases, because in a sequence of resistances the largest will dominate. Its effects have recently been given more attention due to the availability of extremely thin, highly permeable Pd-alloy membranes [76, 129, 130].

The net effect of concentration polarization can be assessed by a film effectiveness factor, as it is common practice in the treatment of film diffusion in heterogeneous catalysis [129]. A laminar boundary layer (film) is assumed adjacent to the membrane surface. In steady state, the hydrogen flux through this film (described by Fick's first law) must equal the hydrogen flux through the membrane (described by Sieverts' law). Both expressions contain the unknown hydrogen partial pressure at the retentate side membrane surface, so equating them yields an expression relating this quantity to the known hydrogen partial pressures in the retentate and permeate bulk gas phases. The film effectiveness factor is defined as the effective hydrogen flux, $j_{H_2}^e$, divided by the ideal flux, $j_{H_2}^{id}$, that is, as obtained without depletion of hydrogen towards the membrane surface.

$$\eta = \frac{j_{H_2}^e}{j_{H_2}^{id}} \tag{7.17}$$

With the square root of ratio of the hydrogen partial pressure in permeate versus retentate gas phase, ρ, and the ratio of the mass transfer resistances of film versus membrane, Φ, according to the definition

$$\rho = \sqrt{\frac{p_{H_2}^P}{p_{H_2}^R}} \tag{7.18}$$

$$\Phi = \frac{RT}{k_G} \cdot \frac{2 \cdot \Pi}{\left(\sqrt{p_{H_2}^R} + \sqrt{p_{H_2}^P}\right)} \tag{7.19}$$

Here, k_G is the mass transfer coefficient describing the diffusion resistance and Π represents the membrane permeability. Following [129], an expression can be derived relating the film effectiveness factor to these two dimensionless variables. High values of Φ indicate high mass transfer resistance of the film compared to the membrane and thus strong concentration polarization.

$$\eta = \frac{\sqrt{1 + 2\rho\left(\frac{\Phi}{1+\rho}\right) + \left(\frac{\Phi}{1+\rho}\right)^2} - \frac{\Phi}{1+\rho} - \rho}{1 - \rho} \tag{7.20}$$

Note that the definition of Φ according to Eq. 7.19 differs slightly from the one in Ref. [129], but offers the advantage that the curves for different values of ρ all approach a common asymptote at large values of Φ, that is, $\rho = 1/\Phi$. This is visible from Figure 7.8

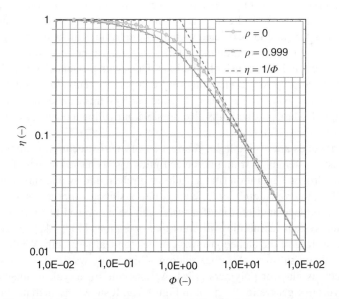

Figure 7.8 *Film effectiveness factor η according to Eq. (7.20) as a function of Φ for the two extreme cases: zero hydrogen partial pressure in the permeate side ($\rho = 0$) and almost identical partial hydrogen pressure on both sides ($\rho = 0.999$)*

showing the curves for the two extremes, that is, $\rho = 0$ and $\rho = 0.999$ in the usual double logarithmic plot.

The benefit of Figure 7.8 or Eq. 7.20 is that the influence of the mass transfer on the effective hydrogen flux can be determined, once the hydrogen permeance of the membrane is known and the feed/retentate side mass transfer coefficient k_G can be estimated. Note that the same formalism can also be applied to treat a possible diffusion limitation inside the support whenever this is placed in contact with a gas mixture and also on the permeate side if sweep gas is used to remove the permeated hydrogen. Furthermore, the film effectiveness factor is a local property as it also depends on the hydrogen partial pressure ratio, which varies along the flow direction due to the combined effects of reaction and separation. Generally, concentration polarization increases with decreasing membrane thickness (large Π), decreasing driving force (large ρ), decreasing Reynolds number (low k_G) and increasing temperature (Π profits more than k_G) [76, 130, 131].

Third, no less important aspect for technical reformers is the supply of heat to drive the endothermic reaction. In conventional systems, gas-fired burners are used to heat the outside surface of the reformer tubes. For large tubes used in world-scale reformer plants, the reaction is strongly heat transfer limited, which increases the fuel consumption by the burners. And also at smaller scale, the consequences of concentration and temperature gradients for system performance can be remarkable pointing out the need for optimal reactor design.

The advantages of microstructured reactors regarding high performance in indirect heat exchange and high mass transport rate are well documented in the literature [132] and have also been proved for catalytic reactions even in prototype operations [17, 133]. Both are connected and arise from the tiny lateral dimensions of microchannels and the large wall surface area per volume associated with it. These features make microstructured reactors attractive also for incorporating membranes, as they will both minimize concentration polarization effects and heat transfer limitations. Thus, with proper design, high space-time yields should become possible enabling very compact systems. In addition, due to the modular format, the scalability is generally good and systems can be easily adapted to the capacity required. Mass production is also possible, providing a perspective for economic fabrication costs [134, 135].

Different methods are available for fabrication of microstructured membrane reactors. Silicon processing stemming from microelectronics has been used to manufacture MEMS for portable hydrogen generation (in the following referred to as micromembrane device) [136–140]. Bulk micromachining of channels into a silicon wafer and surface micromachining of layers deposited on top of a wafer can be combined to fabricate even complex integrated systems combining different unit operations [138, 141]. With appropriate methods, the surface of the etched grooves can be atomically smooth. While this requires high-quality silicon substrates, surface micromachining allows the use of less expensive substrates such as glass or polymers [136]. An example for a typical integrated MEMS device for hydrogen generation is given in Figure 7.9. It shows a schematic view of a catalytic MEMS micromembrane device. The device consists of two layers, each containing one flow channel. The feed side contains the membrane support, a pattern of 4-μm-diameter holes, as well as resistive heaters and temperature sensors [142]. A dense 200-nm-thin Pd-Ag alloy film was coated on the silicon wafer via electron-beam deposition. With the MEMS-type membrane reactor, methanol partial oxidation experiments were conducted on

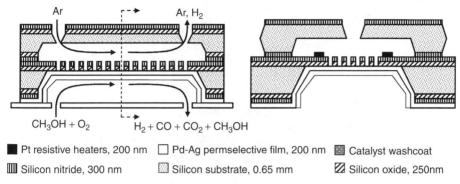

Figure 7.9 *Schematic view of a catalytic MEMS micromembrane device for hydrogen generation, side view (left), cross section (right),* (Source: *After Ref. [143]*)

a complex oxide methanol reforming catalyst wash-coated in the microchannels demonstrating the proof of concept for portable hydrogen generation. Methanol conversion of up to 63% was obtained at 475°C and atmospheric pressure. A sweep gas was applied at the permeate side, resulting in an hydrogen recovery of 47% [143].

So far, these ultra-thin Pd-based membranes ruptured at a pressure difference of around 400–500 kPa at room temperature and the transmembrane pressure was limited to ~70 kPa by the poor mechanical properties of the support structure [138, 144, 145]. Recently, a micromembrane device for hydrogen purification with a silicon support with enhanced stability was tested by Deshpande *et al.*, which tolerated operating pressures of up to 200 kPa resulting in a H_2 flux of 50 mol/m²s at 350°C [146]. The increased pressure tolerance originates from an additional device layer of 50 μm. Chip-like reactors such as this present a degree of miniaturization and membrane integration, which allows watt and sub-watt fuel cells to be fuelled. Yet, up to date, no reports were given on a device performance after stacking several MEMS-type membrane devices to a larger compact membrane reformer unit. Although thin membranes combining high flux and acceptable selectivity have been prepared and progress was demonstrated in the last decade, challenges, such as sealing of metal membranes with a silicon-based device and membrane stability at elevated pressures and temperatures, remain, and, therefore, such systems are limited to portable applications [137].

A different approach [14, 16] is the microstructured technology originating from plate-type heat exchangers and is based on thin metal foils, which carry channels in the micrometre range that are coated with a thin catalyst layer or packed with a fixed bed of catalyst in the channels. With highly precise joining techniques, such as laser welding, a planar stack of these microchannel foils together with membrane films and micromachined supports, all made from metals, results in a modular, scalable and very compact unit. Following this concept, for example, microchannel foils having a channel depth of 200 μm and a 10-μm-thick catalyst coating on the bottom would offer a ratio of the Pd-membrane surface area to catalyst volume as high as 100.000 m²/m³, provided that each microstructured catalyst foil is in contact with the Pd membrane. This would call for an extremely active catalyst to deliver hydrogen at a rate comparable to the fast hydrogen

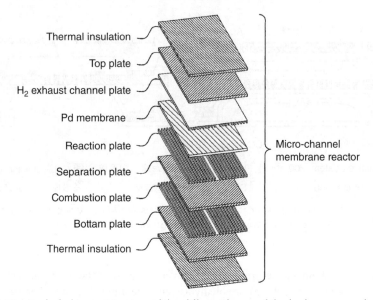

Figure 7.10   *Exploded isometric view of the different layers of the hydrogen membrane reactor,* (Source: After Ref. [148])

removal rate through such a large membrane area. This underlines that the design of these microstructured membrane reactors should be adapted to the process needs in order to neither oversupply Pd-membrane area nor expensive catalytic material. As mentioned, superior heat integration and reduced concentration polarization add to the potentials of microstructured membrane reactor technology.

Systems of this type have been described in the patent literature [147, 148], usually for the production of hydrogen in compact energy packs for mobile and portable devices, as well as uninterrupted power supply units using hydrogen generation from liquid fuels such as methanol, propane and kerosene and stationary fuel cell applications. The system described by Chellappa *et al.* aims for a compact device with a volume of less than 1 l and a mass of less than 1 kg [148]. It is realized either by thin metal plates, containing meso- or microchannels, or porous metal plates for the distribution of fluids. The membrane reactor contains a top plate, a bottom plate and a Pd-based membrane, operating at a maximum temperature of 650°C, see Figure 7.10.

For further minimization of the system, the membrane reactor, the combustion zone for heat integration, the pre-reformation of the hydrocarbons, the fuel heat exchanger and the oxygen heat exchanger for the combustion of the tail gas are combined to one compact integrated system, which is sealed by brazing. The system is equipped with an additional methanation unit to decrease the CO concentration in the permeate as well as with a shut-down procedure, in case of membrane failure. An energy density of 2000 W h/kg and an average power output level of 20 W are reported. The hydrogen production rate from liquid hydrocarbons is 200 ml/min. However, stacking of more than one single microchannel membrane module for a scalable hydrogen production unit has not yet been reported.

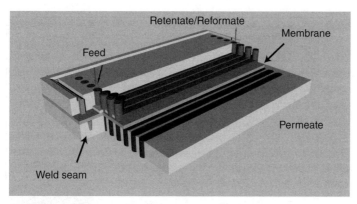

Figure 7.11 *Schematic of a microstructured membrane module consisting of two thin microstructured metal sheets with an integrated palladium-based membrane,* (Source: *After Ref. [129])*

With a similar approach to decrease the distance between the wall or catalyst layer and the membrane surface drastically, all-metallic membrane modules with micromachined plates directly attached to the membrane have been fabricated by KIT [129]. Pd-alloy foils with different thicknesses ranging from 61 μm to 3 μm have been leak-tight integrated in the modules by laser welding (see Figure 7.11). This is considered a very practical approach and represents the first step towards the anticipated compact multi-layered microchannel membrane reformer system.

For moderate transmembrane pressure difference, that is, up to about 650 kPa, and for temperatures not higher than 400°C the microstructured metallic plate having straight channels 200–500 μm wide provided good mechanical support even for the thinnest membranes tested. Moreover, mixed gas permeation experiments revealed that concentration polarization effects are expected to be subordinate [129].

However, application temperatures above 400°C are not possible with the membrane in direct contact with the metal plate because of intermetallic diffusion, and for higher transmembrane pressure, it needs to be stabilized by an additional support. Several solutions have been developed: the preferred concept uses chemically etched metal sieves coated by magnetron sputtering with yttria-stabilized zirconia (YSZ) to form a diffusion barrier layer. The microsieves are 500 μm thick and have almost perfectly circular holes of 120 μm diameter with 100 μm spacing resulting in a void fraction of about 40%, the YSZ barrier is about 0.8 μm thick. The coated supports are placed both on the feed/retentate side and on the permeate side of the membrane for maximum protection. An alternative concept uses porous sinter metal plates as a support, which first receives a porous YSZ diffusion barrier layer applied by suspension plasma spraying and then a Pd layer by electroless plating. Figure 7.12 compares the hydrogen flux for the three different concepts. Note that for the calculation of the flux in all cases the free area of the microchannels was used as the permeation area, which gives reasonable results for the permeability of the thin membrane foils [129]. The application of the microsieve supports should lead to a decrease in effective membrane area. However, this is partially compensated by making the region above

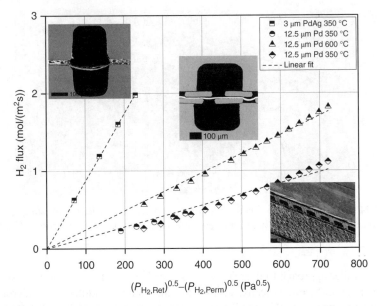

Figure 7.12 *Sieverts' plot of hydrogen permeation measurements with three different microstructured membrane modules. (1) no additional support (squares, circles). (2) porous support (diamonds). (3) etched metal sieves as support (triangles),* (Source: After Ref. [149])

the fins between the microchannels partly accessible to hydrogen permeation. Depending on the quality of the alignment of the microsieves, a 20–30% reduction of the hydrogen flow was observed. On the other hand, the mechanical stability of the laser-welded unit equipped with a 12.5-μm-thick Pd membrane was preserved up to 600°C and 1100 kPa in pure hydrogen permeation experiments and up to 2000 kPa retentate pressure and 550°C in water–gas shift reaction conditions. The bursting pressure was never reached, which confirms the high potential of the module design for application. Integration of even thinner membranes, that is, ultra-thin $Pd_{77}Ag_{23}$ foils prepared by SINTEF, is underway and should enable further improved hydrogen fluxes.

In a more sophisticated system, the arrangement of the microchannels was designed in such a way that enables stacking of several membrane modules for higher gas throughput and hydrogen production rates adapted to the process needs (see Figure 7.13). Single microchannel membrane modules are joined with laser welding, whereas several modules can be stacked to a modular membrane reactor connecting the modules through graphite rings.

Each microchannel membrane module is coated with a highly active reforming catalyst and contains a section for the pre-reforming of the hydrocarbons and a section where the reforming zone is in direct contact with the hydrogen-separating palladium membrane (see Figure 7.13b). The system shown in Figure 7.13 is electrically heated, but heat integration by catalytic oxidation of the reformate, or additional methane, is also possible by adding combustion modules.

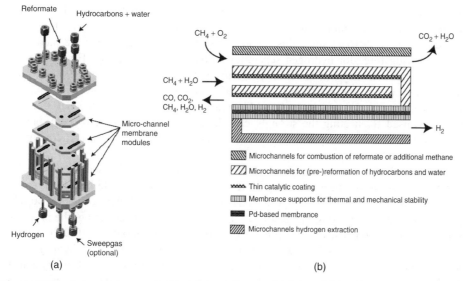

Figure 7.13 *(a) compact reactor design based on stacked microchannel membrane modules for scalable hydrogen production; (b) schematic illustration of the channel configuration in the microstructured membrane module, (Source: After Ref. [149])*

7.7 Integration of Pd-Based Membranes and Fuel Cells

Integration of Pd-based membrane technology and PEMFC for stationary application seems more straightforward than for transportation and portable systems. Small fuel cell units for household CHP systems usually plugged into the natural gas grid have been installed in large quantities in Japan [10]. For such CHP systems, a Pd-based membrane fuel processor has been suggested by various studies [150, 151]. According to calculations [151], a system solution based on membrane reactor technology yields net electrical efficiency as high as 43%, about 10% greater than the values obtained by conventional SR or ATR reactors, and an overall efficiency of 80–95% when generated heat is included [152]. The simpler Pd-based membrane fuel processor appears, therefore, as attractive for these small units with a net electrical power output in the range of 0.5–10 kW [152].

High efficiency requires high utilization of the fuel, and therefore, the fuel cell systems must be designed to utilize unconverted fuel from the stack and retentate outlets. Mendes *et al.* compared HYSYS process simulation by a conventional ethanol steam reforming process for hydrogen production that included a steam reforming reactor, high- and low-temperature WGS reactors and a CO preferential oxidizer reactor to reduce CO to acceptable levels upfront a PEM fuel cell stack, with a simpler process layout where the three last reactors were replaced by a membrane reactor consisting of 19 tubular Pd-Ag membranes performing simultaneously the WGS reaction and hydrogen separation [153]. The membrane retentate stream and unconverted fuel vented from the fuel cell stack were sent to a post-combustion unit that generated heat for the ethanol steam reformer and the evaporation of liquids prior to the reactor. The membrane process showed a higher total efficiency in terms of power output than the conventional process. The optimal condition for the membrane process was for minimum steam/ethanol ratio, $R = 3$, and highest pressure (20 bar) since this favours high hydrogen partial pressure, and, thus, a high driving force for hydrogen separation in the membrane reactor. However, a low steam/ethanol ratio may lead to coking problems of the catalyst, in particular for non-noble metal catalysts, and the membrane surface [97], which will reduce conversion and flux. Lowering the reformer temperature could increase the overall system efficiency since less energy is used for heating the strongly endothermic steam reforming process, but this must be balanced against potential side reactions such as methanation and coking that could make additional steam necessary.

Also for PEMFC used for portable power, the adaption of Pd-based membrane fuel processors has shown clear advantages over conventional small-scale hydrogen generation systems [154–156]. Wilson reported on a membrane-based fuel processor for portable power applications by using methanol as fuel [155]. Instead of reforming, the processor uses methanol decomposition to enable the by-product, carbon monoxide (CO), to be used as the heat source. The resulting self-heated processor produces 85% of the hydrogen content of the fuel, while a 20 W autonomous power source demonstrates a fuel energy density >1.5 Wh/g (electrical), nearly twice as high as for microreformer power sources [155].

High-pressure hydrogen storage is currently the dominating technology for PEMFC applications in the transport sector. A considerable compression work is thus required at the hydrogen filling station. By way of example, assuming that hydrogen is produced at atmospheric pressure and thereafter compressed for storage to 425 bars using electricity from the grid (assuming transmission grid electricity efficiency of 32.5%), close to 40% of the fuel's lower heating value (LHV) will be consumed [87]. To avoid compression, on-board hydrogen production from liquid fuels and renewable sources seemingly appears as an alternative solution [157]. De Falco has described a process and module of 40 cm in diameter and 1.52 cm length, containing four membrane tubes packed with a reforming catalyst for on-board production of 64.7 Nl/min hydrogen from ethanol for light vehicles using a 4-kW PEMFC stack [158]. Depending on the output power, the integrated membrane–fuel cell-driven electrical engine may reach an energy efficiency above 40% compared to that of internal combustion engines of 20–25% (see Table 7.3). It should be noted that applying water-bioethanol solutions from fermentation as primary fuel, the energy demanding distillation and purification required for mixing bioethanol with gasoline is avoided [159].

Table 7.3 *Comparison of green house gas emissions of gasoline/diesel engines and membrane ethanol reformer + PEMFC system power generator*

Model	GHG emission (g_{CO_2}/km)
Audi A2 1.2 TDI (diesel)	86
Toyoya Yaris 1.4 (diesel)	113
Fiat Stilo 1.2 16V (gasoline)	149
Ford Ka 1.3 (gasoline)	150
Membrane ethanol reformer + PEMFC	30–80

Source: After Ref. [158].

Several challenges must be met for on-board systems, for instance, at start-up, which requires heating of the reactor, and operation during variable load of the PEMFC. A combined fuel system with pure hydrogen as a buffer [25, 160] and a main liquid fuel could increase the dynamics and efficiency of the integrated fuel processor-PEMFC system. Gas cleaning systems, for example, a ZnO bed or liquid fuel desulphurization process, may be required for bio-fuels and add complexity to on-board hydrogen production where space and weight limitations are often stringent.

7.8   Final Remarks

During the past two decades, the research on Pd-alloy membranes has led to a technology matureness that appears ready for up-scaling for some applications. These are typically being applications with operation temperature less than about 550°C and where no detrimental impurities are present (e.g., sulphur). It has been demonstrated beyond doubt that hydrogen production from primary fuels as hydrocarbons, oxygenates and ammonia is advantageously performed in simple tubular membrane reactor designs where pure hydrogen is extracted from the reaction. Hydrogen production rates and yields generally increase with feed pressure and reduced temperature, even if reaction thermodynamics predict otherwise. However, to improve production rates and economics, research must continue so that the full advantage of the Pd-alloy membrane reactor concept is utilized. Such research should lead to membrane reactors integrating thin membranes within a smart reactor design that reduces concentration polarization to a minimum and heat management to a maximum. These boundary conditions are placing stringent requirements on membrane stability, catalyst activity, sealing technology, support materials and module design. The ongoing research on microstructured membrane reactors is a step in this direction towards low-cost and high-productivity units for hydrogen production.

In addition, more research is required for optimal integration of membrane reactors with fuel cells. Very few technologies are cost-effective over a wide range of operation conditions, and this applies for both fuel cells and membrane reactors. Both technologies work better as long as the fuel/feed side hydrogen partial pressure is kept high. Integrated membrane reactor–fuel cell systems may, however, advantageously take use of unconverted fuel in the retentate as a source for heat generation in endothermic conversion of primary fuels and any required evaporation process. Tuning of the operation therefore require modelling tools and understanding of each component's behaviour under different operating conditions in order to design and operate the integrated system in the most efficient way. For Pd-based membrane reactor technology to become an important contributor to the foreseen broad deployment of fuel cells, further research on system integration is required.

Acknowledgements

The financial support from the Research Council of Norway (RCN) through the following programs, RENERGI (Project No: 190779/S60), CLIMIT (Contract No. 215666/E20), and the Helmholtz Research School for Energy-Related Catalysis is gratefully acknowledged.

References

1. Hwang, J.J. (2013) Sustainability study of hydrogen pathways for fuel cell vehicle applications. *Renewable and Sustainable Energy Reviews*, **19**, 220–229.
2. Chaubey, R., Sahu, S., James, O.O. and Maity, S. (2013) A review on development of industrial processes and emerging techniques for production of hydrogen from renewable and sustainable sources. *Renewable and Sustainable Energy Reviews*, **23**, 443–462.
3. Oertel, M., Schmitz, J., Weirich, W. *et al.* (1987) Steam reforming of natural gas with integrated hydrogen separation for hydrogen production. *Chemical Engineering Technology*, **10**, 248–255.
4. Uemiya, S. (1999) State-of-the-art of supported metal membranes for gas separation. *Separation & Purification Reviews*, **28**, 51–85.
5. Beavis, R. (2011) The EU FP6 CACHET project - final results. *Energy Procedia*, **4**, 1074–1081.
6. Middleton, P., Hurst, P. and Walker, G. (2005) in *Carbon dioxide capture for storage in deep geological formations - results from the CO₂ capture project; capture and separation of carbon dioxide from combustion sources* (ed D.C. Thomas), Elsevier, Naperville.
7. Tanksale, A., Beltramini, J.N. and Lu, G.M. (2010) A review of catalytic hydrogen production processes from biomass. *Renewable and Sustainable Energy Reviews*, **14**, 166–182.
8. Dicks, A. and Larminie, J. (2003) *Fuel Cell Systems Explained*, 2nd edn, John Wiley Sons Inc., Chichester, United Kingdom.
9. The Industry Review 2011, Fuel Cell Today (2011).
10. Carter, D., Ryan, M. and Wing, J. (2012) *The Industry Review 2012*, Fuel Cell Today.
11. Haussinger, P., Lohmuller, R. and Watson, A.M. (2000) *Ullmann's Encyclopedia of Industrial Chemistry*, Wiley-VCH Verlag GmbH & Co, KGaA.
12. Spath, P.L., Mann, M.K. (2001) National Renewable Energy Laboratory (NREL).
13. Ritter, J.A. and Ebner, A.D. (2007) State-of-the-art adsorption and membrane separation processes for hydrogen production in the chemical and petrochemical industries. *Separation Science and Technology*, **42**, 1123–1193.
14. Kolb, G. (2013) Review: Microstructured reactors for distributed and renewable production of fuels and electrical energy. *Chemical Engineering and Processing: Process Intensification*, **65**, 1–44.
15. Kolb, G., Hessel, V., Cominos, V. *et al.* (2006) Microstructured fuel processors for fuel-cell applications. *Journal of Materials Engineering and Performance*, **15**, 389–393.
16. Pfeifer, P. (2012) Application of catalysts to metal microreactor systems, in *Chemical Kinetics* (ed V. Patel), InTech, pp. 325–344.
17. Jahnisch, K., Hessel, V., Lowe, H. and Baerns, M. (2004) Chemistry in Microstructured Reactors. *Angewandte Chemie International Edition*, **43**, 406–446.
18. Baier, T. and Kolb, G. (2007) Temperature control of the water gas shift reaction in microstructured reactors. *Chemical Engineering Science*, **62**, 4602–4611.

19. Myrstad, R., Eri, S., Pfeifer, P. *et al.* (2009) Fischer-Tropsch synthesis in a microstructured reactor. *Catalysis Today*, **147**, 301–304.

20. Cook-Chennault, K.A. (2008) Powering MEMS portable devices - a review of non-regenerative and regenerative power supply systems with special emphasis on piezoelectric energy harvesting systems. *Smart Materials and Structures*, **17**, 043001.

21. Chou, S.K., Yang, W.M., Chua, K.J. *et al.* (2011) Development of micro power generators GÇô A review. *Applied Energy*, **88**, 1–16.

22. Hu, J., Wang, Y., VanderWiel, D. *et al.* (2003) Fuel processing for portable power applications. *Chemical Engineering Journal*, **93**, 55–60.

23. Paglieri, S.N. and Way, J.D. (2002) Innovations in palladium membrane research. *Separation and Purification Methods*, **31**, 1–169.

24. Ockwig, N.W. and Nenoff, T.M. (2007) Membranes for hydrogen separation. *Chem. Rev.*, **107**, 4078–4110.

25. Lu, G.Q., Diniz da Costa, J.C., Duke, M. *et al.* (2007) Inorganic membranes for hydrogen production and purification: A critical review and perspective. *Journal of Colloid and Interface Science*, **314**, 589–603.

26. Nenoff, T.M., Spontak, R.J. and Aberg, C.M. (2006) Membranes for hydrogen purification: an important step toward a hydrogen-based economy. *MRS Bulletin*, **31**, 735–741.

27. Chiappetta, G., Clarizia, G. and Drioli, E. (2006) Design of an integrated membrane system for a high level hydrogen purification. *Chemical Engineering Journal*, **124**, 29–40.

28. Grashoff, G.J., Pilkington, C.E. and Corti, C.W. (1983) Purification of hydrogen. *Platinum Metals Review*, **27**, 157–169.

29. McKinley, D.L. US Patent 3,439,474, 22-4-1969.

30. McKinley, D.L. US Patent 3,350,845, 7-11-1967.

31. Dittmeyer, R., Hollein, V. and Daub, K. (2001) Membrane reactors for hydrogenation and dehydrogenation processes based on supported palladium. *Journal of Molecular Catalysis A: Chemical*, **173**, 135–184.

32. Collins, J.P., Schwartz, R.W., Sehgal, R. *et al.* (1996) Catalytic dehydrogenation of propane in hydrogen permselective membrane reactors. *Industrial and Engineering Chemistry Research*, **35**, 4398–4405.

33. Tosti, S., Bettinali, L. and Violante, V. (2000) Rolled thin Pd and Pd/Ag membranes for hydrogen separation and production. *International Journal of Hydrogen Energy*, **25**, 319–325.

34. Holleck, G.L. (1970) Diffusion and solubility of hydrogen in palladium and palladium-silver alloys. *The Journal of Physical Chemistry*, **74**, 503–511.

35. Faroldi, B., Carrara, C., Lombardo, E.A. and Cornaglia, L.M. (2007) Production of ultrapure hydrogen in a Pd-Ag membrane reactor using Ru/La_2O_3 catalysts. *Applied Catalysis A: General*, **319**, 38–46.

36. Hughes, D.T. and Harris, I.R. (1978) A comparative study of hydrogen permeabilities and solubilities in some palladium solid solution alloys. *Journal of the Less Common Metals*, **61**, 9–21.

37. Iyoha, O., Enick, R., Killmeyer, R. and Morreale, B. (2007) The influence of hydrogen sulfide-to-hydrogen partial pressure ratio on the sulfidization of Pd and 70 mol% Pd-Cu membranes. *Journal of Membrane Science*, **305**, 77–92.
38. Kulprathipanja, A., Alptekin, G.O., Falconer, J.L. and Way, J.D. (2005) Pd and Pd-Cu membranes: inhibition of H_2 permeation by H_2S. *Journal of Membrane Science*, **254**, 49–62.
39. Gao, H.Y., Lin, J.Y.S., Li, Y.D. and Zhang, B.Q. (2005) Electroless plating synthesis, characterization and permeation properties of Pd-Cu membranes supported on ZrO_2 modified porous stainless steel. *Journal of Membrane Science*, **265**, 142–152.
40. Kulprathipanja, A., Alptekin, G.O., Falconer, J.L. and Way, J.D. (2004) Effects of water gas shift gases on Pd-Cu alloy membrane surface morphology and separation properties. *Industrial and Engineering Chemistry Research*, **43**, 4188–4198.
41. Roa, F., Way, J.D., McCormick, R.L. and Paglieri, S.N. (2003) Preparation and characterization of Pd-Cu composite membranes for hydrogen separation. *Chemical Engineering Journal*, **93**, 11–22.
42. Roa, F., Block, M.J. and Way, J.D. (2002) The influence of alloy composition on the H_2 flux of composite Pd-Cu membranes. *Desalination*, **147**, 411–416.
43. Gade, S.K., Payzant, E.A., Park, H.J. *et al.* (2009) The effects of fabrication and annealing on the structure and hydrogen permeation of Pd-Au binary alloy membranes. *Journal of Membrane Science*, **340**, 227–233.
44. Okazaki, J., Tanaka, D.A.P., Tanco, M.A.L. *et al.* (2008) Preparation and hydrogen permeation properties of thin Pd-Au alloy membranes supported on porous alpha-alumina tube. *Materials Transactions*, **49**, 449–452.
45. Yakabe, H. (2012) Operations of a 40 Nm^3/h-class Membrane Reformer System at Tokyo Gas, Presentation at the International Joint Workshop on Palladium Membrane Technology - Rome, Italy, 12-14 November.
46. Peters, T.A., Kaleta, T., Stange, M. and Bredesen, R. (2012) Hydrogen transport through a selection of thin Pd-alloy membranes: Membrane stability, H_2S inhibition, and flux recovery in hydrogen and simulated WGS mixtures. *Catalysis Today*, **193**, 8–19.
47. O'Brien, C.P., Howard, B.H., Miller, J.B. *et al.* (2010) Inhibition of hydrogen transport through Pd and Pd47Cu53 membranes by H_2S at 350C. *Journal of Membrane Science*, **349**, 380–384.
48. Scura, F., Barbieri, G., De Luca, G. and Drioli, E. (2008) The influence of the CO inhibition effect on the estimation of the H_2 purification unit surface. *International Journal of Hydrogen Energy*, **33**, 4183–4192.
49. Chen, C.H. and Ma, Y.H. (2010) The effect of H_2S on the performance of Pd and Pd/Au composite membrane. *Journal of Membrane Science*, **362**, 535–544.
50. Peters, T.A., Kaleta, T., Stange, M. and Bredesen, R. (2013) Development of ternary PdAgTM alloy membranes with improved sulphur tolerance. *Journal of Membrane Science*, **429**, 448–458.
51. Tsai,C.Y., Tam, S.Y. (2013) US20120325087 A1.
52. Hydrogen from Coal Program (2010) Research Development and Demonstration Plan for the period 2010-2016, US Department of Energy.

53. Shu, J., Grandjean, B.P.A. and Kaliaguine, S. (1994) Methane steam reforming in asymmetric Pd- and Pd-Ag/porous SS membrane reactors. *Applied Catalysis A: General*, **119**, 305–325.
54. Abdollahi, M., Yu, J., Liu, P.K.T. *et al.* (2012) Ultra-pure hydrogen production from reformate mixtures using a palladium membrane reactor system. *Journal of Membrane Science*, **390–391**, 32–42.
55. Tong, J. and Matsumura, Y. (2006) Pure hydrogen production by methane steam reforming with hydrogen-permeable membrane reactor. *Catalysis Today*, **111**, 147–152.
56. Tong, J.H., Matsumura, Y., Suda, H. and Haraya, K. (2005) Thin and dense Pd/CeO2/MPSS composite membrane for hydrogen separation and steam reforming of methane. *Separation and Purification Technology*, **46**, 1–10.
57. Gallucci, F., Paturzo, L., Fama, A. and Basile, A. (2004) Experimental study of the methane steam reforming reaction in a dense Pd/Ag membrane reactor. *Industrial and Engineering Chemistry Research*, **43**, 928–933.
58. Kikuchi, E., Nemoto, Y., Kajiwara, M. *et al.* (2000) Steam reforming of methane in membrane reactors: comparison of electroless-plating and CVD membranes and catalyst packing modes. *Catalysis Today*, **56**, 75–81.
59. Lægsgaard Jørgensen, S., Nielsen, P.E.H. and Lehrmann, P. (1995) Steam reforming of methane in a membrane reactor. *Catalysis Today*, **25**, 303–307.
60. Uemiya, S., Sato, N., Ando, H. *et al.* (1990) Steam reforming of methane in a hydrogen-permeable membrane reactor. *Applied Catalysis*, **67**, 223–230.
61. Chang, H.F., Pai, W.J., Chen, Y.J. and Lin, W.H. (2010) Autothermal reforming of methane for producing high-purity hydrogen in a Pd/Ag membrane reactor. *International Journal of Hydrogen Energy*, **35**, 12986–12992.
62. Gallucci, F., Van Sint Annaland, M. and Kuipers, J.A.M. (2008) Autothermal reforming of methane with integrated CO_2 capture in a novel fluidized bed membrane reactor. Part 2: Comparison of reactor configurations. *Topics in Catalysis*, **51**, 146–157.
63. Lin, W.H., Hsiao, C.S. and Chang, H.F. (2008) Effect of oxygen addition on the hydrogen production from ethanol steam reforming in a Pd-Ag membrane reactor. *Journal of Membrane Science*, **322**, 360–367.
64. Ferreira-Aparicio, P., Benito, M., Kouachi, K. and Menad, S. (2005) Catalysis in membrane reformers: a high-performance catalytic system for hydrogen production from methane. *Journal of Catalysis*, **231**, 331–343.
65. Paturzo, L., Gallucci, F., Basile, A. *et al.* (2003) An Ru-based catalytic membrane reactor for dry reforming of methane--its catalytic performance compared with tubular packed bed reactors. *Catalysis Today*, **82**, 57–65.
66. Ferreira-Aparicio, P., RodrÆguez-Ramos, I. and Guerrero-Ruiz, A. (2002) On the applicability of membrane technology to the catalysed dry reforming of methane. *Applied Catalysis A: General*, **237**, 239–252.
67. Ferreira-Aparicio, P. and Ma, Y.H. (2001) in *Studies in Surface Science and Catalysis, Spillover and Mobility of Species on Solid Surfaces* (ed A. Guerrero-Ruiz), Elsevier, pp. 461–468.
68. Matsumura, Y. and Tong, J. (2008) Methane steam reforming in hydrogen-permeable membrane reactor for pure hydrogen production. *Topics in Catalysis*, **51**, 123–132.

69. Lin, Y.M., Liu, S.L., Chuang, C.H. and Chu, Y.T. (2003) Effect of incipient removal of hydrogen through palladium membrane on the conversion of methane steam reforming: Experimental and modeling. *Catalysis Today*, **82**, 127–139.

70. Li, H., Pieterse, J.A.Z., Dijkstra, J.W. *et al.* (2011) Performance test of a bench-scale multi-tubular membrane reformer. *Journal of Membrane Science*, **373**, 43–52.

71. Kurokawa, H., Shirasaki, Y. and Yasuda, I. (2011) Energy-efficient distributed carbon capture in hydrogen production from natural gas. *Energy Procedia*, **4**, 674–680.

72. Yasuda, I. and Shirasaki, Y. (2007) Development and demonstration of membrane reformer system for highly-efficient hydrogen production from natural gas. *Materials Science Forums*, **539–543**, 1403–1408.

73. Kurokawa, H., Yakabe, H., Yasuda, I. *et al.* (2014) The inhibition effect of CO on hydrogen permeability of Pd-Ag membrane applied in a microchannel module configuration. *International Journal of Hydrogen Energy*, **39**, 17201–17209.

74. Zhang, J., Liu, D., He, M. *et al.* (2006) Experimental and simulation studies on concentration polarization in H_2 enrichment by highly permeable and selective Pd membranes. *Journal of Membrane Science*, **274**, 83–91.

75. Hara, S., Sakaki, K. and Itoh, N. (1999) Decline in Hydrogen Permeation Due to Concentration Polarization and CO Hindrance in a Palladium Membrane Reactor. *Industrial and Engineering Chemistry Research*, **38**, 4913–4918.

76. Peters, T.A., Stange, M., Klette, H. and Bredesen, R. (2008) High pressure performance of thin Pd-23%Ag/stainless steel composite membranes in water gas shift gas mixtures; influence of dilution, mass transfer and surface effects on the hydrogen flux. *Journal of Membrane Science*, **316**, 119–127.

77. Gallucci, F., Van Sint Annaland, M. and Kuipers, J.A.M. (2008) Autothermal reforming of methane with integrated CO_2 capture in a novel fluidized bed membrane reactor. Part 1: Experimental demonstration. *Topics in Catalysis*, **51**, 133–145.

78. Gallucci, F., Van Sint Annaland, M. and Kuipers, J.A.M. (2010) Theoretical comparison of packed bed and fluidized bed membrane reactors for methane reforming. *International Journal of Hydrogen Energy*, **35**, 7142–7150.

79. Gallucci, F., Van Sint Annaland, M. and Kuipers, J.A.M. (2010) Pure hydrogen production via autothermal reforming of ethanol in a fluidized bed membrane reactor: A simulation study. *International Journal of Hydrogen Energy*, **35**, 1659–1668.

80. Múnera, J.F., Carrara, C., Cornaglia, L.M. and Lombardo, E.A. (2010) Combined oxidation and reforming of methane to produce pure H_2 in a membrane reactor. *Chemical Engineering Journal*, **161**, 204–211.

81. Faroldi, B.M., Lombardo, E.A. and Cornaglia, L.M. (2011) Ru/La_2O_3-SiO_2 catalysts for hydrogen production in membrane reactors. *Catalysis Today*, **172**, 209–217.

82. Alves, H.J., Bley Junior, C., Niklevicz, R.R. *et al.* (2013) Overview of hydrogen production technologies from biogas and the applications in fuel cells. *International Journal of Hydrogen Energy*, **38**, 5215–5225.

83. Balat, H. and KIrtay, E. (2010) Hydrogen from biomass - Present scenario and future prospects. *International Journal of Hydrogen Energy*, **35**, 7416–7426.

84. Lopez, E., Divins, N.J. and Llorca, J. (2012) Hydrogen production from ethanol over Pd-Rh/CeO_2 with a metallic membrane reactor. *Catalysis Today*, **193**, 145–150.

85. Montane, D., Bolshak, E. and Abello, S. (2011) Thermodynamic analysis of fuel processors based on catalytic-wall reactors and membrane systems for ethanol steam reforming. *Chemical Engineering Journal*, **175**, 519–533.

86. De Falco, M. and Gallucci, F. (2010) Ethanol steam reforming heated up by molten salt CSP: Reactor assessment. *International Journal of Hydrogen Energy*, **35**, 3463–3471.

87. Papadias, D.D., Lee, S.H.D., Ferrandon, M. and Ahmed, S. (2010) An analytical and experimental investigation of high-pressure catalytic steam reforming of ethanol in a hydrogen selective membrane reactor. *International Journal of Hydrogen Energy*, **35**, 2004–2017.

88. Tosti, S., Basile, A., Borelli, R. *et al.* (2009) Ethanol steam reforming kinetics of a Pd-Ag membrane reactor. *International Journal of Hydrogen Energy*, **34**, 4747–4754.

89. Yu, C.Y., Lee, D.W., Park, S.J. *et al.* (2009) Study on a catalytic membrane reactor for hydrogen production from ethanol steam reforming. *International Journal of Hydrogen Energy*, **34**, 2947–2954.

90. Gallucci, F., De Falco, M., Tosti, S. *et al.* (2008) Co-current and counter-current configurations for ethanol steam reforming in a dense Pd-Ag membrane reactor. *International Journal of Hydrogen Energy*, **33**, 6165–6171.

91. Gallucci, F. and Basile, A. (2008) Pd-Ag membrane reactor for steam reforming reactions: a comparison between different fuels. *International Journal of Hydrogen Energy*, **33**, 1671–1687.

92. Gallucci, F., De Falco, M., Tosti, S. *et al.* (2008) Ethanol steam reforming in a dense Pd-Ag membrane reactor: a modelling work. *Comparison with the traditional system, International Journal of Hydrogen Energy*, **33**, 644–651.

93. Urasaki, K., Tokunaga, K., Sekine, Y. *et al.* (2008) Production of hydrogen by steam reforming of ethanol over cobalt and nickel catalysts supported on perovskite-type oxides. *Catalysis Communications*, **9**, 600–604.

94. Tosti, S., Basile, A., Borgognoni, F. *et al.* (2008) Low temperature ethanol steam reforming in a Pd-Ag membrane reactor: Part 1: Ru-based catalyst. *Journal of Membrane Science*, **308**, 250–257.

95. Tosti, S., Basile, A., Borgognoni, F. *et al.* (2008) Low-temperature ethanol steam reforming in a Pd-Ag membrane reactor: part 2, Pt-based and Ni-based catalysts and general comparison. *Journal of Membrane Science*, **308**, 258–263.

96. Wieland, I., Melin, I. and Lamm, I. (2002) Membrane reactors for hydrogen production. *Chemical Engineering Science*, **57**, 1571–1576.

97. Basile, A., Pinacci, P., Iulianelli, A. *et al.* (2011) Ethanol steam reforming reaction in a porous stainless steel supported palladium membrane reactor. *International Journal of Hydrogen Energy*, **36**, 2029–2037.

98. Cheekatamarla, P.K. and Finnerty, C.M. (2006) Reforming catalysts for hydrogen generation in fuel cell applications. *Journal of Power Sources*, **160**, 490–499.

99. Tosti, S., Zerbo, M. and Basile, A. (2013) V. Calabr+, F. Borgognoni, A. Santucci, Pd-based membrane reactors for producing ultra pure hydrogen: Oxidative reforming of bio-ethanol. *International Journal of Hydrogen Energy*, **38**, 701–707.

100. Santucci, A., Annesini, M.C., Borgognoni, F. *et al.* (2011) Oxidative steam reforming of ethanol over a Pt/Al$_2$O$_3$ catalyst in a Pd-based membrane reactor. *International Journal of Hydrogen Energy*, **36**, 1503–1511.

101. Lin, W.H., Liu, Y.C. and Chang, H.F. (2010) Autothermal reforming of ethanol in a Pd-Ag/Ni composite membrane reactor. *International Journal of Hydrogen Energy*, **35**, 12961–12969.

102. Iulianelli, A., Longo, T., Liguori, S. *et al.* (2009) Oxidative steam reforming of ethanol over Ru-Al$_2$O$_3$ catalyst in a dense Pd-Ag membrane reactor to produce hydrogen for PEM fuel cells. *International Journal of Hydrogen Energy*, **34**, 8558–8565.

103. Diaz Alvarado, F. and Gracia, F. (2010) Steam reforming of ethanol for hydrogen production: Thermodynamic analysis including different carbon deposits representation. *Chemical Engineering Journal*, **165**, 649–657.

104. Gallucci, F., Basile, A., Tosti, S. *et al.* (2007) Methanol and ethanol steam reforming in membrane reactors: An experimental study. *International Journal of Hydrogen Energy*, **32**, 1201–1210.

105. Iulianelli, A., Longo, T. and Basile, A. (2008) Methanol steam reforming reaction in a Pd-Ag membrane reactor for CO-free hydrogen production. *International Journal of Hydrogen Energy*, **33**, 5583–5588.

106. Basile, A., Parmaliana, A., Tosti, S. *et al.* (2008) Hydrogen production by methanol steam reforming carried out in membrane reactor on Cu/Zn/Mg-based catalyst. *Catalysis Today*, **137**, 17–22.

107. Joensen, F. and Rostrup-Nielsen, J.R. (2002) Conversion of hydrocarbons and alcohols for fuel cells. *Journal of Power Sources*, **105**, 195–201.

108. Harold, P., Nair, B. and Kolios, G. (2003) Hydrogen generation in a Pd membrane fuel processor: assessment of methanol-based reaction systems. *Chemical Engineering Science*, **58**, 2551–2571.

109. Lin, K.H., Lin, W.H., Hsiao, C.H. *et al.* (2012) Hydrogen production in steam reforming of glycerol by conventional and membrane reactors. *International Journal of Hydrogen Energy*, **37**, 13770–13776.

110. Chang, A.C.C., Lin, W.H., Lin, K.H. *et al.* (2012) Reforming of glycerol for producing hydrogen in a Pd/Ag membrane reactor. *International Journal of Hydrogen Energy*, **37**, 13110–13117.

111. Iulianelli, A., Seelam, P.K., Liguori, S. *et al.* (2011) Hydrogen production for PEM fuel cell by gas phase reforming of glycerol as byproduct of bio-diesel. The use of a Pd-Ag membrane reactor at middle reaction temperature. *International Journal of Hydrogen Energy*, **36**, 3827–3834.

112. Wang, X., Wang, N., Li, M. *et al.* (2010) Hydrogen production by glycerol steam reforming with in situ hydrogen separation: a thermodynamic investigation. *International Journal of Hydrogen Energy*, **35**, 10252–10256.

113. Iulianelli, A., Longo, T. and Basile, A. (2008) CO-free hydrogen production by steam reforming of acetic acid carried out in a Pd-Ag membrane reactor: the effect of co-current and counter-current mode. *International Journal of Hydrogen Energy*, **33**, 4091–4096.

114. Basile, A., Gallucci, F., Iulianelli, A. *et al.* (2008) Acetic acid steam reforming in a Pd-Ag membrane reactor: the effect of the catalytic bed pattern. *Journal of Membrane Science*, **311**, 46–52.

115. Rakib, M.A., Grace, J.R., Lim, C.J. *et al.* (2010) Steam reforming of propane in a fluidized bed membrane reactor for hydrogen production. *International Journal of Hydrogen Energy*, **35**, 6276–6290.

116. Rakib, M.A., Grace, J.R., Lim, C.J. and Elnashaie, S.S.E.H. (2010) Steam reforming of heptane in a fluidized bed membrane reactor. *Journal of Power Sources*, **195**, 5749–5760.

117. Chen, Z. and Elnashaie, S.S.E.H. (2005) Autothermal CFB Membrane Reformer for Hydrogen Production from Heptane. *Chemical Engineering Research and Design*, **83**, 893–899.

118. Rahimpour, M.R. (2009) Enhancement of hydrogen production in a novel fluidized-bed membrane reactor for naphtha reforming. *International Journal of Hydrogen Energy*, **34**, 2235–2251.

119. Chen, Y., Xu, H., Wang, Y. and Xiong, G. (2006) Application of Coprecipitated Nickel Catalyst to Steam Reforming of Higher Hydrocarbons in Membrane Reactor. *Chinese Journal of Catalysis*, **27**, 772–777.

120. Miyamoto, M., Hayakawa, C., Kamata, K. *et al.* (2011) Influence of the pre-reformer in steam reforming of dodecane using a Pd alloy membrane reactor. *International Journal of Hydrogen Energy*, **36**, 7771–7775.

121. Di Carlo, A., Dell'Era, A. and Del Prete, Z. (2011) 3D simulation of hydrogen production by ammonia decomposition in a catalytic membrane reactor. *International Journal of Hydrogen Energy*, **36**, 11815–11824.

122. Garcia-Garcia, F.R., Ma, Y.H., Rodriguez-Ramos, I. and Guerrero-Ruiz, A. (2008) High purity hydrogen production by low temperature catalytic ammonia decomposition in a multifunctional membrane reactor. *Catalysis Communications*, **9**, 482–486.

123. Rahimpour, M.R. and Asgari, A. (2008) Modeling and simulation of ammonia removal from purge gases of ammonia plants using a catalytic Pd-Ag membrane reactor. *Journal of Hazardous Materials*, **153**, 557–565.

124. Rahimpour, M.R. and Asgari, A. (2009) Production of hydrogen from purge gases of ammonia plants in a catalytic hydrogen-permselective membrane reactor. *International Journal of Hydrogen Energy*, **34**, 5795–5802.

125. Abashar, M.E.E. (2002) Integrated catalytic membrane reactors for decomposition of ammonia. *Chemical Engineering and Processing*, **41**, 403–412.

126. Rahimpour, M.R., Mottaghi, H.R. and Barmaki, M.M. (2010) Hydrogen production from urea wastewater using a combination of urea thermal hydrolyser-desorber loop and a hydrogen-permselective membrane reactor. *Fuel Processing Technology*, **91**, 600–612.

127. Bredesen, R., Peters, T.A., Stange, M. and Vicinanza, N. (2011) H. J. Venvik, in *Membrane Engineering for the Treatment of Gases* (eds E. Drioli and G. Barbieri), RSC Publishing, Cambridge, United Kingdom, pp. 40–86.

128. Mahecha-Botero, A., Boyd, T., Gulamhusein, A. *et al.* (2008) Pure hydrogen generation in a fluidized-bed membrane reactor: Experimental findings. *Chemical Engineering Science*, **63**, 2752–2762.

129. Boeltken, T., Belimov, M., Pfeifer, P. *et al.* (2013) Fabrication and testing of a planar microstructured concept module with integrated palladium membranes. *Chemical Engineering and Processing: Process Intensification*, **67**, 136–147.

130. Caravella, A., Barbieri, G. and Drioli, E. (2009) Concentration polarization analysis in self-supported Pd-based membranes. *Separation and Purification Technology*, **66**, 613–624.

131. Li, H., Dijkstra, J.W., Pieterse, J.A.Z. *et al.* (2010) Towards full-scale demonstration of hydrogen-selective membranes for CO_2 capture: Inhibition effect of WGS-components on the H_2 permeation through three Pd membranes of 44 cm long. *Journal of Membrane Science*, **363**, 204–211.

132. Hessel, V. and Noel, T. (2012) *Ullmann's Encyclopedia of Industrial Chemistry*, Wiley-VCH Verlag GmbH & Co. KGaA.

133. Kolb, G. and Hessel, V. (2004) Micro-structured reactors for gas phase reactions. *Chemical Engineering Journal*, **98**, 1–38.

134. Men, Y., Kolb, G., Zapf, R. *et al.* (2008) A complete miniaturized microstructured methanol fuel processor/fuel cell system for low power applications. *International Journal of Hydrogen Energy*, **33**, 1374–1382.

135. Bieberle-Hutter, A., Santis-Alvarez, A.J., Jiang, B. *et al.* (2012) Syngas generation from n-butane with an integrated MEMS assembly for gas processing in micro-solid oxide fuel cell systems. *Lab Chip*, **12**, 4894–4902.

136. Mitsos, A. and Barton, P.I. (2009) *Microfabricated power generation devices*, Wiley-VCH.

137. Holladay, J.D., Wang, Y. and Jones, E. (2004) Review of developments in portable hydrogen production using microreactor technology. *Chemical Reviews*, **104**, 4767–4790.

138. Karnik, S.V., Hatalis, M.K. and Kothare, M.V. (2003) Towards a palladium micro-membrane for the water gas shift reaction: microfabrication approach and hydrogen purification results. *Journal of Microelectromechanical Systems*, **12**, 93–100.

139. Pattekar, A.V. and Kothare, M.V. (2004) A microreactor for hydrogen production in micro fuel cell applications. *Journal of Microelectromechanical Systems*, **13**, 7–18.

140. Lindroos, V., Tilli, M., Lehto, A. and Motooka, T. (2010) *Handbook of silicon-based MEMS materials and technologies*, Elsevier.

141. de Jong, J., Lammertink, R.G.H. and Wessling, M. (2006) Membranes and microfluidics: a review. *Lab Chip*, **6**, 1125–1139.

142. Wilhite, B.A., Schmidt, M.A. and Jensen, K.F. (2004) Palladium-Based Micromembranes for Hydrogen Separation: Device Performance and Chemical Stability. *Industrial and Engineering Chemistry Research*, **43**, 7083–7091.

143. Wilhite, B.A., Weiss, S., Ying, J.Y. *et al.* (2006) High-Purity Hydrogen Generation in a Microfabricated 23 wt% Ag-Pd Membrane Device Integrated with 8:1 LaNi0.95Co0.05O3/Al$_2$O$_3$ Catalyst. *Advanced Materials*, **18**, 1701–1704.

144. Tong, H.D., Berenschot, J.W.E., de Boer, M.J. *et al.* (2003) Microfabrication of palladium-silver alloy membranes for hydrogen separation. *Journal of Microelectromechanical Systems*, **12**, 622–629.

145. Tong, H.D., Gielens, F.C., Gardeniers, J.G.E. *et al.* (2005) Microsieve supporting palladium-silver alloy membrane and application to hydrogen separation. *Journal of Microelectromechanical Systems*, **14**, 113–124.

146. Deshpande, K., Meldon, J.H., Schmidt, M.A. and Jensen, K.F. (2010) SOI-Supported microdevice for hydrogen purification using palladium–silver membranes. *Journal of Microelectromechanical Systems*, **19**, 402–409.

147. Goerke, O., Pfeifer, P., Schubert, K. 2006 European Patent, EP1669323.

148. Chellappa, A.S., Powell, M.R., Call, C.J. (2008) WO2008024089.

149. Boeltken, T., Lee, S., Kreuder, H., Pfeifer, P., Peters, T.A., Bredesen, R., Dittmeyer, R. (2013) Microstructured reactors with integrated palladium membranes for hydrogen generation, Proceedings of the 11th International Conference on Catalysis in Membrane Reactors, Porto, Portugal.

150. Roses, L., Manzolini, G. and Campanari, S. (2010) CFD simulation of Pd-based membrane reformer when thermally coupled within a fuel cell micro-CHP system. *International Journal of Hydrogen Energy*, **35**, 12668–12679.

151. Campanari, S., Macchi, E. and Manzolini, G. (2008) Innovative membrane reformer for hydrogen production applied to PEM micro-cogeneration: Simulation model and thermodynamic analysis. *International Journal of Hydrogen Energy*, **33**, 1361–1373.

152. http:www.fuelcelltoday.com/about-fuel-cells/applications (2013).

153. Mendes, D., Tosti, S., Borgognoni, F. *et al.* (2010) Integrated analysis of a membrane-based process for hydrogen production from ethanol steam reforming. *Catalysis Today*, **156**, 107–117.

154. Damle, A.S. (2009) Hydrogen production by reforming of liquid hydrocarbons in a membrane reactor for portable power generation - Experimental studies. *Journal of Power Sources*, **186**, 167–177.

155. Wilson, M.S. (2009) Methanol decomposition fuel processor for portable power applications. *International Journal of Hydrogen Energy*, **34**, 2955–2964.

156. Damle, A.S. (2008) Hydrogen production by reforming of liquid hydrocarbons in a membrane reactor for portable power generation - model simulations. *Journal of Power Sources*, **180**, 516–529.

157. Han, J., Kim, I.-S. and Choi, K.S. (2000) Purifier-integrated methanol reformer for fuel cell vehicles. *Journal of Power Sources*, **86**, 223–227.

158. De Falco, M. (2011) Ethanol membrane reformer and PEMFC system for automotive application. *Fuel*, **90**, 739–747.

159. Manzolini, G. and Tosti, S. (2008) Hydrogen production from ethanol steam reforming: energy efficiency analysis of traditional and membrane processes. *International Journal of Hydrogen Energy*, **33**, 5571–5582.

160. Capobianco, L., Del Prete, Z., Schiavetti, P. and Violante, V. (2006) Theoretical analysis of a pure hydrogen production separation plant for fuel cells dynamical applications. *International Journal of Hydrogen Energy*, **31**, 1079–1090.

8

From Biomass to SNG

Luca Di Felice[1] *and Francesca Micheli*[2]

[1] *Chemical Process Intensification, Department of Chemical Engineering
and Chemistry, Eindhoven University of Technology, Den Dolech 2, 5612AD,
Eindhoven, The Netherlands*
[2] *Chemical Engineering Department, University of L'Aquila,
Monteluco di Roio (AQ), 67040, L'Aquila, Italy*

8.1 Introduction

The use of biomass as a renewable energy source increased in the last decades up to approximately 10% of the global energy supply; 35% of this amount is used in industrialized countries mainly for industrial applications within heat and power sector and road transportation [1]. Among other options, attention is growing on production of substitute (or synthetic) natural gas from biomass (substitute natural gas (SNG) or bio-SNG), which offers the advantage of high conversion efficiency – reported to be in the range 64–70% of chemical energy output of CH_4 compared to chemical energy input of wood [2] – and an already existing gas distribution and end-use infrastructure (from natural gas grid to fuel station for vehicles). Moreover, an inherent carbon dioxide separation step is integrated in the SNG production process to obtain a highly concentrated CH_4 stream, while CO_2 is available for further storage; from this point of view, bio-SNG production offers the advantage of a carbon negative balance without additional separation costs.

There are two main options to produce the so-called green natural gas, that is, a gaseous energy carrier produced from biomass [3]: through anaerobic digestion of organic materials at low temperature or gasification of biomass and subsequent methanation of the product gas. While the first route is referred to produce the so-called biogas (containing 50–75% of CH_4), bio-SNG is related to the latter option, in which a gaseous fuel containing up to 95% vol CH_4 can be obtained.

Process Intensification for Sustainable Energy Conversion, First Edition.
Edited by Fausto Gallucci and Martin van Sint Annaland.
© 2015 John Wiley & Sons, Ltd. Published 2015 by John Wiley & Sons, Ltd.

Currently, the attention is focused on the synthesis of both liquid biofuels – mainly FT-diesel products and methanol – and gaseous synthetic fuels such as dimethyl ether (DME) and substitute natural gas (SNG, which is addressed in this chapter); the main parameters (pressure, temperature, type of catalyst, H_2/CO ratio) governing the different synthesis pathways of these fuels are reported in reference [4].

Among all the developing bioenergy technologies recently listed by the Intergovernmental Panel on Climate Change, the conversion of syngas into SNG by a biomass gasification process is estimated to have the production costs of 6 to 12/GJ USD$_{2005}$; this is in line with that of other gaseous biofuels such as H_2 and is sensibly lower compared to available alternatives such as Integrated Gasification Combined Cycle (IGCC) [5].

On the other hand, between 2004 and 2006, total SNG production costs have been estimated to be in the range of 10–30 €/GJ for capacities of 10–1000 MW$_{th}$, which increases by decreasing the plant size. Taking into consideration a natural gas price of 6 €/GJ, the amount of subsidy required for the implementation of SNG production facilities (1.2–8.9 €ct/kWh) has been estimated to be comparable with the range of financial support for producing electricity from biomass (6.0–9.7 €ct/kWh), suggesting that this technology can be afforded [6, 7].

In this chapter, a brief description of the main bio-SNG facilities and projects in Europe is presented as well as the main process units (gasification, gas cleaning and methanation) integrated for bio-SNG production. Therefore, a case study for bio-SNG production is modeled by using the CHEMCAD 6.3.1.4168 software. Two process technologies, a fixed (adiabatic case) or fluidized (isothermal) bed methanation reactors are considered, while three different product gas compositions from "real" biomass gasification data are fed as input syngas for the modeled system. Finally, CH_4 yield and chemical efficiency of the different cases are compared and discussed.

8.2   Current Status of Bio-SNG Production and Facilities in Europe

Industrial-scale methanation reactors integrated in biomass gasification plants have not been commercially exploited yet, and, as a matter of fact, there is only one commercial SNG production plant currently in operation in North Dakota (Great Plains Synfuels Plant since 1984), where lignite coal is gasified and 4.8×10^6 m^3/day CH_4 are produced, according to a technology review in the field [6]. However, in October 2013, Topsøe announced the start-up of the largest single-train SNG plant in the world (in China), with a yearly production of 1.4×10^9 Nm^3 SNG based on coal gasification; the plant uses Haldor Topsøe catalysts and process technology, the so-called TREMP™ [8]. Although not directly related to bio-SNG production, it is worth mentioning here that also Audi is planning the first industrial-scale plant for conversion of H_2 (produced by water electrolysis, which in turn is powered by wind and solar renewable energies) and CO_2 (from high-purity exhausted flow of an adjacent bio-methane plant) into SNG [9].

There are essentially two main demonstration/pilot-scale facilities in Europe that are actively involved in the research area of bio-SNG production; they are associated with the 8 MW$_{th}$ biomass combined heat and power (CHP) gasification plant in Güssing (Austria), and the 800 kW$_{th}$ pilot-scale MILENA gasifier in Petten (The Netherlands).

Since 2002, the Energy Research Centre of the Netherlands (ECN) began preliminary study to investigate the feasibility of SNG production from biomass (wood, sewage sludge, and lignite) via indirectly heated gasification. Both a 30 kW$_{th}$ and an 800 kW$_{th}$ allothermal biomass gasifiers (the so-called MILENA gasification technology) are used at ECN as syngas generators for the studies on SNG production from biomass. By implementing data obtained from the Milena gasifier and relevant cleaning/upgrading techniques, simulation data had shown an expected composition of bio-SNG to be 93.4% CH_4, 4.6% N_2, and 1.2% H_2 [10].

The Paul Scherrer Institute (Switzerland) worked on the demonstration of a 10 kW$_{th}$ fluidized bed methanation reactor (Comflux technology), and, based on the obtained results (1000 h of stable operation for the methanation unit without catalyst deactivation), a scale-up to 1 MW$_{th}$ has been erected in collaboration with other commercial (Biomassekraftwerk, CTU, and Repotec GmbH) and academic (TU Vienna) partners in Güssing, Austria. The SNG Güssing concept has been developed in the frame of the "Bio-SNG" project, between 2006 and 2009, aimed at demonstrating the production, upgrading, and the final utilization of synthetic natural gas. Moreover, simulated exit gas composition from the methanation unit has been found to meet the requirements of different national directives (Austria, Germany, Sweden, Denmark, Switzerland) regulating the feeding into the natural gas grid, for example, a N_2 content <5% and CH_4 content >96%. Composition of bio-SNG gas from real plant operations is in line with the requirements. The same composition data were obtained for car fueling station SNG [11].

It is worth mentioning here that other ongoing projects are in an early stage of implementation. By using the same gasification technology of Güssing, Gothenburg biomass gasification (GoBiGas) aims at extracting biogas by thermal gasification of forest residues and biodiesel. Phase 1 was planned to be finished by 2013, with the construction of a demonstration 20 MW gas production plant. The goal of phase 2 – planned for completion in 2016 – will be the construction of a commercial plant with a planned gas production of 80–100 MW. The objective is to reach a biomass conversion into bio-SNG of 65% and an overall energy efficiency of more than 90% [12].

GDF Suez is leading the so-called Gaya project, with the aim of validating, at a preindustrial scale, a portfolio of technological solutions to develop an innovative industry for bio-methane production. The industrial-scale stage of the project is scheduled to start in 2016, while realization is foreseen in 2020. A demonstration plant (biomass treatment, gasification, and methanation) was commissioned in 2011 and will be constructed near Lyon (France). Gasification and methanation units are based on the technology and experience acquired in the Biomass CHP plant located in Güssing [13, 14].

8.3   Bio-SNG Process Configuration

Following the Güssing and ECN models, a general flow diagram for bio-SNG process generation is shown in Figure 8.1. The produced syngas is first cleaned from impurities in different "cold" removal steps for acid gases, tar, and ammonia abatement (see Section 3.2). The cleaned gas can undergo two possible pathways, in which CO_2 capture and water-gas

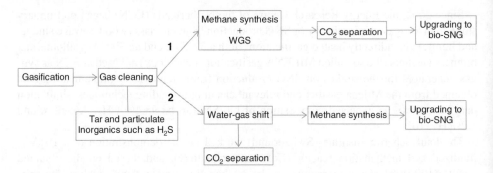

Figure 8.1   *Schematic presentation of the integrated bio-SNG production chain*

shift are carried out at different positions in the line-up, either before or after the methanation step. As a matter of fact, the Güssing plant is based on simultaneous WGS and methanation with subsequent CO_2 capture by amine scrubbing, while the ECN model is reported with a separated, upstream WGS unit followed by CO_2 separation. The latter configuration is also reported by TREMP, the Topsøe Recycle Energy-efficient Methanation Process [15]. The CO_2 removal is tuned in such a way that the remaining CO_2 reacts with the surplus of H_2 in the gas to form CH_4 and H_2O [16].

Both the Dutch and the Austrian/Swiss facilities have reported data on gas composition at inlet/outlet of different process units, and the collected information is reported in Table 8.1 [7, 11]. ECN data refer to the $30\,kW_{th}$ laboratory-scale MILENA concept, which is described elsewhere [17]. The reported values of gas composition refer to steam and nitrogen-free conditions. As it can be seen, the H/C ratio of the two types of (dry and ash free) biomass fuels used is almost the same for the ECN and Güssing (about 1.5) plants. This ratio slightly changes after the conversion from biomass to syngas (about 1.25 for the former, 1.8 for the latter installation); at this stage, the H_2/CO ratio of the Güssing plant is higher, while the CH_4 content is lower compared with the ECN data. The H/C ratio noticeably increases up to 4 after methanation and upgrading steps, which is the expected value

Table 8.1   *Reported inlet/outlet gas composition from Güssing and ECN studies at different units of the process line-up (adapted from [7, 10])*

	Biomass		Syngas		SNG (Before upgrading)		SNG (After upgrading)	
	ECN	Güssing	ECN	Güssing	ECN	Güssing	ECN (*)	Güssing (*)
C (wt%)	31.7	31.1	33.2	25.9	24.8	–	20.3	20
H (wt%)	46.9	47	41.3	47.1	43.3	–	77.6	80
O (wt%)	21.4	21.8	25.5	27	31.9	–	2.14	<0.1
CO (vol%)	–	–	46	25.2	0.54	–	<0.1	0
CO_2 (vol%)	–	–	10.6	22	53.2	–	5.2	<0.1
H_2 (vol%)	–	–	21.3	38.4	14.2	–	1.9	0.7
CH_4 (vol%)	–	–	16.6	10.9	29.2	–	92.9	99.3
Others $> C_1$ (vol%)	–	–	6.6	3.5	3.3	–	0	0

*Calculated from a modeling study.

when CH_4 dominates the gas composition at the end of the process. This is clearly an effect of combined water-gas shift, methanation, and C dilution by CO_2 capture. It should be noted that the reported final compositions for both facilities refer to modeled data, although a CH_4 composition higher than 95% has been observed experimentally in the Güssing plant. On the other hand, upgrading of the raw product SNG, that is, water and CO_2 removal, was not included in the ECN laboratory-scale facility for practical considerations.

As a whole, it has been reported that 40% of the carbon contained in the biomass fuel is converted into SNG, an equal amount of carbon is separated as CO_2, and the remaining 20% is lost as flue gas from the process [18].

8.3.1 The Gasification Step

Gasification is the first step of biomass conversion toward SNG and is probably the most critical one, being pointed out as the most expensive unit of the process (21% of total equipment cost) [18]. It is a reliable bioenergy route; the estimated production costs for a commercial system range between 9 and 35 USD_{2005}/GJ for small-scale plants (2.2–13 MW); however, only few industrial-scale operational units have been found in Europe, and the annual biomass gasification capacity has been estimated to be about 15 PJ [5]. On the other hand, the World Gasification Database listed only 12 gasification plants (out of 160) using biomass as feedstock, and these plants are small scale compared to fossil fuel operations, as mentioned by Minchener [19].

The gasification process of biomass is a thermochemical conversion step capable of producing a syngas, which can be used either as a combustible for heat and power generation or as a synthesis gas for subsequent utilization; the latter option – which is addressed in this chapter – requires a specific feeding gas composition (e.g. hydrogen to carbon monoxide ratio) and an extensive gas cleaning and upgrading. A suitable product gas distribution and the gas purification steps involved are often achieved by chemical conversion processes.

The produced gas contains a mixture of H_2, CO, CO_2, H_2O, and CH_4, coming mainly from the combination of hydrocarbons reforming and the reversible water-gas shift reactions.

$$C_nH_x + nH_2O \rightarrow nCO + \left(n + \frac{x}{2}\right)H_2 \tag{R1}$$

$$CO + H_2O \leftrightarrow CO_2 + H_2 \tag{R2}$$

Optionally, N_2 (up to 50% in the product gas) can be present if O_2 from air is used as (co)gasification agent without the implementation of Air Separation Unit (ASU) technology; however, N_2 content higher than 3% should be avoided [10] in order to prevent dilution of the Bio-SNG, which results in a reduced size of the facility, and, eventually, high heat loss in the flue gas.

Although a detailed discussion on available gasification technologies is beyond the scope of this paper, it is worth mentioning that both ECN and Güssing plants have developed an indirectly heated (allothermal) biomass gasification process, based on the dual fluidized bed technology (DFB; see also [20]). This configuration is advantageous because the highly endothermic steam gasification reactions are coupled with combustion of char-tar that is generated in the gasifier; the heat produced during combustion is therefore transferred to the gasifier by the particles bed circulating between two interconnected fluidized beds,

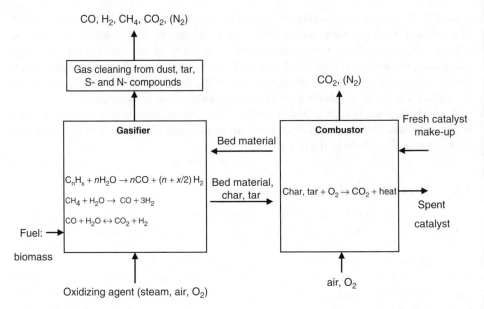

Figure 8.2 *General scheme of a biomass gasification process involving two interconnected fluidized bed*

as schematized in Figure 8.2. An overall efficiency to SNG of 67% (on an LHV basis, including electricity consumption and pretreatment, but excluding transport of biomass) has been reported when indirect gasification technology is used: this is higher than that for entrained flow (54%) and for conventional circulating fluidized bed (58%) technologies [10, 21].

As mentioned before, ECN has already developed both laboratory-scale ($30\,kW_{th}$) and pilot-scale ($800\,kW_{th}$) gasifiers coupled with an OLGA tar removal unit; the pilot plant has ran for more than $500\,h$ of operation with results compatible for scale-up and commercial applications (96% of availability of the gasifier). The design of a $12\,MW_{th}$ plant for waste gasification is under development, while more ambitious projects ($100\,MW_{th}$) are under evaluation [16]. On the other hand, the Güssing plant – with its $8\,MW_{th}$ installation – is a demo scale of the concepts developed in the $100\,kW_{th}$ pilot plant at University of Vienna and is in operation since 2001 (counting more than $9700\,h$ of operation only in the year 2004 [22]); both Austrian facilities support a huge number of EU projects in the field of fluidized bed systems for advanced fuel conversion technologies, among others GRACE, CLC GAS POWER, CACHET, AER GAS II, BiGPower (FP6), Fecundus, UNIQUE, GreenSyngas, Innocuous (FP7)[23].

8.3.2 Gas Cleaning

In order to upgrade a gasification gas stream into bio-SNG, the key step is purification by removing trace levels of pollutants before the catalytic conversion of the gas into CH_4. The main gas impurities of the product gas are tar and inorganic compounds such as H_2S, NH_3 and HCl. Tar is a complex mixture of condensable aromatic hydrocarbons, which can

Table 8.2 *Reported levels of S-, Cl-, N- impurities from Güssing and ECN studies before cleaning process (raw), and cleaning performance at ECN only (cleaned), adapted from [7, 23]*

			ECN		Güssing
Biomass content	N		0.2		0.2
	S		0.02		0.01
	Cl		0.004		–
Product gas		Raw		Cleaned	Raw
H_2S	ppmV	116		nd	150
Other S-compounds	ppmV	5		$\approx 10^{-2}$	nr
HCl	ppmV	0.7		nd	5
NH3	ppmV	1300		0.06	1000–2000
Benzene	g/Nm^3	23.7		0.08	nr
Toluene	g/Nm^3	2.9		0.35	nr
Naphthalene	g/Nm^3	6.5*		nd	1–2
Tar (PAH larger than naphthalene)	g/Nm^3	nr		nr	0.02–0.05

nr: Not reported.
nd: Not detectable (below detection limits).
*Total tar content expressed as naphthalene equivalent to 1140 ppmv of tar.

condense and clog pipelines, gas engines, and turbines. HCl and H_2S derive from biomass. 90% of sulfur is generally transformed in H_2S and the rest in COS or thiophene; chlorine is partially transformed in HCl and the rest is generally found in ashes.

Both Güssing and the ECN facilities have published detailed data on the impurities content of the syngas produced before the cleaning step, as summarized in Table 8.2, with cleaning performance achieved by the ECN facility [7, 11]. Focusing on the Dutch cleaning system technology, the following gas cleaning path has been reported [7, 24]:

- Tar abatement involves the so-called OLGA technology for energy-efficient tar abatement (78% energy efficiency); this system – based on organic scrubbing liquid – also removes dust and contaminants such as thiophenes and dioxins. The OLGA is operating downstream a hot gas filter at the exit of the gasifier; the gas enters at $T > 400°C$ – to avoid condensation of tar and water – and leaves at a temperature of 60–100°C.
- Sulfur compounds (H_2S, thiophene, COS) are further removed (at 300°C), from levels of 100–200 ppm to $\ll 1$ ppm, by using HDS catalysts and solid oxide sorbents (ZnO).
- NH_3 and HCl are removed by water scrubbing.
- An additional catalytic hydrocarbon removal step can be added to avoid soot formation downstream the methanation unit.

The composition/purity of the product gas from gasification is critical for the bio-SNG synthesis: for example, soot formation has been found to be the catalyst deactivation factor at ECN [7], while deactivation was due to sulfur deposition on the catalyst surface for the $10 \, kW_{th}$ pilot-scale tests COSYMA in Güssing [25].

A detailed description of available technologies and equipment for gas cleaning is beyond the scope of this chapter. Interesting reviews in the scientific literature can be considered for additional studies, including catalytic tar conversion [26–31] and H_2S and NH_3 abatement technologies [31].

8.3.3   The Synthesis Step

The methanation reaction has been extensively studied as a gas purification step in the synthesis of NH_3 and as an alternative to CO combustion for H_2 purification in PEM fuel cells [32]. It consists essentially of the catalyzed hydrogenation of carbon monoxide to CH_4:

$$CO + 3H_2 \leftrightarrow CH_4 + H_2O \quad \Delta H_R^0 = -206 \, kJ/mol \tag{R3}$$

Since CO_2 is also present in the product gas, direct CO_2 methanation can be used as CH_4 synthesis step:

$$CO_2 + 4H_2 \leftrightarrow CH_4 + 2H_2O \quad \Delta H_R^0 = -165 \, kJ/mol \tag{R4}$$

where reaction (R4) is a linear combination of (R3) and the reverse of water-gas shift (R2), with $\Delta H_R^0 = -41 \, kJ/mol$.

$$CO + H_2O \leftrightarrow CO_2 + H_2$$

Because a low-temperature range is needed for methanation reaction, carbon formation via Boudouard reaction can also be taken into account:

$$2CO \leftrightarrow C + CO_2 \quad \Delta H_R^0 = -173 \, kJ/mol \tag{R5}$$

Both reactions R3 and R4 are exothermic, suggesting that methanation is thermodynamically favored at low temperature (typically, 300–400°C); moreover, the large decrease in mole number suggests that high CH_4 yield is predicted at high pressure. Figure 8.3 shows the thermodynamic equilibrium calculation for the system of reactions R2–R3 at different temperature and pressure. The software CHEMCAD 6.3.1.4168 has been used for the calculation.

An excellent overview on methanation technology development from 1950 to 2009 is given by Kopicinky *et al.* [33]. Basically two different options are used for methanation processes: fixed bed (Conoco/BGC, Lurgi, TREMP, Linde, HICOM, RMProcess technologies) or fluidized bed reactors (e.g. Bureau of mine, BGR, Comflux). The former process consists of multiple reactors with gas recycling and intermediate cooling steps, that is, 4–6 reactors are used in the RMProcess, 2 in Lurgi, 3 (or more) in TREMP. For this technology, one of the main challenges is temperature control, and high pressure of operation even increases the specific heat production per volume of CH_4 produced. For example, the TREMP technology (within the ADAM and EVA projects) shows an inlet and outlet temperature range of 300°C and 600°C, respectively, for the first adiabatic reactor; this range decreases to 350–450°C inlet–outlet temperature in the second reactor, and becomes 250–350°C for the third one (at 30 bar). The CH_4 content increases from 28.12 to 37.44% in the first vessel and to 47.28 up to 82.95 (containing 12% N_2 and 3% H_2) in the second and third reactors.

Catalyst sintering and deactivation problems can occur due to high-temperature reactions, and heat removal is a focal point of the process; moreover, high temperature decreases the methanation conversion and CH_4 selectivity. To overcome such problems, isothermal

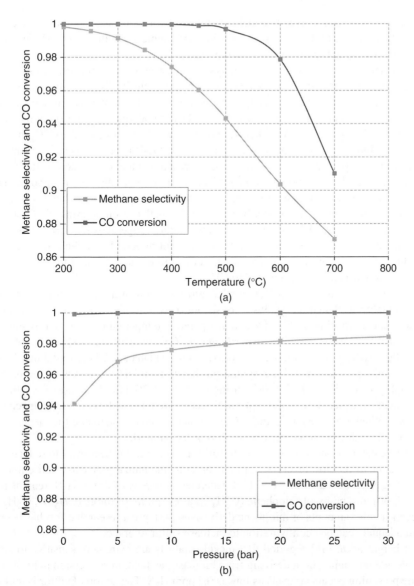

Figure 8.3 *Thermodynamic equilibrium of methanation and water-gas shift reaction system as a function of (a) temperature and (b) pressure*

fluidized bed reactors have been developed: a temperature profile between 350 and 400°C is reported for a multiple-feed-inlet fluidized bed reactor [56].

8.4   Catalytic Systems

The choice of a catalyst is an important step in the process design, and its optimization has to take into account both deactivation by biomass contaminants and high-temperature sintering problems in the fixed bed methanation configuration. As an example, a methanation catalyst lifetime of the order of 4 years has been achieved in the previously mentioned coal-based SNG plant in operation in North Dakota [8].

Nickel is the most commonly used catalyst in industrial process because of its low cost, good catalytic activity, and low capacity of growing chain in Fischer-Tropsch [34]. It is known to be active in catalytic CO_x methanation since the first catalytic tests in 1902 by Sabatier and Sanderens [35]. It has been widely studied on different supports such as ZrO_2 [36] and La_2O_3 [37]. A higher selectivity toward CO compared to CO_2 has been observed by Inui *et al.* by using Ni-La_2O_3 and Ni-La_2O_3-Ru, which has been attributed to the stronger adsorption strength of CO during the reaction, blocking the active sites for CO_2 [38]. Yaccato *et al.* [39] studied the identification of zirconia- and ceria-supported noble metal catalyst through combinatorial and high-throughput approaches. They tested around 500 samples and selected the most advantageous for methanation in terms of CO_x activity, selectivity, and reverse WGS reaction in a temperature range between $T = 300–400°C$ at atmospheric pressure. They concluded that CO_2 is not directly hydrogenated but is firstly shifted to CO (with 1 order of magnitude faster reaction rate compared to the direct CO_2 methanation path) and then converted to CH_4.

Hoekman *et al.* [40] studied CO_2 methanation reaction over Haldor Topsøe commercially available methanation catalysts consisting of Ni and NiO on an alumina substrate with total nickel loading of 20–25% and an operating temperature range of 190–450°C in an extruded ring-shaped catalyst. Approximately 60% conversion of CO_2 was observed at $T = 300–350°C$ and stoichiometric CO_2/H_2 ratio. Aldana *et al.* [41] found that Ni over ceria-zirconia (prepared by sol–gel synthesis) shows an initial CO_2 activity of almost 80%, with a CH_4 selectivity of 97.3%, decreasing down to 84.7% after 90 hours of reaction. By IR operando analysis, they found that for Ni-ceria-zirconia catalysts the main mechanism for CO_2 methanation does not require CO as reaction intermediate and the mechanism is based on CO_2 adsorption on weak basic sites of the support.

Ni-based catalysts are very sensitive to sulfur poisoning: it is necessary to reduce sulfur concentration to very low levels (<1 ppm) in order to maintain a good catalyst lifetime [42]. Since syngas produced from wood fuel generates hundreds of ppmv H_2S, tens of ppmv COS, 10 or more ppmv of thiophene (C_4H_4S), a few ppmv of thiols (CH_3SH, C_2H_5SH), and even smaller amounts of thiophene derivatives [43], great research effort has been put to find adequate catalyst configuration to increase sulfur tolerance.

Hausberger *et al.* [44] reported that nickel catalysts are extremely sensitive to sulfide and chloride poisoning, both deriving from biomass gasification conversion to bio-syngas. Sulfide poisoning occurs at levels as low as 0.3 ppm H_2S. The action of sulfur is undoubtedly through the reaction of hydrogen sulfide with nickel, according to the nickel sulfide formation:

$$Ni + H_2S \leftrightarrow NiS + H_2 \qquad (R6)$$

Depending on the partial pressure ratio P_{H_2}/P_{H_2S}, nickel can be poisoned due to its reaction with sulfur or regenerated (when the ratio is higher than the equilibrium value) [45].

Czekaj *et al.* studied sulfur poisoning on Ni/Al_2O_3 catalyst used for the methanation process. Using density functional theory calculations (StoBe) with cluster model and non-local functional (RPBE) approach, they reported that several poisoning agents such as

COS, H_2S, or hydrogen thiocarbonates are stable on both Ni clusters and alumina support [46].

Ni-based methanation catalysts have been found to be significantly poisoned by H_2S within 2–3 days exposure to a reaction mixture containing 1–10 ppm H_2S. The deactivation extent has been found to increase by increasing H_2 content and reaction temperature. The adopted Ni-Mo and Ni-Co bimetallic catalysts did not show any significant improvement in thioresistance properties [47]. Rabou and Bos [43] focused on gas cleaning for the production of SNG from biomass and studied the selective adsorption of thiophene by hydrodesulphurization of commercial Al_2O_3-supported CoMoO catalysts with a nominal composition of 3–5% CoO and 10% MoO_3, CoO/Al_2O_3 (1–10%), and MoO/Al_2O_3 (10–20%), at different operating conditions. Thiophene is removed for 98–99% at $T = 450$–475°C and GHSV = 40–65 h^{-1} and to 93% when $T = 425$°C and GHSV = 70 h^{-1}. CoMoO sample absorbed H_2S completely for about 50 h and with a sulfur uptake of 1.5–2% on weight basis of the active material (about 0.2% relative to the total weight). Tests performed show that a DHS guard bed allows a total conversion of benzene and ethylene on a pre-reforming bio-SNG unit for almost 50 h, then the breakthrough starts, and total deactivation is achieved after few days (200 h); on the other hand, removal of hydrocarbons from Ni catalyst is not efficient without any HDS pretreatment. The HDS catalysts also promote the water-gas shift and hydrogenation reactions.

As alternative to Ni-based catalysts, Jiang *et al.* [48] studied CeO_2-Al_2O_3 supports prepared with impregnation, deposition precipitation, and solution combustion methods to prepare the MoO_3/CeO_2-Al_2O_3 catalyst to investigate sulfur-resistant properties in methanation reactions. Methanation tests were carried out flowing syngas with a $H_2/CO = 1.0$ ratio containing 1.2 vol% H_2S, 20 vol% N_2, at 550°C and 3 MPa. They found that catalysts prepared by deposition precipitation have the highest conversion (65%) when thermally treated at temperature $T = 600$°C; conversion decreased to 60% after 30 h, and CH_4 and CO_2 selectivity were about 50% and 45%, respectively. They explained this behavior was due to a CeO_2 layer that does not interact with MoO_3 and improve its dispersion. More expensive noble metal methanation catalysts are also reported in literature; their activity was found to decrease in the following order: Ru>Rh>Pt>Ir~Pd~Re [49]. König *et al.* successfully studied an integrated process using Ru/Al_2O_3 (2 wt% Ru) catalyst. The authors show the possibility of removing sulfur compounds directly during the methanation reactions, circumventing the need of a low-temperature scrubbing. Through their experimental tests, combining mass spectrometry and modulated excitation XAS data, they found that a Ru-sulfide phase is formed, deactivating the metal site; however, during oxidation (catalyst regeneration), SO_2, RuO_2, and $Al_2(SO_4)_3$ phases are formed, the metal is re-activated and sulfur is released from the support.

8.5 The Case Study

As presented in this chapter, methanation is exploited in commercial and demonstration applications by using both fixed bed reactors with intermediate cooling (e.g. TREMP technology) or fluidized bed reactors (ECN, Güssing); for both, the temperature transfer and hotspots control are critical factors. Path 1 of Figure 8.1 has been exploited and implemented in the simulations, a CO_2 capture step being conceived downstream the methanation section.

In this case study, a simplified scheme is modeled using CHEMCAD 6.3.1.4168 to convert a syngas produced from a (real) biomass gasification process into bio-SNG. Two different thermal configurations of the methanation section have been studied: an isothermal case, which is used to approach the fluidized bed thermodynamic behavior, and an adiabatic case in which the large inlet and outlet temperature differences are representative of the fixed bed configuration. For the isothermal case, the temperature is fixed at 380°C (in the range of the experimental values published by different fluidized bed methanation reactors, see Section 3.3), while for the adiabatic case three reactors in series with intermediate cooling at an initial temperature of 300°C (for the first two) and 200°C (for the last one) are considered. A schematic representation of the simulated process is given in Figure 8.4; the unit numbers are detailed in Legend 1.

A comparison between two different thermodynamic approaches, the equilibrium reaction (EREA) and the GIBBS reactors, is also proposed. The Gibbs free energy minimization approach consists of calculating the product composition at which the Gibbs free energy is a minimum and is independent on the path (reactions) involved. On the other hand, in the equilibrium calculations a reaction network is assumed. CHEMCAD provides thermodynamic data for both approaches. The process is schematized in Figure 8.4 for the adiabatic case; the different process units and flows are described with more details in the following sections.

8.5.1   The Feeding Composition

As previously discussed in this chapter (see Table 8.1), the syngas composition coming from a biomass gasification process can be different depending on process conditions (gasification temperature, steam/oxygen/biomass ratios, biomass feedstock, etc.) and can have an effect on CH_4 yield and conversion efficiency. Three different compositions (Table 8.3) have been used in this study; all of them are obtained from the data reported in the frame of the EU project UNIQUE (www.uniqueproject.eu) [50]. A theoretical syngas flow rate for $1MW_{th}$ installation (referred to the biomass thermal power) is calculated for different feed compositions in these simulations.

The first biomass product gas derives from a bench-scale fluidized bed facility consisting of an electrically heated single reactor vessel (see Ref. [51] for more details), where steam was used as gasification agent and almond shells were fed as fuel (8–10 g/min fed). A nickel-based catalytic ceramic filter candle for hot cleaning was housed in the freeboard of the gasification reactor to perform continuous gasification tests integrated by tar catalytic reforming and particulate abatement; olivine and Fe-olivine were used as catalytic bed material, and a biomass/steam ratio equal to 1 was adopted for the tests.

The second composition derives from a bench-scale facility at ENEA Trisaia (Italy) and makes use of a $10\,kw_{th}$ internal circulating bubbling fluidized bed (ICBFB) gasifier. The same bed material and biomass particles were used in these tests; however, oxygen required for autothermal conditions was added as gasification agent, the adopted biomass/steam weight ratio was 2/1, and no filter candle was used for these tests.

The third composition was obtained from a theoretical scale-up of the above-mentioned test facility ($1\,MW_{th}$ prototype).

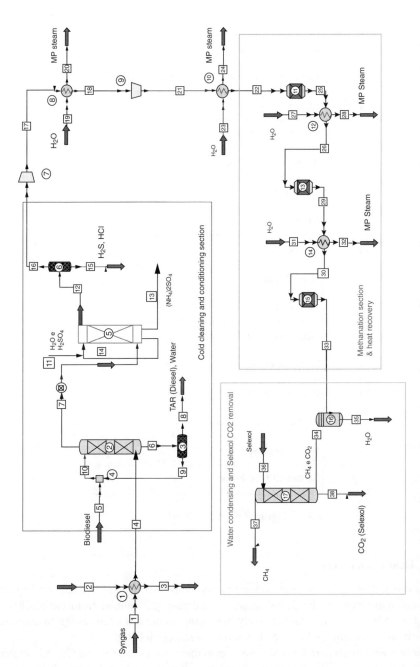

Figure 8.4 *Process simulation of SNG production from a biomass gasification product gas. Legend 1: Process units identification*

Number	Corresponding unit
1	Heat exchanger
2	Biodiesel tar scrubber
3	Component separator water-tar-biodiesel/biodiesel
4	Stream mixer
5	Ammonia scrubber
6	AC separation column and guard bed
7	Compressor
8	Heat exchanger
9	Compressor
10	Heat exchanger
11	Methanation reactor I
12	Heat exchanger
13	Methanation reactor II
14	Heat exchanger
15	Methanation reactor III
16	Water condenser
17	Selexol CO_2 adsorber

Figure 8.4   *(continued)*

8.5.2   Heat Exchangers

In the process, there are three heat exchangers for the isothermal and five heat exchangers for the adiabatic case; in the former case, the entering gas coming from the gasifier is cooled from 850 to 300°C. This relatively high temperature has been chosen to avoid tar condensation, according to the ECN data reported earlier (400°C).

Other heat exchangers are located downstream the compressors (7) and (9) and as intermediate cooling stages between two fixed-bed methanation reactors (for the adiabatic case only). Their exit stream temperature (300°C) and pressure (10 bar) are fixed, and the minimum water flow rate (at $T = 25$°C) required to cool down the gaseous stream is used.

Table 8.3 *Biomass syngas compositions used for the reported simulations, streams I, II, and III*

Composition	I	II	III
Molar flow (kmol/h)	15.3	10.9	13.1
Gas yield, Nm^3 dry/kg daf	1.75	1.25	1.5
Tar content, g/Nm^3 dry	0.5	6	0.5
H_2 (vol% dry gas, N_2 free)	56.5	31.7	33.5
CO_2 (vol% dry gas, N_2 free)	22.1	35.6	33.5
CO (vol% dry gas, N_2 free)	17.4	25.2	25
CH_4 (vol% dry gas, N_2 free)	4.1	7.6	8
HCl (ppm)	50	50	n.d.a.
H_2S (ppm)	50	50	
NH_3 (ppm)	500	500	500

Table 8.4 *Gaseous inlet/outlet temperature difference (ΔT) and water flow rate required for stream II*

Heat exchange unit	Minimum water flow rate (kmol/h)	ΔT of gases (°C)
1	4.8	550
8	1.1	156
10	1.1	255
12	3.3	369
14	3.0	355

The inlet/outlet temperature difference of the compressed gases and the minimum amount of water flow rate required for stream II are reported in Table 8.4.

8.5.3 Scrubber Tar Removal

A biodiesel scrubber for tar removal is placed downstream the heat exchanger. Naphthalene has been chosen as tar model compound, which is removed in the "cold" ($T = 300°C$) cleaning section. The choice of naphthalene as the most representative aromatic hydrocarbon of the complex tar mixture is supported by many studies in the field (see for instance [51]). On the other hand, the model compound 1-1' bicyclohexyl has been used to simulate the biodiesel mixture, being representative of physical and chemical properties of the commercially produced biodiesel [52, 53].

The CHEMCAD SCDS column is used to model the scrubber with a multistage vapor–liquid equilibrium approach. The number of stages of this unit has been fixed to 4 in order to achieve a cleaning efficiency of 99.15% (moles naphthalene eliminated/moles of naphthalene entering the scrubber). CHEMCAD provides thermodynamic data for vapor–liquid equilibrium of the binary system of biodiesel–naphthalene. The gas enters the scrubber at 148°C and exits at 95°C; for liquids, a recycle is carried out where water and hydrocarbon are separated by the CHEMCAD component separator CSEP (3): the stream is split into a recirculation (at $T = 25°C$) and a waste-stripped tar and H_2O stream (at 50°C) with a molar ratio of 90:10, respectively. The total amount of biodiesel used corresponds to about 1 kmol/h per 0.08 kmol/h naphthalene removed.

8.5.4   Ammonia Absorber

Ammonia is removed by a 1M H_2SO_4 water solution scrubber: the liquid solution entering from the top of the tower (a SCDS column settled as packed column mass transfer simulation model) is continuously fed by a make-up quantity corresponding to the amount needed for the ammonia removal. At the bottom of the column gaseous ammonia enters at $T = 95°C$, it dissolves into the acid solution, diffuses and rapidly reacts with the H^+ ions via ammonia protonation following thermodynamics of electrolyte non-random two liquid (Electrolyte NRTL) approach.

To run the SCDS column simulation, the number of stages and the packing height are required by ChemCAD software, hence the packed column height and the number of transfer units (NTU) have to be manually calculated. The packing height and the NTU are evaluated using the assumptions and parameters detailed below as reported by Westerterp et al. [54]:

- the gas phase is considered ideal;
- mass transfer coefficients of ammonia in gas and liquid phases are equal to $k_G = 15.7 \times 10^{-3}$ m/s and $k_L = 0.37 \times 10^{-3}$ m/s [55].
- Ammonia solubility in aqueous solution is very high ($m \approx 10^3$ 1/l) and the mass transfer resistance for ammonia can be considered only in the gas phase ($k_L/(m \cdot k_G) << 1$) [57].

To evaluate the packaging height, the mass balance along the infinitesimal column height needs to be written, allowing to study the differential variation of ammonia concentration; neglecting longitudinal dispersion it can be written as:

$$dC_{NH_3} \cdot Q_{syngas} = -dz \cdot S \cdot a \cdot J_{NH_3} \tag{8.1}$$

where C_{NH_3} is ammonia concentration in the feed stream [vol%], Q_{syngas} is the syngas flow rate [m³/s], z is the height [m], S is the cross sectional area of the column [m²], a the effective specific surface for mass transfer estimated at 50% of specific surface of Raschig rings (220) [m²/m³] [55] and J_{NH_3} is the ammonia flux [mol/(m²·s)].

Assuming the resistance in mass transfer is only in the gas phase, we can assume:

$$J_{NH_3} = k_G \cdot C_{NH_3,eq} \tag{8.2}$$

The substitution of Eq. (8.2) in Eq. (8.1), and the evaluation of the latter at the exit of the column allows to find the length of the column L.

$$L = \frac{Q_{syngas} * \ln(C_{NH_3,in}/C_{NH_3,eq})}{k_g \cdot a \cdot S} = 2.4 \tag{8.3}$$

and the number of transfer unit, used in the simulation as number of stages:

$$NTU = -\ln \frac{C_{NH_3}}{C_{NH_3,eq}} = 13.5 \tag{8.4}$$

The obtained values (*L* and NTU) are inserted into CHEMCAD SCDS column window to run the simulation and can be adjusted if the software requires (i.e. recirculation does not converge).

CHEMCAD component separator CSEP is used to separate ammonium sulfate from the unreacted 1M H_2SO_4 water solution that is recycled at the top of the column, where fresh acid water solution is also fed (0.0025 kmol/h of H_2SO_4 and 0.180 kmol/h of H_2O). A recirculation of 1M H_2SO_4 water solution is carried out to control the temperature rise inside the packed column (see also [55]).

A total water flow rate of 10 kmol/h and 0.18 kmol/h of H_2SO_4 enter the column at $T = 20°C$, clean syngas leaves the column at $T = 40°C$, and the bottom temperature is $T = 70°C$, the recycle solution is cooled down to 20°C and mixed with the fresh liquid feed. Ammonia is almost completely removed, and only 1.63×10^{-8} kmol/h leave the system in the cleaned syngas.

8.5.5 HCl and H_2S Removal

A simple component separator *CSEP* is used for the complete removal of HCl and H_2S. The exiting top stream temperature (16) is set to $T = 150°C$ and the bottom temperature to $T = 50°C$.

8.5.6 Compression Section

The syngas exiting the cold cleaning section is compressed up to 30 bar to increase the CH_4 selectivity (see Figure 8.3). For this purpose, a multistage compression consisting of two compressors (7–9) with intermediate cooling (8–10) is used in the simulation, with an efficiency of 0.75 and a polytropic model type with ideal cp/cv = 1.337 (for the first compressor) and 1.303 (for the second one). The temperature and pressure values of the gas exiting the compressors are shown in Table 8.5.

8.5.7 Separation Section: H_2O and CO_2 Removal

The outlet methanation stream contains a high percentage of water and carbon dioxide that need to be removed in order to obtain a concentrated CH_4 stream. In the CHEM-CAD simulation the stream leaving the methanation section enters into a multipurpose flash unit operating at constant pressure ($P = 30$ bar) and ambient temperature ($T = 25°C$), where water removal takes place by condensation.

On the other hand, CO_2 is removed by an SCDS column, where the number of stages has been set to 20 and the feeding solvent flow rate to 8 kmol/h at $T = 10°C$ and $P = 30$ bar. Selexol process for dynamic absorption/desorption (already implemented in the CHEM-CAD library) has been selected to perform this step.

Table 8.5 Inlet/outlet temperature and pressure data for the pre-methanation compression section, composition II

	Tin	Tout	Pout [bar]
Compressor I (7)	50	367	10
Compressor II (9)	200	410	30

Methane exits from the system in a compressed form and is available for different uses and further conversions. Obtained upgraded concentrations are discussed in the following section, where the methanation unit is presented.

8.5.8   Methanation Section Case 1: Adiabatic Fixed Bed with Intermediate Cooling

A methanation section consisting of three adiabatic reactors is used to simulate the fixed bed with the intermediate cooling configuration (reactors 11, 13, 15, see Figures 8.4 and 8.5).

Clean, dry, and compressed syngas enters the first methanation reactor (11) at $P = 30$ atm and $T = 300°C$; the stream exiting the methanation reactors (11 and 13) is cooled down (heat exchangers 12–14) in order to lower the temperature of the gas entering the subsequent reactor. The inlet and outlet temperatures and stream compositions for reactors 11, 13, and 15 are reported in Table 8.6, for stream II and EREA reactor thermodynamics.

A comparison between the EREA and the GIBBS approaches in terms of CH_4 mass flow rate entering the reactor (11) and exiting from the reactor (15), the obtained $\%CH_4$ vol. before and after the upgrading step is shown in Table 8.7 for the three composition streams adopted.

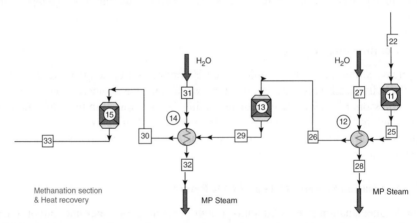

Figure 8.5 *Process simulation of SNG production from a biomass gasification product gas: zoom on the adiabatic fixed bed reactors with the intermediate cooling section*

Table 8.6 *Inlet/outlet gas composition and temperature for the simulated adiabatic configuration of the methanation section, stream II, reactors 11, 13, and 15*

	R11		R13	R15
	Inlet	Outlet	Outlet	Outlet
T	300	694.5	597	431
CH_4 (kmol/h)	0.82	1.61	2.03	2.32
H_2 (kmol/h)	3.44	0.86	0.36	0.06
CO (kmol/h)	2.72	2.14	0.96	0.09
CO_2 (kmol/h)	3.86	3.65	4.41	4.99
H_2O (kmol/h)	0	0.99	0.65	0.37
H_2/CO (kmol/h)	1.26	0.403	0.375	0.65

Table 8.7 *Inlet (unit 11) and outlet (unit 15) CH$_4$ mass flow rate, and CH$_4$ vol% before and after CO$_2$ and steam separation, for the simulated adiabatic configuration, stream II. T$_{(15)inlet}$ = 200°C*

Stream	CH$_4$ entering/exiting the adiabatic EREA methanation section (kg/h)		CH$_4$ entering/exiting adiabatic GIBS methanation section (kg/h)		CH$_4$ %		Upgraded CH$_4$ vol%	
	IN	OUT	IN	OUT	EREA	GIBBS	EREA	GIBBS
Composition I	10.05	53.02	10.05	53.08	33.2	33.3	85.0	85.35
Composition II	13.20	37.34	13.20	37.65	29.6	29.7	93.7	94.3
Composition III	18.30	48.03	18.30	48.10	31.8	31.9	94.5	95.0

For each stream composition, both EREA and GIBBS reactor simulations give similar results but slightly higher values have been found for the latter, indicating that the combination of methanation and water-gas shift only may underestimate CH$_4$ production rate.

The effect of steam addition on H$_2$/CO ratio and on CH$_4$ production is reported in Table 8.8 for stream II, using the EREA method. Steam has been ideally added before the first methanation reactor; a comparison of the outlet H$_2$/CO ratio for reactors 11, 13, and 15 with varying injected streams (from 0 to 1.5 H$_2$O/CO ratio) is reported in Table 8.9. As an effect of water–gas shift equilibrium enhancement, the outlet H$_2$/CO ratio at the exit of the first methanation reactor (11) is more than two (H$_2$O/CO = 1) or three times (H$_2$O/CO = 1.5) the base case where no steam is fed to the unit (11). The same effect is

Table 8.8 *Effect of steam addition on H$_2$/CO ratio and obtained final (upgraded) gas composition at the inlet/outlet methanation reactors, stream II*

		R11 inlet	R11 outlet	R13 outlet	R15 outlet
H$_2$O/CO = 0	H2/CO	1.3	0.4	0.4	0.5
	vol% CH$_4$	7.55	17.34	24.04	29.52
	vol% CO	25.03	23.04	11.37	1.15
	vol% H$_2$	31.65	9.26	4.26	0.76
	vol% CO$_2$	35.52	39.30	52.22	63.50
	vol% H$_2$O	0.00	10.66	7.70	4.71
H$_2$O/CO = 1	H$_2$/CO	1.26	1.08	1.80	12.9
	vol% CH$_4$	6.03	13.50	18.93	22.20
	vol% CO	20.01	11.92	3.35	0.06
	vol% H$_2$	25.31	12.83	6.16	0.66
	vol% H$_2$O	20.01	25.17	26.54	28.94
	vol% CO$_2$	28.40	36.33	44.66	48.01
H$_2$O/CO = 1.5	H$_2$/CO	1.26	1.49	3.0	10.03
	vol% CH$_4$	5.48	12.22	17.17	19.30
	vol% CO	18.18	8.77	1.86	0.08
	vol% H$_2$	22.99	13.12	5.75	1.50
	vol% H$_2$O	27.34	31.11	34.09	36.42
	vol% CO$_2$	25.80	34.56	40.89	42.44

Table 8.9 Inlet/outlet gas composition for the isothermal case, stream II

	R11	
	Inlet	Outlet
T	380	380
vol% CH_4	7.6	30.1
vol% H_2	31.7	0.4
vol% CO	25.2	0.4
vol% CO_2	35.5	64.5
vol% H_2O	0	4.6
H_2/CO	1.26	1.0

also evident at the exit of the second methanation unit (13), with an increase in the H_2/CO ratio from 0.38 to 1.8 (feeding $H_2O/CO = 1$) and 3 (feeding $H_2O/CO = 1.5$), and of the third one (15) where the final H_2/CO ratio becomes 20 times higher when the H_2O/CO feeding ratio increases from 0 to 1.5. The steam addition does sensibly affect the final CH_4 percentage in the exiting gas flow: an optimum value is found for the H_2O/CO ratio of 1 (96.8% CH_4, 2.9% H_2 on dry and CO_2 free basis). An excess of H_2 (7.2% on dry and CO_2 free basis) is found in the third case (H_2O/CO ratio $= 1.5$), while the amount of CO exiting has always been lower than 1% when steam is added.

8.5.9 Methanation Section Case 2: Isothermal Fluidized Bed

The inlet/outlet composition of the isothermal reactor configuration is shown in Table 8.9 for stream II. In this case, the adopted relatively low temperature of the inlet/outlet streams (380°C) favors the equilibrium of both methanation and water–gas shift in one single reactor and the addition of steam has not been considered.

A comparison between the EREA and GIBBS approaches in terms of %CH_4 vol entering the single isothermal reactor is shown in Table 8.10 for the three composition streams adopted. Similar trends – but slightly higher CH_4 yield and content – are found for the isothermal technology compared with the adiabatic case. The obtained composition

Table 8.10 Inlet and outlet CH_4 mass flow rate, and CH_4 vol% before and after CO_2 and steam separation, for the simulated isothermal configuration, stream II

Stream	CH_4 entering/exiting isothermal EREA methanation reactor (kg/h)		CH_4 entering/exiting isothermal GIBBS methanation reactor (kg/h)		CH_4 (%)		Upgraded CH_4 (vol%)	
	IN	OUT	IN	OUT	EREA	GIBS	EREA	GIBS
Composition I	10.05	54.8	10.05	54.8	35.1	35.2	96.1	96.1
Composition II	13.2	37.7	13.2	37.7	30.1	30.1	97.3	97.4
Composition III	18.3	48.4	18.3	48.4	32.2	32.3	97.5	97.6

Table 8.11 CO and H_2 vol% in the final upgraded methane EREA reactor simulation T = 380°C; P = 30 bar, fluidized bed isothermal configuration

Composition	Final vol% CO	Final vol% H_2	Final vol% CH_4
Composition I	0.27	3.6	96.1
Composition II	1.30	1.35	97.3
Composition III	1.03	1.5	97.5

Table 8.12 Chemical efficiency of SNG produced from different bio-syngas compositions

Gasification stream	Chemical efficiency (Isothermal configuration)	Chemical efficiency (Adiabatic configuration)
Composition I	0.84	0.82
Composition II	0.58	0.58
Composition III	0.74	0.74

(Table 8.11) indicates H_2 content within 1.5–3.5%, while CO content is lower than 1.3%, in the range of the simulation results from ECN and Güssing models (Table 8.1).

8.6   Chemical Efficiency

The chemical efficiency of the methanation process has been calculated as the ratio between the thermal power derived from CH_4 and the installed one (1 MW_{th}), as shown in Table 8.12. Comparable values are found for the adiabatic and isothermal configurations under the adopted simulation conditions; it has to be pointed out that the syngas composition quality can produce large differences in efficiency performance at same simulation conditions. Stream I – coming from a steam gasification process – has been found to generate a process chemical efficiency in the range of 0.82–0.84. This high chemical efficiency value, however, does not take into account the need of an external heat source to thermally sustain the endothermic steam gasification process.

8.7   Conclusions

In this chapter, the biomass-to-SNG process has been discussed in detail. Reference data from two main facilities working in Europe in the field, at ECN (The Netherlands) and Güssing (Austria) have been collected and discussed in the first part of the chapter. A literature survey on catalysts and deactivation phenomena has been proposed. A case study has been presented where the effect of different key process parameters (mainly adiabatic/isothermal conditions, syngas composition, and steam addition) has been evaluated. A high ($\geq$96%) CH_4 concentration and chemical efficiency (0.6–0.84) is obtainable using both adiabatic and isothermal technological solutions, under the adopted simulation conditions.

References

1. E. Vakkilainen, K. Kuparinen, H. Jussi (2013) Large Industrial Users of Energy Biomass, IEA Bioenergy Report.
2. Van Der Drift, A., Vreugdenhil, B.J. (2010) Comparing the options to produce SNG from biomass pp. 3–7.
3. Bio-SNG (Synthetic Natural Gas) and Gasification Technologies, (n.d.).
4. Hofbauer, H. (2009) Gas production for polygeneration strategies, in: 1st International Conference on Polygeneration Strategies, Vienna.
5. Renewable energy sources and climate change mitigation: special report of the Intergovernmental Panel on Climate Change (2012).
6. Mozaffarian, M., Zwart, R.W.R., Boerrigter, H., Deurwaarder, E.P. (2004) Biomass and Waste-related SNG Production Technologies Technical, economic and ecological feasibility.
7. Deurwaarder, E.P., Production of Synthetic Natural Gas (SNG) from Biomass Development and operation of an integrated bio-SNG system, (n.d.).
8. http://www.topsoe.com/Media/News/2013/281013.aspx
9. http://www.audi.com/com/brand/en/company/corporate_responsibility/product/audi_e-gas_new_fuel.html, (n.d.).
10. Van Der Meijden, C.M. (2010) Development of the MILENA Gasification Technology for the Production of Bio-SNG the MILENA, Ph.D. Thesis. Eindhoven University of Technology Library, ISBN: 978-90-386-2363-4.
11. Rehling, B., Hofbauer, H., Rauch, R. and Aichernig, C. (2011) BioSNG—process simulation and comparison with first results from a 1-MW demonstration plant. *Biomass Conversion and Biorefinery*, **1**, 111–119.
12. http://gobigas.goteborgenergi.se/En/Start, (n.d.).
13. http://www.repotec.at/index.php/gaya-rd.html, (n.d.).
14. https://www.ecn.nl/fileadmin/ecn/units/bio/Biomassa/Syngas_and_SNG/Gasification_2010/GAYA_project_A_unique_demonstration_platform_in_Europe.pdf
15. From solid fuels to substitute natural gas (SNG) using TREMP^{TM}, http://www.netl.doe.gov/technologies/coalpower/gasification/gasifipedia/pdfs/tremp-2009.pdf%20.
16. Van Der Meijden, C.M., Konemann, J.W., Sierhuis, W., Van Der Drift, A., Rietveld, B. (2013) Wood to Bio-Methane demonstration project in the Netherlands, in: ECN Report Number: ECN-M–13-009.
17. Van Der Drift, A., Van Der Meijden, C.M., Boerrigter, H. (2005) MILENA gasification technology for high efficient SNG production from biomass, 17–21.
18. C.C.P. Partnership, D. University (2009) Synthetic natural gas (SNG): technology, environmental implications, and economics.
19. Minchener, A.J. (2005) Coal gasification for advanced power generation. *Fuel*, **84**, 2222–2235.
20. Corella, J., Toledo, M. and Molina, G. (2007) A review on dual fluidized-bed biomass gasifiers. *Industrial and Engineering Chemistry Research*, **46**, 6831–6839.
21. Van Der Meijden, C.M., Veringa, H. and Rabu, L. (2010) The production of synthetic natural gas (SNG): a comparison of three wood gasification systems for energy balance and overall efficiency. *Biomass and Bioenergy*, **34**, 302–311.

22. Aichernig, C., Tremmel, H., Voigtlaender, K., Koch, R., Lehner, R., Guessing, B.C.H.P., Steam gasification of biomass at chp plant guessing – status of the demonstration plant.

23. http://www.vt.tuwien.ac.at/chemical_process_engineering_and_energy_technology/

24. Zwart, R.W.R., Van Der Drift, A., Bos, A. *et al.* (2009) Oil-based gas washing - flexible tar removal for high efficient production of clean heat and power as well as sustainable fuels and chemicals. *Progress and Sustainable Energy*, **28**, 324–335.

25. Seiffert, M., Rönsch, S., Schmersahl, R., Majer, S., Pätz, C., Kaltschmitt, M., et al. (2009) D 1. 4 Final Project Report.

26. Abu El-Rub, Z., Bramer, E.A. and Brem, G. (2004) Review of catalysts for tar elimination in biomass gasification processes. *Industrial and Engineering Chemistry Research*, **34**, 6911–6919.

27. Dayton, D. (2002) A review of the literature on catalytic biomass tar destruction, NREL/TP–510–32815.

28. Sutton, D., Kelleher, B. and Ross, J.R.H. (2001) Review of literature on catalysts for biomass gasification. *Fuel Processing Technology*, **72**, 155–173.

29. Devi, L., Ptasinski, K.J. and Janssen, F.J.J.G. (2003) A review of the primary measures for tar elimination in biomass gasification processes. *Biomass and Bioenergy*, **24**, 125–140.

30. Li, C. and Suzuki, K. (2009) No TitlTar property, analysis, reforming mechanism and model for biomass gasification–an overview. *Renewable and Sustainable Energy Reviews*, **13**, 594–604.

31. Torres, W., Pansare, S.S. and GoodwinJr, J.G. (2007) Hot gas removal of tars, ammonia, and hydrogen sulfide from biomass gasification gas. *Catalysis Reviews*, **49**, 407–456.

32. Zyryanova, M., Snytnikov, P., Gulyaev, R. *et al.* (2013) Performance of Ni/CeO_2 catalysts for selective CO methanation in hydrogen-rich gas. *Chemical Engineering Journal*, **238**, 189–197.

33. Kopyscinski, J., Schildhauer, T.J. and Biollaz, S.M.A. (2010) Production of synthetic natural gas (SNG) from coal and dry biomass – a technology review from 1950 to 2009. *Fuel*, **89**, 1763–1783.

34. Ernst, B., Hilaire, L. and Kiennemann, A. (1999) Effects of highly dispersed ceria addition on reducibility, activity and hydrocarbon chain growth of a Co/SiO_2 Fischer–Tropsch. *Catalysis Today*, **50**, 413–427.

35. Sabatier, P. and Senderens, J.B. (1902) No Title. *Comptes Rendus De l'Académie Des Sciences Paris*, **134**, 514.

36. Takenaka, S., Shimizu, T. and Otsuka, K. (2004) Complete removal of carbon monoxide in hydrogen-rich gas stream through methanation over supported metal catalysts. *International Journal of Hydrogen Energy*, **29**, 1065–1073.

37. Choudhury, M.B.I., Ahmed, S., Shalabi, M. and Inui, T. (2006) Preferential methanation of CO in a syngas involving CO_2 at lower temperature range. *Applied Catalysis A: General*, **314**, 47–53.

38. Inui, T., Funabiki, M. and Takegami, Y. (1980) Simultaneous methanation of CO and CO_2 on Supported Ni-based composite catalysts. *Industrial & Engineering Chemistry Product Research and Development*, **19**, 385–388.

39. Yaccato, K., Carhart, R., Hagemeyer, A. *et al.* (2005) Competitive CO and CO$_2$ methanation over supported noble metal catalysts in high throughput scanning mass spectrometer. *Applied Catalysis A: General*, **296**, 30–48.

40. Hoekman, S., Broch, A., Robbins, C. and Purcell, R. (2010) CO$_2$ recycling by reaction with renewably-generated hydrogen. *International Journal of Greenhouse Gas Control*, **4**, 44–50.

41. Aldana, U., Ocampo, F., Kobl, K. *et al.* (2013) Catalytic CO$_2$ valorization into CH$_4$ on Ni-based ceria-zirconia. Reaction mechanism by operando IR spectroscopy. *Catalysis Today*, **215**, 201–207.

42. Speight, L.S.J. and Loyalka, S. (2007) *Handbook of Alternative Fuel Technologies*, CRC Press Editor.

43. Rabou, P. and Bos, L. (2012) High efficiency production of substitute natural gas from biomass. *Applied Catalysis B: Environmental*, **111–112**, 456–460.

44. Hausberger, A.L., Atwood, K. and Knight, C.B. (1975) No Title. *Advances in Chemistry Series*, **146**, 47.

45. Franko, B.R. and Gruber, G. (1975) Survey of methanation chemistry and processes, in *Methanation of Synthesis Gas*, A.C. Society (Ed.), Washington DC.

46. Czekaj, I., Struis, R., Wambach, J. and Biollaz, S. (2011) Sulphur poisoning of Ni catalysts used in the SNG production from biomass: computational studies. *Catalysis Today*, **176**, 429–432.

47. Bartholomew, C.H. and Jarvi, G.A. (1979) Sulfur poisoning of nickel methanation catalysts: I. in situ deactivation by H$_2$S of nickel and nickel bimetallics. *Journal of Catalysis*, **60**, 257–269.

48. Jiang, M., Wang, B., Yao, Y. *et al.* (2013) A comparative study of CeO$_2$-Al$_2$O$_3$ support prepared with different methods and its application on MoO$_3$/CeO$_2$-Al$_2$O$_3$ catalyst for sulfur-resistant methanation. *Applied Surface Science*, **285**, 267–277.

49. Solymosi, F. and Erdöhelyi, A. (1980) Hydrogenation of CO$_2$ to CH$_4$ over alumina-supported noble metals. *Journal of Molecular Catalysis*, **8**, 471–474.

50. Bench scale tests of gasification and hot gas cleaning and conditioning in one reactor vessel, UNIQUE Project G. A. No. 211517, n.d.

51. Rapagnà, S., Gallucci, K., Di Marcello, M. *et al.* (2012) First Al$_2$O$_3$ based catalytic filter candles operating in the fluidized bed gasifier freeboard. *Fuel*, **97**, 718–724.

52. Phuphuakrat, T., Namioka, T. and Yoshikawa, K. (2011) Absorptive removal of biomass tar using water and oily material. *Bioresource Technology*, **102**, 543–549.

53. Blasi, A., Fiorenza, G., Viola, E., Braccio, G., Biomass Pilot plant for hydrogen production: design of a novel biodiesel scrubber.

54. Westerterp, A., van Swaaij, R. and Beenackers, W.P.M. (1984ISBN 0 471 90183 0, n.d) *Chemical Reactor Design and Operation*, Wiley.

55. Heine, S. (2010) *Process Integration Opportunities for Synthetic Natural Gas (SNG) Production and Thermal Gasification of Biomass*, Chalmers University of Technology.

56. Schlesinger, M.D., Demeter, J.J. and Greyson, M. (1956) Catalyst for producing methane from hydrogen and carbon monoxide. *Industrial and Engineering Chemistry*, **48**, 68.

57. Byron Bird, R., Stewart, W.E. and Edwin, N. (2008) Lightfoot, transport phenomena, in *Definition of Transfer Coefficient in Two Phases*, Wiley student edition, 2nd edition, pp. 688–689.

9

Blue Energy: Salinity Gradient for Energy Conversion

Paolo Chiesa, Marco Astolfi and Antonio Giuffrida
Department of Energy, Politecnico di Milano, via Lambruschini 4, 20156,
Milano, Italy

9.1 Introduction

Since long time, humankind have been exploiting the gravitational potential energy associated with water drops to drive machines like grain mills, looms and, more recently, to produce electric power. Similar to a difference in elevation, potential energy is also associated with a difference in salt concentration so that electric power can eventually be produced by exploiting the salinity gradient between freshwater of rivers and seawater. The potential energy implied in this salinity difference is of the same order of a hundred meter high water drop. Since the hydropower accounted for about 13% of the world electricity production in 2011, it derives that potential contribution of this source could be huge.

Unfortunately, exploiting the salinity gradient for power production is not so easy as waterfalls. The present technology advancement is still far from any commercial application. Demonstration of the most studied and experimented technologies is just at a laboratory scale, prototypes of few kilowatt capacity. Nevertheless, the technology has extremely attractive characteristics related to a wide geographic distribution of the source and the possibility of continuous operation in opposition to random availability of sun and wind power.

This chapter presents the different technologies for power generation from salinity gradient, which have been so far proposed in the technical literature pointing out the theoretical operating principles, the possible plant configurations and identifying the development gap to bridge so as to achieve technical and economical maturity.

Process Intensification for Sustainable Energy Conversion, First Edition.
Edited by Fausto Gallucci and Martin van Sint Annaland.
© 2015 John Wiley & Sons, Ltd. Published 2015 by John Wiley & Sons, Ltd.

9.2   Fundamentals of Salinity Gradient Exploitation

The Gibbs free energy gradient is the driving force for chemical species migration in a system with concentration gradient. The Gibbs free energy is also called chemical potential and its value is minimized when the system reaches an equilibrium state. In particular, if the difference of Gibbs energy between the final and the initial state (called ΔG_{mix}) is negative, the process is spontaneous [1] and ΔG_{mix} represents the work that can be theoretically obtained by a sequence of reversible processes between the two states [2]. A system with two volumes of aqueous solution at different salinity is an example of this principle. The chemical potential of two separate solutions is greater than the potential of the solution resulting from their mixing [3] and the resulting ΔG_{mix} can be calculated as

$$\Delta G_{mix} = G_b - (G_c + G_d) \tag{9.1}$$

where G (J/mol) is the Gibbs free energy and subscripts c, d, and b are for concentrated, diluted and brackish solutions, respectively.

The Gibbs free energy can be expressed by Eq. 9.2 in terms of the activity of each chemical species:

$$G = \sum_i \mu_i n_i \tag{9.2}$$

where μ_i and n_i are the chemical potential (J/mol) and the number of moles (mol) in solution for the i-component, respectively. According to [4], the chemical potential can be expressed as the sum of different contributions as reported in Eq. 9.3.

$$\mu_\downarrow i = \mu_\downarrow i^\uparrow 0 + v_\downarrow i \, \Delta p + RT \, \ln(\gamma_\downarrow i \, x_\downarrow i) + |z_\downarrow i| F \Delta \varphi \tag{9.3}$$

where:

μ_i^0 (J/mol) is the molar free energy of the pure species at reference temperature and pressure

Δp (Pa) is the pressure difference between system and reference state

v_i (m³/mol) is the partial molar volume of i-component

R is the gas constant (8.314 J/mol K)

T (K) is the absolute temperature

x_i is the molar fraction of i-component in solution

γ_i is the activity coefficient of the i-component in solution

z_i (equivalent/mol) is the ion valence

F is the Faraday constant (96485 C/mol)

$\Delta \varphi$ (V) is the electrical potential difference.

Considering ideal solutions[1] and assuming an ideal mixing process at the same pressure of the reference state with no transfer of electrical charges, Eq. 9.3 reduces to Eq. 9.4

$$\mu_i = \mu_i^0 + RT \, \ln(x_i) \tag{9.4}$$

Substituting Eq. 9.4 into Eq. 9.2 and simplifying μ_i^0 terms, equation (Eq. 9.5) for the calculation of Gibbs free energy released during mixing is obtained:

$$\Delta G_{mix} = \sum_i \{(n_{i,c} + n_{i,d})RT \, \ln(x_{i,b}) - [n_{i,c}RT \, \ln(x_{i,c}) + n_{i,d}RT \, \ln(x_{i,d})]\} \tag{9.5}$$

---

[1] In ideal solutions, activity coefficients $\gamma_i = 1$.

The final form of ΔG_{mix} is obtained by substitution of $n_i = c_i V$ and $V_b = V_c + V_d$:

$$\Delta G_{\text{mix}} = \sum_i \{c_{i,b} V_b RT \ \ln(x_{i,b}) - [c_{i,c} V_c RT \ \ln(x_{i,c}) + c_{i,d} V_d RT \ \ln(x_{i,d})]\} \tag{9.6}$$

where c_i is the molar concentration of the i-component in solution (mol/l).

If the concentrations of initial and final solutions are relatively low, the molar fraction of water approaches unity and the contribution of free energy of mixing can be neglected. At low concentrations, furthermore the volumetric and mole contribution of salt is small compared to water so the mole fraction of salt can be expressed as $x_s \approx c_s V_w$, where V_w is the molar volume of pure water. Finally, considering that NaCl is a strong electrolyte salt, the multiple ionic species contribution can be taken into account with a coefficient $v = 2$. Using the approximate formulation for ΔG_{mix} reported in Eq. 9.7, errors lower than 10% are obtained with respect to values calculated with Eq. 9.6 for salinity lower than 7M. Discrepancies are higher when higher salinity solutions are considered.

$$\Delta G_{\text{mix}} = vRT[c_{s,b} V_b \ \ln(c_{s,b}) - c_{s,c} V_c \ \ln(c_{s,c}) - c_{s,d} V_d \ \ln(c_{s,d})] \tag{9.7}$$

The maximum work available by reversibly mixing two equal volumes of solutions at different salinity is a linear function of the temperature and logarithmic function of the two concentrations. In order to give some value of the achievable theoretical work, contours for ΔG_{mix} are reported in Figure 9.1 as a function of the two solution salinities. It is possible

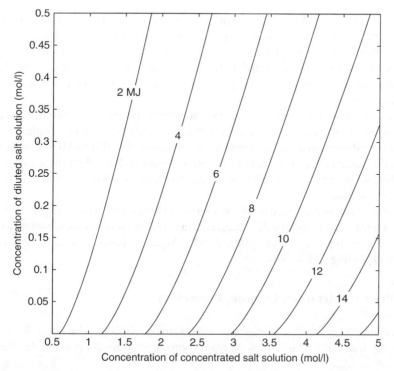

Figure 9.1 *Theoretical work achievable by reversibly mixing two 1 m³ volumes of solutions having different NaCl concentrations at 20°C*

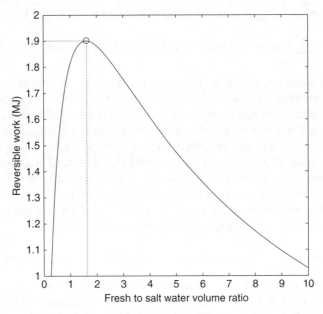

Figure 9.2   *Theoretical available work by mixing different volumes of concentrated and diluted solutions. The overall volume is set to 2 m³ for all the cases while temperature is equal to 20°C*

to extract up to 16.9 MJ by mixing 1 m³ of pure water with 1 m³ of a 5M[2] solution of NaCl. Concentrations with values above 1M can be found just in few hypersaline lakes while seawater has an average concentration of 35 kg/m³ (around 0.6M). By mixing 1 m³ of seawater and 1 m³ of river water (0.01M) at temperature of 20°C, the reversible work is 1.8 MJ, still an extremely interesting value considering the huge potential availability of this energy source.

All the considerations reported so far refer to equal volumes of concentrated and diluted solutions. Figure 9.2 shows the reversible work producible by mixing, according to different ratios, freshwater and seawater with a concentration of 0.01M and 0.6M, respectively. The total volume of the two solutions is kept constant at 2 m³. With these assumptions, the highest work can theoretically be generated by choosing a freshwater to saltwater ratio slightly higher than 1.

The energy values presented in this section were obtained by considering reversible processes. Realistic work output will be presented in the following sections according to design specs and operating limits of the different technologies proposed for electric power generation by exploiting salinity gradient.

9.3   Pressure Retarded Osmosis Technology

This section focuses on the pressure retarded Osmosis (PRO) technology. Because this technology may somewhat borrow its main components from commercial desalinization

---

[2] Molarity identifies the salinity concentration of a solution expressed in mol/l. 5 M indicates a solution containing 5 kmol of NaCl (equal to 292.3 kg) per cubic meter.

and hydraulic power plants, it has come to a more advanced development stage than the other ones proposed for power generation from salinity gradient, and it is currently the most viable option to take advantage of this renewable source. For this reason, the PRO technology is treated at a deeper level than alternative systems inside this chapter.

9.3.1 Operating Principles

PRO technology is based on an osmotic process that allows water spontaneously moving across a semipermeable (i.e. permeable to water but not to salt) membrane placed in between two solutions at different salinity concentration, as shown in Figure 9.3 forward osmosis, FO). Water flux from more to less concentrated solution can be reduced by applying a hydraulic pressure over the more concentrated solution according to the PRO concept (Figure 9.3b). When the hydraulic pressure difference equals the osmotic pressure difference, no water permeates across the membrane (as in example of Figure 9.3c). Osmotic pressure difference for ideal solutions at low concentration can be evaluated according to Eq. 9.8 by the van't Hoff equation:

$$\Delta\Pi = iRT(M_1 - M_2) \tag{9.8}$$

where:

i is a dimensionless dissociation constant (equal to 2 for NaCl, meaning that one mole of salt dissociated in two moles of ions)
R is the gas constant (8314 J/kmol-K)
T is the absolute temperature
M is the salinity concentration expressed in kmol/m^3 (molarity)

Assuming 0.01 and 0.6 molarity for freshwater and seawater, respectively, the resulting pressure difference is 28.7 bar at 20°C, equivalent to an hydrostatic head of 293 m.

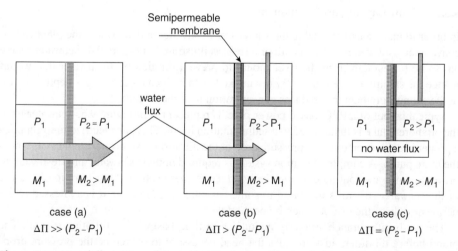

case (a)   case (b)   case (c)
$\Delta\Pi \gg (P_2 - P_1)$ $\Delta\Pi > (P_2 - P_1)$ $\Delta\Pi = (P_2 - P_1)$

Figure 9.3 *Visual illustration of the water flux induced by a salinity gradient in presence of a semipermeable membrane*

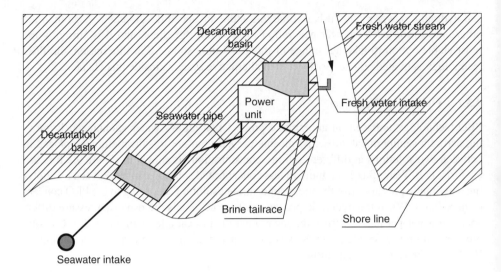

Figure 9.4   *Typical layout of a PRO technology-based power plant*

When the pressure difference is lower than the osmotic pressure (as in the case depicted in Figure 9.3b), water is transferred from a low-pressure, less concentrated solution to a high-pressure chamber containing a more concentrated solution. Mechanical power is finally generated by exploiting the extra flow permeated into the high-pressure room to drive a hydraulic turbine.

In reverse osmosis (RO) instead, the hydraulic pressure applied to the saltwater side is so high to exceed the osmotic pressure difference and water permeates from the high- to the low-concentration membrane side. This operating concept is largely applied in desalinization plants.

9.3.2   Plant Layout and Components

In order to make easily available the sources of freshwater and saltwater, the plant can be conveniently located near the mouth of a river as illustrated in Figure 9.4. Seawater intake has to be far enough from the river mouth to prevent dilution by freshwater that would reduce the salinity concentration. Water intake has to be located at a proper depth because of the salinity gradient that reduces concentration near the surface.

Typical layout of a PRO-based power plant is represented in Figure 9.5. Seawater from the offshore intake is first moved to a sedimentation basin. Then, an electric pump supplies a proper head to overcome the pressure drop of the filtration system and avoid cavitation in the main pump. A complex filtration system is required both on seawater and freshwater to prevent membrane fouling that negatively affects its permeation. After filtration, pressure of the seawater stream is increased to some bar (stream S2) before it is addressed to the permeate side of the osmotic membrane module.

The freshwater branch similarly includes a settling basin and a filtration system. The pump before the filters just provides the head necessary to overcome the pressure drops along the freshwater circuit.

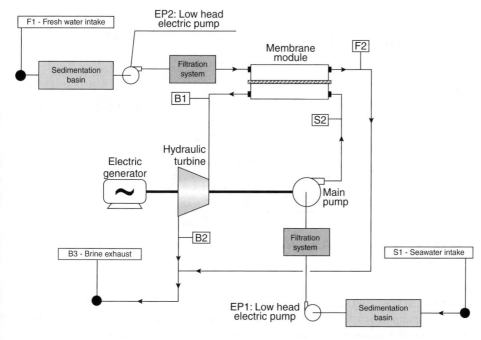

Figure 9.5   *Process flow diagram of a PRO technology-based power unit*

Provided that the pressure difference between the two sides of the membrane is set to a value lower than the difference in their osmotic pressure, a water flux establishes across the membrane from the low- to the high-pressure side (i.e. from the low to the high salt concentration side). Therefore, the water flow rate on the high-pressure side of the membrane increases. The seawater stream diluted by pure water in the membrane module (brackish stream B1) is then sent to a hydraulic turbine connected to the same shaft of the main pump. Since the turbine head equals the pump head (apart from the pressure losses along the circuit and the small head provided by the EP1 pump) and the turbine flow rate is higher than the pump flow rate because of the water permeated across the membrane, a net mechanical power output is available at the shaft. A generator connected to the shaft finally converts mechanical to electric power.

The freshwater flow rate at the membrane module inlet is kept significantly higher than the flow permeated across the membrane so as to limit the increase of salinity concentration (resulting from water permeation) that would consequently reduce the osmotic pressure difference on the two membrane sides. The excess flow rate released from the membrane module (stream F2) is eventually mixed to the brine stream B2 at the turbine outlet and then they are released into a tailrace located downstream the freshwater intake.

A higher plant efficiency could be achieved by replacing the main pump with a positive displacement energy recovery device. Also known as pressure exchanger, this volumetric machine can transfer the pressure energy to a low-pressure liquid stream by depressurizing a high-pressure stream having the same volume flow rate. According to the analysis carried out in [5], the direct work exchange between the two fluids (simply separated by a

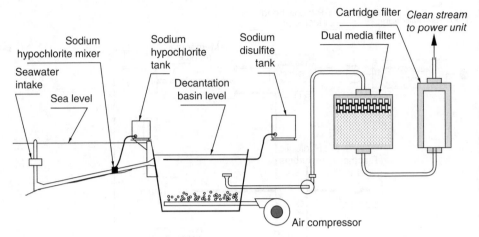

Figure 9.6 *Outline of a possible water filtration process included in a PRO technology-based power station*

moving wall) occurring in the pressure exchanger allows to increase the energy recovery efficiency compared to the solution previously considered, which pays for higher hydraulic losses inside the turbomachines and friction losses on the rotating shaft.

Since the pressure exchanger operates on streams having the same volume, just a fraction of the high-pressure brine stream exiting the membrane module is fed to the device while the remainder (equivalent to the flow rate permeated across the osmosis membrane) is sent to a supplementary hydraulic turbine, whose power output practically represents the plant net effect.

A proper filtration system is essential to prevent membrane fouling that reduces the plant electric output and shorten the membrane lifetime. Depending on the actual characteristics of freshwater and seawater, a number of treatment processes and filtering devices are placed along the path from the water intake to the membrane module. Besides effectively reducing the amount of entrained particles, it is important the filtration system is designed to maintain low-pressure losses that negatively affect the plant power output. The following section describes the seawater purification process usually adopted in commercial desalinization plants based on reverse osmosis (RO) membranes (Figure 9.6), which have seawater filtration requirements similar to the power plant considered here. For the latter class, a similar treatment is however necessary for the freshwater stream too.

Water is first made biologically inactive to avoid formation of algae by adding sodium hypochlorite (NaClO), which is a cheaper solution than the UV ray treatment in the case of considerable flow rate. Sodium hypochlorite is directly injected along the suction water pipe to allow a proper dilution and an effective treatment in the following decantation pool. Dirty water is then clarified in a sedimentation basin, a solution preferable to centrifugal cleaner because of the lower head losses. Provided that membranes are very sensitive to chlorine poisoning, an effective removal treatment is required which can be performed in activated carbon filters or, as shown in Figure 9.6, by adding an oxidant additive, which can neutralize the reducing chlorine. Sodium disulfite ($NaHSO_3$) is commonly added with a significant excess (300–400%) to ensure a complete chlorine precipitation on the basin bottom. Online sampling systems are useful to add chemicals in proportion to the amount of pollutants in the water preventing overconsumption or insufficient introduction of reagents.

Since traces of oils could build up on the porous wall of the membrane preventing permeation, a complete water deoiling is required. It is achieved in a two-stage process: flotation by air blown in the settling pool by a compressor can reduce oil content down to 5 ppm. Deoiling is then completed in the following mechanical filters. Water ventilation inside the settling basin is also useful to oxidize metal ions that coagulate in particles retained in the filters.

Mechanical filtration is first performed in dual media components featuring a first layer of anthracite to capture the final oil traces, followed by a bed of quartz sand of proper granulometry to prevent silica entrainment by water. Dual media filters can retain particles larger than few microns and are followed by cartridge filters. They represent the last filtration stage before the membrane and are suited to remove fine particles having size larger than about 1 μm.

The membrane module is surely the most critical component in the development of the PRO technology for power generation. Semipermeable membranes can theoretically be derived from those currently in use in commercial desalinization plants based on reverse osmosis. They usually have an asymmetric structure being composed of two different layers: (i) an active thin layer (whose thickness is typically lower than 1 μm) in charge of preventing salt migration while allowing water permeation, (ii) a thicker porous support able to offer the mechanical strength required to withstand the high pressure difference between the two sides. Suitable materials for the filtering layer of the PRO membranes are the same in use for reverse osmosis:

- Cellulose acetate, one of the earlier materials exploited in water depuration, is usually produced in flat sheets suitable to be arranged in spiral wound modules. Cellulose acetate–based membranes are cheap and present high resistance to chlorine and high stability to dirty water but they show lower permeability and are less tolerant of organic solvents than other materials.
- Aromatic polyamides can be produced in hollow fibers with very fine diameter that allow an extremely high filtering surface-to-module volume ratio and good mechanical strength to pressure difference. The weak point of this material is a very low tolerance to chlorine that entails additional requirements of adding reagents in the depuration section.

Active skin, exposed to high-pressure side, is supported on a porous layer of polymeric material about one tenth of millimetre thick. The latter is in turn deposited on top of a non-woven fabric layer of the same thickness that gives the membrane the desired mechanical strength.

Even although first experiments aiming to prove power generation by PRO technology were actually based on commercial reverse osmosis membranes developed for desalinization, pretty soon researchers realized that specifically designed membranes were required to fully exploit the potential of the PRO technology. RO membranes are subjected to a hydraulic pressure difference higher than the osmotic one requiring thicker support that entails a higher resistance to water flux. The extremely high salt rejection necessarily required in RO membranes for desalinization, which is usually achieved at the expense of a reduction in water permeation, is not essential in PRO technology where high permeation is instead desired to reduce the plant cost.

In contrast to the RO case, the PRO membrane permeates water from the low to the high concentration sides increasing the salt concentration in the water soaking the support (in contact with the low concentration side) and reducing the effective salinity gradient on the two sides of the active skin. This mechanism, termed concentration polarization, may severely reduce the water flux across the membrane and, to a lesser degree, exists also on the highly concentrated side given that the pure water flux dilutes the solution on the active skin surface leading to a reduction of the effective salinity gradient on the two sides of the active skin.

9.3.3 Design Criteria and Optimization

Performance analysis of a PRO system can be carried out by solving the salt and water mass balance and the energy balance for every component included in the plant. A detailed characterization of the membrane module is crucial for an accurate prediction of the plant performance. An adequate discretization is therefore required [6–10] since local flux is substantially affected by local conditions and local polarization phenomena.

A correct performance prediction of a PRO plant requires several assumptions regarding membrane permeability and pressure drop of the stream along the modules, efficiencies of the turbine, the pumps and the energy recovery device, length and geometry of the ducts and pressure drops introduced by the filtration system. Ambient conditions and salinity concentration for both freshwater and seawater have to be known as well.

Once these parameters have been fixed and provided that the availability of freshwater is usually the limiting factor that sets the plant capacity, three design parameters are basically required to completely define the plant operating conditions. The parameters that can be conveniently adopted for such a purpose are (i) the hydraulic pressure on the membrane salt side (p_{hyd}), (ii) the concentration of the brine discharged from the membrane (C_b) and (iii) the ratio between the volumetric flow rate of freshwater and seawater (V_f/V_s). In principle, these design parameters should be selected by a techno-economic optimization so as to provide the lowest cost of electricity or the minimum plant cost specific to power output.[3] However, due to the lack of information related to the cost for the water treatment section (which is strongly site dependent), turbomachinery and even for the membrane, literature analyses usually discard this methodology in favor of a more general approach [7, 10–12] that adopt the power density (i.e. the power generated per unit membrane area) as the objective function of the optimization procedure. It is important to emphasize that this approach can lead to acceptable results just in case the membrane cost largely prevails over the aggregate cost of all the other components included in the plant.

Figure 9.7 reports the effects of the three main design parameters identified on power density. Trends are qualitative and have been obtained by sequentially varying one variable at constant values of the other two.

Increasing the hydraulic pressure p_{hyd} reduces the flow crossing the membrane for assigned boundary conditions (membrane area, freshwater and seawater flow rate). On the other hand, since p_{hyd} is the pressure at turbine inlet, as its value increases, more work per permeated mass unit can be produced by the turbine. The optimal p_{hyd} value is a trade-off

---

[3] In case of renewable plants relying on a free energy source (like sun, wind, waterfall besides salinity gradient), the lowest cost of the electricity is generally obtained in correspondence to the minimum specific plant cost.

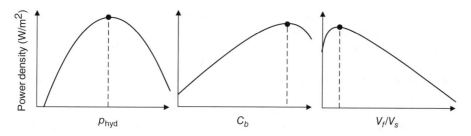

Figure 9.7 *Effects of the main plant design parameters on power density*

between these two opposing effects and usually power density is maximized for a pressure about half of the osmotic pressure difference.

Small values of brine concentration C_b are the result of a large flow rate across the membrane that contributes to dilute seawater flowing on the permeate side. The ensuing large difference between the brine and the seawater flow rate entails a high power production. A small value of C_b leads to a low average salinity difference on the two membrane sides, resulting in a little average flux across the membrane and hence a huge module area. The optimal C_b value maximizes the power density deriving from these conflicting effects. For real operating conditions, it is usually in the range of 16–23 g/l meaning that the flow permeated across the membrane accounts for 34–54% of the brine flow rate processed in the turbine.

For an assigned membrane surface, low values of the V_f/V_s ratio favor a reduction of the consumption of freshwater auxiliaries without significant changes in the gross power output. However, for very small values of the V_f/V_s ratio and nonnegligible salinity of the inlet freshwater, the salt concentration at the freshwater side outlet significantly increases, reducing the average salinity gradient between the membrane sides and consequently the flow permeated across the membrane. This condition should be avoided because it entails a strong drop in power density. Due to these reasons, the V_f/V_s ratio for maximum power density typically results around 0.5.

In summary, a careful evaluation of the opposite effects of these three parameters should be performed in the design of a PRO plant with the ultimate aim to minimize the resulting cost of electricity and guarantee efficient operations both at the nominal point and even in off-design conditions.

9.3.4   Technology Review

PRO has seen significant progress in the past few years, mainly thanks to the development of better and better membranes. However, experimental activity has been so far mostly limited to laboratory-scale testing of commercially available RO and FO membranes [6, 7, 13, 14] or prototype lab-cast membranes [12, 15, 16]. In particular, experiments have been reported where commercial FO spiral-wound modules were used: the maximum power densities reported as inferred from experimental measurements, using seawater-equivalent as the concentrated stream, are 0.5 W/m² in a commercially available spiral-wound module [14] and 3.5 W/m² with a prototype, lab-cast TFC membrane [12].

There has been significant progress made recently in the fabrication of FO membranes. While these were not specifically tested for PRO operation (e.g. permeation rates under pressurized conditions), their estimated potential performance may be calculated based on the experimentally determined characteristics, namely, the water and salt permeabilities of the filtering layer and the mass transfer resistance of the support to diffusive transport. Membrane permeability is expressed in m/s·Pa and measures the water volume flow rate permeated across a membrane area unit under a 1 Pa osmotic pressure difference between the membrane sides. High values of water permeability denote the membrane capability of avoiding resistance to water flux.

Membrane selectivity quantifies instead the ratio between water and salt permeability. Extremely low salt permeability (and consequently high selectivity) is ultimately required in RO desalinization applications. In PRO plants for power generation, this feature is not much essential considering that salt permeation has just the secondary effect of reducing the difference in concentration on the two sides of the membrane and consequently the average flux.

Ramon *et al.* [17] compiled a list of commercial and prototype osmotic membranes and evaluated the resulting theoretical power densities according to their characteristics. The analysis showed that prototype lab-cast thin-film composite membranes based on a selective polyamide active layer can achieve water permeability in the range of 5–7 m/s·Pa, which result in theoretical power density in the range of 5–6 W/m^2 with seawater feed and over 15 W/m^2 in plant fed with 1.1M brine discharged from RO-based desalinization plants. Theoretical power densities achievable with commercial membranes are below 3 W/m^2 with seawater feed.

Although the reported values can be useful to rank the different membranes, they cannot be regarded as expected performance of real plants because they have been calculated neglecting dilution effects, pressure losses and hydraulic machine inefficiency. According to the author's evaluation, by accounting for these effects, the power density of a real plant easily reduces to 10–25% of these theoretical values. In these conditions, the corresponding electric energy produced is in the range 15–25% of the reversible work reported in Figure 9.2. These figures suggest that, accounting for performance penalty due to the aforementioned effects, a plant based on current commercial membrane is still pretty far from a 4 W/m^2 power density quoted in [18] as the threshold to produce energy at sustainable cost in a PRO technology-based plant.

Ultimately, an important characteristic of any PRO membrane is its ability to withstand the applied hydraulic pressure in the feed stream. Since the maximum power density is achieved when this pressure is about half the osmotic pressure, the higher the concentration difference, the higher the pressure to be applied for power density maximization. For example, when RO brine (1.1M, twice the concentration of seawater) is contacted with 0.02M wastewater, the optimum hydraulic operating pressure would be around 24 bar, but it is not clear whether the currently available membranes would be able to mechanically withstand such a large pressure [17].

9.3.5   Pilot Testing

Apart from laboratory tests, the most significant experience with a pilot plant has been carried out by Statkraft with the first prototype PRO installation in Tofte, Norway, inaugurated on November 24, 2009 [19]. This plant utilizes 10 l of freshwater and 20 l of saltwater per second and generates a power output in the range 2–4 kW according to the operating conditions. Planned improvements in membrane features should increase the power output to about 10 kW.

In the selection of the Tofte site, various compromises had to be made in terms of proximity to Statkraft facilities, utilization of existing building and access to seawater and freshwater. In order to secure good freshwater quality, a pretreatment plant based on a 50 μm filtration system was installed. As a matter of fact, rivers in the Tofte area contain certain amounts of natural organic matter and silt with contents that may vary considerably during the year, even though the relatively low temperature (less than 10°C) is important in reducing the risk of biofouling. The maintenance of the plant is based on treatment with sodium hypochlorite at regular intervals combined with back flushing of the membrane. After the first year of operation, the plant was dismantled and some components were cut open for visual inspection. The membrane envelopes seemed in good condition both on the filtering layer side exposed to seawater and on the support side exposed to freshwater, suggesting that pretreatments of the seawater and freshwater were sufficient to achieve an effective depuration and maintain the membranes in good operating condition.

The whole membrane module is composed of six racks, each of them consisting of eleven pressure vessels, operated in parallel, designed to house one spiral wound membrane each. The pressure vessels were originally equipped with cellulose acetate membranes specifically developed for PRO. During the first period, the plant operation has been optimized and the performance has been monitored, resulting in a power density lower than 0.5 W/m². Next, thin film composite membranes developed for PRO were installed in early 2011. So far the measured power density has reached nearly 1 W/m², which is a major improvement compared to the cellulose acetate membranes originally installed. Based on this preliminary experience, Statkraft plans to build a full-scale 25 MW osmotic power plant by 2015 [20].

9.4   The Reverse Electrodialysis Technology

This is a second interesting technology proposed to exploit salinity gradient for direct electric power generation. As for the PRO technology, the reverse electrodialysis (RED) concept has been proved by experimental laboratory-scale plants to be capable of generating a net power output. A detailed description of this technology is presented in this section.

9.4.1   Operating Principles and Plant Layout

RED is a renewable energy–based sustainable technology used to generate energy by mixing water streams with different salinity, as first proved by Pattle with his pioneering work [21]. Figure 9.8 onceptually schematizes an energy conversion system based on such a technology, where ion-selective membranes separate relatively diluted (e.g. river water) and concentrated (e.g. seawater) solutions. Opposite to electrodialysis, where an applied voltage induces ions migration, in RED the driving force is the concentration difference between the feed streams, which originates an electrochemical potential gradient and

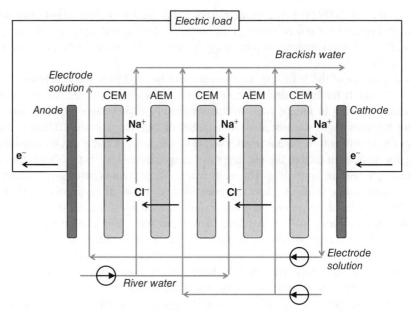

Figure 9.8 *Schematic representation of RED technology (AEM is an anion exchange membrane, CEM a cation exchange membrane)*

induces an electric voltage across the membranes. Driven by the concentration difference, negatively and positively charged ions, that is, anions and cations, diffuse across the membranes with opposing-charge functional groups, creating an ionic flux. A full-scale RED-based system is composed of multiple cell pairs, each consisting of a dilute feed channel, concentrate feed channel and corresponding anion/cation exchange membranes. Anion exchange membrane (AEM) is a semipermeable membrane designed to conduct anions while it is impervious to cations or water. On the contrary, cation exchange membranes (CEMs) are permeable to just cations.

In particular, for a NaCl solution, Na^+ ions permeate through the cation exchange membrane in the direction of the cathode, and Cl^- ions permeate through the anion exchange membrane in the direction of the anode.

Such a membrane stack terminates with electrodes at each end, which convert the ionic flux into an electric current by red-ox (reduction-oxidation) reactions occurring between the electrodes and a hexacyanoferrate aqueous solution in a bulk of NaCl (typically, $K_4Fe(CN)_6$: 0.05 mol/l, $K_3Fe(CN)_6$: 0.05 mol/l, NaCl: 0.25 mol/l [22]). As a result, an electron can be transferred from the anode to the cathode via an external electric circuit. This electrical current and the potential difference over the electrodes can be used to generate electrical power, when an external load is connected to the circuit.

As it can be noticed from Figure 9.8, there is actually a net Na^+ ions transfer across the stack from the anode to the cathode. The electrode solution is therefore recirculated between the electrode compartments to bring the sodium back to the anode.

The key advantage of RED-based systems for power generation is its direct salinity gradient to electricity conversion, which makes the plant layout simpler in comparison with

PRO-based systems, which entail intermediate conversions to pressure energy and then to mechanical work.

In reverse electrodialysis, permselective ion-exchange membranes, which separate a concentrated solution from the diluted solution, allow the selective passage of ions and retain the transport of water. A difference in free energy between the concentrated (c) and the diluted (d) solutions causes the driving force for ion migration that will continue until equilibrium is reached. The ideal electric potential $\Delta\phi$ originated by the salinity gradient is proportional to the difference in free energy between diluted and concentrated solutions. Neglecting any pressure difference between the solutions, $\Delta\phi$ can be calculated by rearranging Eqs. 9.2 and 9.3 as follows:

$$\Delta\phi = \frac{G_c - G_d}{F} = \frac{R \cdot T}{|z_{Na}| \cdot F} \ln(x_{Na,c}) + \frac{R \cdot T}{|z_{Cl}| \cdot F} \ln(x_{Cl,c})$$
$$- \frac{R \cdot T}{|z_{Na}| \cdot F} \ln(x_{Na,d}) - \frac{R \cdot T}{|z_{Cl}| \cdot F} \ln(x_{Cl,d}) \tag{9.9}$$

Since $|z_{Na}| = |z_{Cl}| = 1$ and $|x_{Na}| = |x_{Cl}| = x$ for NaCl solutions, Eq. 9 an reduce to the Nernst equation for an aqueous monovalent electrolyte:

$$\Delta\phi = \frac{2 \cdot R \cdot T}{F} \ln\left(\frac{x_c}{x_d}\right) \tag{9.10}$$

Assuming molarities equal to 0.01 and 0.6 for freshwater and seawater, respectively, and 20°C temperature, the resulting $\Delta\phi$ is 206 mV.

Even if this electric potential difference is very limited to achieve significant electric power, several cell pairs can be stacked as illustrated in Figure 9.8 to increase the voltage applied to the external load. In summary, $\Delta\phi$ can be considered the electric potential difference across the electrodes of a reversible cell composed of a cation-exchange membrane, a compartment filled with a concentrated salt solution, an anion-exchange membrane and a compartment filled with a diluted salt solution. In real operations, a reduction of the cell voltage is expected provided that the stack offers a resistance to the passage of ionic current. Assuming an overall resistance R (Ω) of the cell, the actual voltage across the cell is $\Delta V = \Delta\phi - R \cdot I$

$$\Delta V = \Delta\phi - R \cdot I \tag{9.11}$$

where I (A) is the current circulating in the cell. The power generated by the cell is therefore

$$W = \Delta V \cdot I = \Delta V \cdot \frac{\Delta\phi - \Delta V}{R} \tag{9.12}$$

The power density of the cell is defined as the power generated (W) divided by the total area of the membranes (2·A, provided that two membranes, an AEM and a CEM, are included in the cell):

$$w = \frac{\Delta V \cdot (\Delta\phi - \Delta V)}{2 \cdot A \cdot R} \tag{9.13}$$

At the maximum power point, identified by the condition $\frac{\partial W}{\partial(\Delta V)} = 0$, ΔV is equal to $\Delta\phi/2$ and I is equal to $\Delta\phi/(2 \cdot R)$. Introducing the area resistance ($\Omega \cdot m^2$) of the whole cell (r), the resulting maximum power density is

$$w_{MAX} = \frac{1}{2 \cdot R} \cdot \frac{\Delta\phi^2}{4}$$
(9.14)

9.4.2 RED Technology Review

Scientific interest on RED is relevant. Investigation focusing on RED as a power generation technology started in the 1950s when Pattle assembled a small stack with a maximum electromotive force of 3.1 V [21]. However, due to the high internal resistance associated with the stack, the resulting power output was relatively low. Better results have been obtained as the quality of the membranes has improved. Research has been conducted using membranes developed for electrodialysis, since ion exchange membranes have not been specifically designed for the RED process.

The most important ion-exchange membrane properties are permselectivity and electrical conductivity. The former characterizes the ability of the material to transport only certain ionic species while remaining impermeable to others. The latter measures the propensity of the material to avoid resistance to the passage of ionic current.

Recent advances in polymers and materials science have resulted in significant improvements for ion-exchange membranes [23]. Most of the commercially available membranes are homogenous,that is, composed just of the ion-exchange polymer. Low-performance heterogeneous membranes are also available. They are fabricated by mixing ion-exchange polymer with an inert polymeric carrier. While they achieve permselectivity comparable to homogeneous membranes, heterogeneous membranes are three to four times less ion conductive. This may represent an issue because the assessment performed in [24], regarding a variety of commercially available anion and cation exchange membranes, concluded that power density is more sensitive to changes in membrane conductivity than in permselectivity. More specifically, the current passage achievable by present state-of-the-art homogeneous membranes is enough for acceptable power density, while heterogeneous membranes should still be improved by reducing their electrical resistance, for example, by making them thinner. However, the resistance imposed to ions transfer in the channels of the dilute solution is even more important in limiting the process performance. According to the calculations reported in [17],[4] for standard stack design parameters and operating conditions, benefits induced by an increase of the turbulence in the dilute channels are more significant than those due to an increase of the membrane conductivity. This trend is shown in Figure 9.9, where the power density is plotted against the membrane area conductivity, for different cross-flow velocities. For typical operating conditions reported in the literature (i.e. Reynolds number on the order of unity), an increase of the conductivity above $2000 \, \text{S/m}^2$ does not nearly affect the power density because the resistance to the passage of ionic current is mainly associated with the transport of ions in the dilute channel. Much more important is the effect induced by an increase of the Reynolds number that promotes mixing and favors the charge transfer. Ideal curve calculated by assuming completely mixed flow (i.e. without any limitations to ionic transport) in the dilute channels is reported for reference. Reduction of the channel resistance is analogously observed when the channel

---

[4] In this study, a 0.95 average permselectivity has been assumed, in line with the best membranes currently available, The concentrated solution is seawater 0.55M and the dilute solution has a molarity of 5 mM)

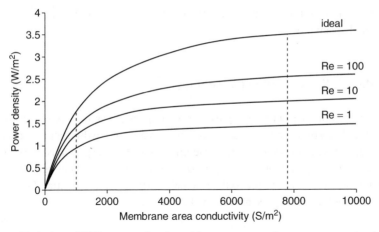

Figure 9.9   *Variation of RED power density with average membrane area conductivity illustrating the effect of external mixing. The vertical dotted and dash-dotted lines mark the typical conductivity of commercial heterogeneous and homogeneous membranes, respectively [17]*

height (i.e. the distance between two membranes) is decreased. As important part of the RED stack, electrode performance has instead been shown to have a minor impact on stack resistance [24].

Coming to experimental activities on real RED stacks, the most significant studies presented in the literature have been carried out by Wetsus (Center for Sustainable Water Technology) in the Netherlands, in the framework of the "Blue Energy" project [25]. Veerman *et al.* [26] compared the performance of six cation–anion membrane pairs reaching a highest power density of 1.2 W/m^2 in a stack of 25 cells,[5] with membranes spaced by 0.2 mm and a functional area of 10×10 cm^2 each, operated with artificial river water and seawater with NaCl concentrations of 1 g/dm^3 and 30 g/dm^3, respectively. Later works by the same research group on a single cell showed that a power density higher than 2 W/m^2 can be achieved by reducing the channel height from 0.2 to 0.1 mm [27]. The gain is less relevant when parasitic pumping power to move water inside the channels is considered provided that the thinner the channel height the higher the pressure drop. A further reduction of the channel height to 0.065 mm, though promising in terms of gross power density, proved ineffective in terms of the net output (stack minus pumping power). This issue becomes more and more important as the size of the stack increases as it was pointed out in [22]. As a first step in the scale-up process [22], the research group at Wetsus also arranged a bigger stack equipped with 50 cells, each measuring 25×75 cm^2 for an overall area of 18.75 m^2. Experiments clearly showed that a key parameter influencing stack power density is the water residence time inside the channels, and power density notably reduces as the residence time increases. Wide membranes are penalized under this aspect because the longer the channel, the higher the flow velocity to maintain the same residence time. This results in large pressure drops, which in turn entail significant pumping power. The outcome was that the maximum net power density for the 18.75 m^2 stack resulted in

---

[5] A single cell is composed of an anion and a cation exchange membrane. Therefore, the stack consists of 50 membranes summing a total active area of 0.5 m^2.

about 40% lower than for the 0.5 m^2 one (0.4 vs. 0.7 W/m^2), assuming the same membrane materials and spacing (0.2 mm).

9.5   Other Salinity Gradient Technologies

Along with PRO and RED technologies, two more possible salinity gradient–based technologies have been proposed: reverse vapor compression and hydrocratic generator. These other concepts have not raised the same scientific interest of the previously presented PRO and RED, and the experimental activity accomplished so far is not enough, in the opinion of the authors of this note, to prove their actual feasibility and potential. Descriptions of these two technologies are presented in the following.

9.5.1   Reverse Vapor Compression

Reverse vapor compression (RVC) produces power by exploiting the vapor pressure difference between two solutions with different salinities. As a matter of fact, it does not require semipermeable or cation/anion-exchange membranes as PRO or RED technologies, and, therefore, it results much less demanding in terms of water filtration and maintenance. This appealing characteristic is however penalized by a really low-power density, which entails huge devices for a relatively limited power production with costs of electricity strongly affected by the cost of the components.

9.5.1.1   Theoretical Background

The vapor pressure of an aqueous solution depends on its salt concentration. Curves of the saturation pressure as a function of temperature are reported in Figure 9.10 for freshwater, seawater, and brine with a salinity of 0, 3.45%, and 28%, respectively [28]. Vapor pressure increases quickly with temperature, even though the trend is affected by salinity. Considering solutions at the same temperature, an extremely small vapor pressure difference can be appreciated between freshwater and seawater, but a wider gap can be observed with reference to a high-salinity brine. The reverse vapor compression technology exploits such a pressure difference.

A scheme of the RVC process is reported in Figure 9.11. It includes two vessels, filled with freshwater and saltwater, respectively, and a steam turbine between them. If the solutions are maintained at the same temperature T_e at a pressure low enough to have vapor–liquid equilibrium in each vessel, a pressure difference is established between the vessels. Thus, the vapor can flow from the freshwater to the saltwater chamber through the expander.

Considering Eq. 5.15 and Figure 9.12, working at higher temperature is recommended owing to the divergence of the two saturation lines (the distance between the two lines is only qualitative and the gap is duly enlarged to better illustrate the working principle), but even at 27°C there is a reasonable pressure difference for a freshwater–seawater system [28].

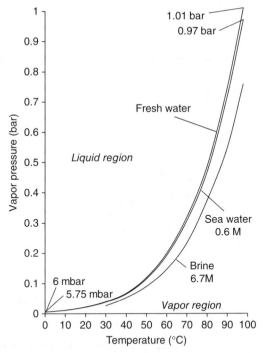

Figure 9.10 *Vapor pressure as a function of temperature for three solutions with different concentrations [28]*

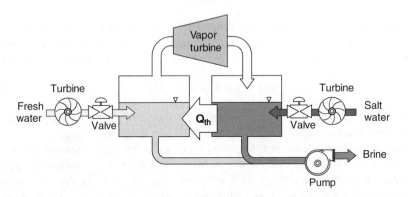

Figure 9.11 *Schematic view of a RVC device*

According to the function of $p_v(T)$ for both saltwater and freshwater, the pressure difference is calculated and the maximum theoretical power results as

$$W_{\text{th,max}} = \bar{v}\Delta p_{v,\text{max}} = \bar{v}(p_{v,f}(T_f) - p_{v,s}(T_s)) = \bar{v}(p_{v,f}(T_e) - p_{v,s}(T_e)) \qquad (9.15)$$

where $\bar{v}$ is the average vapor specific volume at temperature T_e and at a pressure equal to the average between $p_{v,f}$ and $p_{v,s}$.

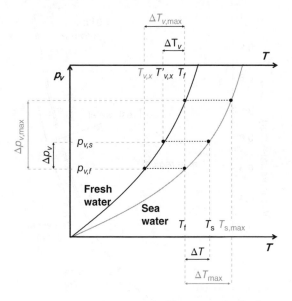

Figure 9.12   *Vapor pressure as a function of temperature for freshwater and seawater*

For an ideal gas, $\bar{v}\Delta p = cp_v\Delta T$, where cp_v is the specific heat at constant pressure. Thus, Eq. 9.15 may be rewritten as

$$W_{\text{th,max}} = cp_v(\Delta T_{v,\text{max}}) = cp_v(T_e - T_{v,x}) \tag{9.16}$$

where the value $T_{v,x}$ is calculated according to the system temperature T_e and the concentration of the two solutions. As reported in Figure 9.12, $T_{v,x}$ is the saturation temperature of freshwater with pressure equal to the vapor partial pressure in the saltwater chamber:

$$T_{v,x} = T_{v,f}[p_{v,s}(T_e)] \tag{9.17}$$

Thus, $\Delta T_{v,\text{max}}$ and $\Delta p_{v,\text{max}}$ are representative for the same phenomena and can be indifferently used to calculate the maximum theoretical specific power $W_{\text{th,max}}$.

Nevertheless, the continuous exploitation of such a pressure difference is not trivial since the process evolves spontaneously to an equilibrium condition. Vapor formation causes the liquid phase to cool down, with a reduction in vapor pressure at the freshwater chamber. In the other chamber, vapor condenses with an increase of temperature and vapor pressure. If no heat is provided to the freshwater chamber and extracted from the saltwater chamber, the small pressure difference that drives the process rapidly vanishes, causing an interruption in the vapor flow.

The solution proposed in literature considers heat transfer from the salt to the freshwater chamber in order for the system to operate continuously. As a result, the theoretical pressure drop cannot be exploited, unless maintaining a temperature difference between the two sides of the conducting wall that divides the two chambers:

$$\Delta T = T_s - T_f \tag{9.18}$$

According to such considerations, the actual specific power output can be calculated as

$$W_r = \bar{v}\Delta p_v = \bar{v}(p_{v,f}(T_f) - p_{v,s}(T_s)) = cp_v(\Delta T_v) = cp_v(T_f - T'_{v,x})\qquad(9.19)$$

where $T'_{v,x} = T_{v,f}[p_{v,s}(T_s)] > T_{v,x}$, leading to a value of Δp_v less than $\Delta p_{v,\max}$, as reported in Figure 9.12.

The optimal temperature difference between the two solutions has to be carefully evaluated in order to avoid huge heat transfer area or extremely poor efficiency. In particular, high-temperature differences between the vessels bring about low Δp values, reflecting on small work output of the expander. On the other hand, a lower temperature difference corresponds to a higher work output, but requires a larger heat exchange surface and an extremely challenging design of the apparatus,that is, a more expensive system. The optimal value of ΔT results from the trade-off between these two demands taking both thermodynamic and techno-economical optimization into account. In order to minimize the cost of electricity, the ratio between net power and heat exchange area can be used as objective function. It is here important adding that limits on acceptable ΔT are really stringent. With reference to Figure 9.12, $\Delta T_{\max}$ corresponding to a zero pressure difference (so no power can be extracted from the system) is around 0.5°C for the usual salinity difference between freshwater and seawater [28] and an ambient temperature of 27°C. The corresponding $\Delta p_{\max,v}$ is about 100 Pa, while the vacuum inside the vessels has to be maintained at an absolute pressure of about 3500 Pa.

Finally, it is important to highlight that freshwater chambers have to be refilled with a freshwater make-up stream to compensate the mass loss due to evaporation. Simultaneously, a purge stream is required to prevent an increase of the salt concentration inside the vessel, which would reduce the salinity difference. Similarly, the vapor flow rate condensed in the saltwater chamber has to be removed. In order to avoid reducing the concentration in the saltwater chamber (because of diluting effect of the condensing vapor), a saltwater stream has to be fed and a corresponding brine purge stream has to be removed from the chamber. Removal of these streams from the vessel represents a problem given that the vessels are kept at a pressure around 3000–4000 Pa. The fluid has to be pumped to atmospheric pressure and the pumping power may significantly affect the plant energy balance. A possible way to reduce this power consumption is by flowing in two turbines the respective streams introduced in the vessels (as illustrated in Figure 9.11), which may produce work by exploiting the pressure difference between the atmosphere and the vessels.

9.5.1.2 Proposed System Layout

A possible design in [28] for the RVC system is shown in Figure 9.13. The heat exchanger has a double spiral arrangement that guarantees a large exchange surface so that significant heat can be transferred under minimal (hundredths or few tenth of centigrade degree) temperature difference. Since no significant pressure difference is applied on the two sides of the spiral, it can be made of thin foil of metal to reduce heat transfer resistance and cost. The spiral is put in slow rotation inside an external vacuum tight vessel to increase liquid-to-wall heat transfer. In each spiral, liquid is in equilibrium with the vapor phase. So the colder freshwater receives heat from the brine and evaporates. Vapor generated inside the spiral on the freshwater side goes out and moves toward the salt side driven by the pressure drop maintained by the salinity gradient. The turbine produces work by exploiting this

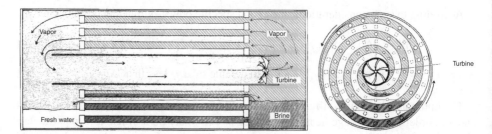

Figure 9.13 *Schematic RVC system with the two chambers divided by a spiral wall of conductive material [28]*

pressure drop. In despite of the minimal pressure gradient, the enthalpy drop in the turbine can reach a few kJ/kg given the huge specific volume of the steam at very low pressure ($38.8 \, m^3/kg$ at $27°C$).

The RVC device does not require accurate filtration of the two feed streams as the membrane-based technologies. Moreover, a small amount of particles and dust may be useful to clean the surfaces of the spiral wound, reducing the fouling resistance. The poor oxygen concentration in the low-pressure chambers limits biofouling, so periodical chemical cleaning is not required. Nevertheless, the effects on the environment should be considered in case of extensive deployment of this technology. The freshwater discharged from the RVC device is slightly cold and poor in oxygen. Difference in temperature is negligible and it does not affect the environment equilibrium, but a re-oxygenation of the fresh stream is required to prevent the damage to aquatic life, especially if huge amount of water is used.

9.5.2 Hydrocratic Generator

"Hydrocrasis" is a term used to define the process of two solutions with different salinity, that is, different osmotic pressures, which are contacted and mixed in absence of any membrane. In particular, the hydrocratic generator (HG) is a machine capable of exploiting the natural mixing of freshwater in a large volume of saltwater for power production. Such a technology has been promoted and studied by Wader LLC [29], a Californian company interested in oceanographic research [30, 31]. According to the experimental activity performed by Wader LLC, the hydrocratic generator allows to enhance the efficiency of desalination and industrial plants, but it can be installed near river estuaries in order to produce electrical power as well.

9.5.2.1 Theoretical Background

The HG operation principle is known and commonly used in sea or salt ponds fish farming. In such facilities, it is crucial to move nutrients from the deep water to the surface, where shellfish and mussel grow. If freshwater is pumped through an upwelling tube, the surrounding saltwater is attracted to the surface. The idea consists of extracting power from this upwelling flow by means of a turbine. The sustainability of the process is strongly affected by the ratio between saltwater and freshwater in the up-tube and by the efficiencies of both pump and turbine.

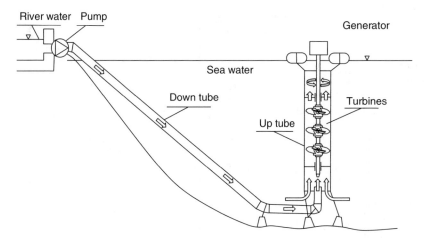

Figure 9.14 *Schematics of some proposed HG-based systems [31]*

9.5.2.2 Proposed System Layout

As reported in a patent by Wader LLC [31] and in Figure 9.14, the HG is mainly formed by four components.

- A down-tube is used to inject the freshwater by means of a pump or exploiting the geodetic difference between the levels of the two reservoirs. Many different configurations are proposed adopting a single tube, several tubes in parallel or positioning such tubes along the up-tube.
- Freshwater and saltwater are mixed together in an up-tube, with a resulting flow to the reservoir surface. The up-tube can present lateral slots in order to favor attraction and mixing of seawater. It is anchored to the sea bottom with chains and concrete blocks.
- One or several submerged turbines are positioned on the top or in the middle of the up-tube. Turbines are proposed in three different configurations: Francis-type turbine, coil turbines and spiral fans.
- A generator is connected to the turbine shaft. Electrical current is then transported at the seaside with cables.

The main advantages of this technology are (i) the relatively simple design of the HG system, (ii) the use of commercially available components, (iii) no requirement for membranes or for heating or cooling the streams. Thus, HG should be cheaper than membrane-based technologies. Moreover, HG technology has limited operational costs, thanks to the low deterioration of components, modest maintenance and a limited biofouling and chemical fouling issues.

Laboratory activities have been performed by Wader LLC with an experimental equipment reported in Figure 9.15a, and quite interesting results have been obtained. In particular, the measured ratio between freshwater and saltwater is greater than 1 [31]. This value greatly exceeds the transfer of kinetic energy from freshwater raising up the tube and the saltwater surrounding it, attesting the feasibility of the system. Other experimental

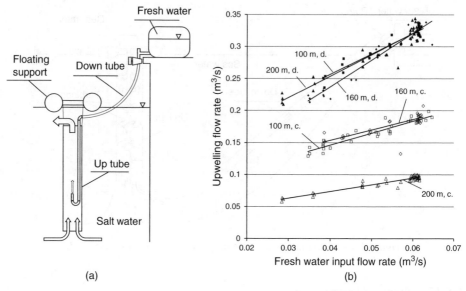

Figure 9.15 *(a) laboratory-scale apparatus for the measurement of the ratio between attracted and injected water [31], (b) upwelling flow results for three different up-tube lengths and two measurement methodologies [30]. Note: In the right figure, label "d" refers to the results obtained by direct flow meter measurements while "c" refers to indirect (concentration) measurements*

campaigns have been performed at the research platform FLIP off San Diego, considering real-size devices and investigating the effect of the up-tube length on upwelling flow [30]. Upwelling flow is measured in both direct (flow meter) and indirect (concentration at the top end of the upwelling tube) ways. The results obtained for three different depths are reported in Figure 9.15b, but inconsistency is found between the two measurement methodologies. The less impressive results seem to confirm the laboratory-scale conclusions with a ratio between freshwater and saltwater in the range of 0.5–2.

9.6   Osmotic Power Plants Potential

Salinity gradient–based power production can be realized in all locations where two water sources with high and low salt concentration, respectively, exist. Typically, such a situation occurs in the estuaries of rivers; so considering the performance of emerging membrane-based technologies and rivers around the world, it is possible to estimate the amount of electric power that could be potentially produced. Of course, a great deal of river resources is concentrated in the equatorial areas, resulting in the most suitable areas for exploiting the salinity gradient for power production. Table 9.1 details the planetary potential; according to the reported data, salinity gradient–based power generators could produce several thousands of TWh of electricity annually [32].

Table 9.1   *Maximum theoretical potential for salinity gradient exploitation (every plant is assumed to generate 2 MW per cubic meter of river water whose entire flow is utilized)*

River	Mouth	Potential (GW)
Mackenzie	Arctic Ocean, Alaska/Canada	23
Mississippi/Missouri	Gulf of Mexico, Louisiana	37
Amazon	Atlantic Ocean	500
Rhine	North Sea, Netherlands	5
Nile	Mediterranean Sea, Egypt	11
Congo	Atlantic Ocean	96
Yenisei	Arctic Ocean, Russia	44
Ganges	Bay of Bengal, Bangladesh/India	32
Meghna	Bay of Bengal, Bangladesh/India	91
Mekong	South China Sea	37
Yangtze	East China Sea	73
Murray/Darling	Southern Ocean, South Australia	2

As detailed in Table 9.1, South America seems to be the geographic area with the most significant potential. However, calculated potentials just refer to theoretical values, preliminarily obtained by taking into account only the difference in salinity between the two sources of freshwater and saltwater. As a matter of fact, the theoretical potential of salinity gradient energy does not depend on a specific conversion technology. Nevertheless, there are severe limitations that significantly restrict the actual potential. In particular, from a technical point of view, not all the potential for the exploitation of the salinity gradient seems to be available. First, the variability in flow rate of the rivers, which depends on the seasons, affects the size of the plant required to achieve reasonable capacity factors. Second, the technology used for power production is a key factor. With reference to the analysis carried out in the previous section about PRO (pressure retarded osmosis) technology, it is clear that just a fraction (indicatively about 20%) of the theoretical potential can be converted to electric energy. Moreover, there are ecological limitations, since it is not possible to consider the whole flow of a river for power production, otherwise the ecosystem of the river mouth would be upset. For this reason, environmental impact studies are required to be as thorough as possible to preserve the ecosystem. Ultimately, considering the most important rivers with their potentially available, that is, enormous, flow rates, the size of a power plant necessarily requires modular units. According to such considerations about the actual potential of the salinity gradient source, the results of a potential analysis proposed by Stenzel and Wagner are reported in Table 9.2 [33]. In order to determine the actually exploitable potential, reference is made to standard compositions of the two water sources (35 g/dm^3 as average salinity for the oceans and 0.13 g/dm^3 as average salinity of river water) and a mixing temperature of 16.1°C, equal to the average surface temperature of the oceans [33]. In particular, focusing on PRO technology, the total global potential sums up to 647 GW$_{el}$ or 5177 TWh$_{el}$/y, which reduces to 65 GW$_{el}$ or 518 TWh$_{el}$/y if an extraction factor equal to 10% for the calculation of the ecological potential is assumed. According to the worldwide use of electrical energy, the maximum contribution of osmotic power plants would be a few percent of the worldwide electrical energy consumption. For other possible

Table 9.2   *Continental discharge data, theoretical potential of salinity gradient energy (index G: Gibbs energy), technical and ecological potential for PRO plants [33]*

Continent	Annual discharge (km³/y)	Mean discharge (m³/s)	Share (%)	Theoretical potential (GW_G)	Theoretical potential (TWh_G/y)	Technical potential (GW_el)	Technical potential (TWh_el/y)	Ecological potential (GW_el)	Ecological potential (TWh_el/y)
Europe	2,752	87,205	7.62	241	2,109	49	395	5	39
Africa	3,511	111,257	9.72	307	2,690	63	503	6	50
Asia	11,603	367,676	32.13	1,015	8,890	208	1,664	21	166
N. America	5,475	173,492	15.16	479	4,195	98	785	10	79
S. America	11,083	351,198	30.69	969	8,492	199	1,589	20	159
Australia*	1,685	53,394	4.67	147	1,291	30	242	3	24
Total	36,109	1,144,222	100	3,158	27,667	647	5,177	65	518

*Including Oceania.

salinity power conversion technologies, for example, reverse electrodialysis, the technical potential might slightly differ from the values in Table 9.2.

9.6.1   Site Criteria for Osmotic Power Plants

The potential values presented in Table 9.2 are based on the consideration of all river systems worldwide. In practice, only a part of the rivers offers suitable conditions for the operation of osmotic power plants, taking practical and economic considerations into account. Salinity distribution depends on the characteristics of the river estuary that can determine the four configurations shown in Figure 9.16.

Partially mixed or well-mixed estuaries are typically located in sites exposed to intense tidal phenomena. In these cases, a high mixing between river and seawaters occurs, resulting in a highly variable salinity profile. In this situation, the salinity gradient of the estuary is horizontal, so the intake of seawater must be placed at a proper distance from the mixing zone. A significant part of the total costs of the power plant is related to seawater intake and transport, often making the project unfeasible. Moreover, long transport distances cause pressure losses due to friction phenomena, with a consequent increase in pumping costs that further reduce the power output. Examples for well-mixed estuaries are the North Sea rivers, Rhine and Elbe, or rivers at the Atlantic coast, for example, the Seine. Due to the necessity of an extensive water transport system, well- and partially mixed estuaries are usually not well suited for the operation of osmotic power plants. Amazon River is another example with a wide brackish water zone of larger than 150 km. Excluding this single river from the worldwide potential considerations, the potential values are reduced by 15%, due to the high share according to the total worldwide discharge.

Salt-wedge estuaries are typical of regions with poor tidal phenomena and are characterized by a very sharp vertical salinity gradient that determines a poor mixing. Some examples of this type of estuaries are the Mississippi River delta (USA) or the Rhone (France). The fjords have a mixing configuration, which is very similar to the salt-wedge estuaries and are characterized by a vertical salinity gradient (fjord-type). In particular, these two types are ideal solutions for the installation of power plants, thanks to the high salinity stability and the limited distance required for seawater intake and transport.

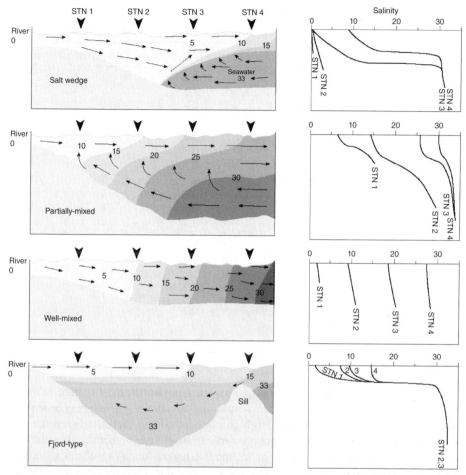

Figure 9.16 *Basic circulation and salinity distribution in salt wedge, partially mixed, well-mixed and fjord-type estuaries. Numbers and shading show salinity values [34]*

Suitable regions regarding the river mouth type are reported in Table 9.3, including regions with low tidal impact where salt-wedge estuaries are assumed to be predominant (e.g. Mediterranean Sea, Gulf of Mexico) or regions with mostly fjord-type estuaries (e.g. Norwegian Sea, New Zealand). For the selected regions as reported in Table 9.3, the ecological potential is calculated by Stenzel and Wagner based on the regional flux data [33]. The classification is made according to the GIWA (Global International Waters Assessment) regions [35] and shows the ecological potential for each selected area. According to the proposed classification (66 regions) and considering the configuration of the estuary, the overall potential compared to the one reported in Table 9.1 is further reduced, since the total ecological potential for the selected regions now sums up to 15 GW$_{el}$ or 122 TWh$_{el}$/y. Another important parameter is the composition of water, with particular interest for the amount of suspended particles and their biological potential. If these parameters are considered too high, an intensive pretreatment will be necessary.

Table 9.3   *Ecological potential for selected GIWA regions [33]*

N°	GIWA regions	Ecological potential	
		GW$_{el}$	TWh$_{el}$/y
1	Arctic	4.5	36
2	Gulf of Mexico	2.0	16
38	Patagonian Shelf	1.7	13
25	Gulf of Alaska	1.3	10
21	Mediterranean Sea	0.8	7
39	Brazil Current	0.7	6
11	Barents Sea	0.7	6
64	Humboldt Current	0.6	5
9	Newfoundland Shelf	0.6	5
63	Tasman Sea	0.5	4
28	East Bering Sea	0.5	4
33	Sea of Japan	0.4	3
29	West Bering Sea	0.2	2
12	Norwegian Sea	0.2	2
14	Iceland Shelf	0.1	1
15	East Greenland Shelf	0.1	1
16	Wet Greenland Shelf	0.1	1
4	Caribbean Islands	0.1	1
	Total	15.2	122

Although the deltas of the major rivers of the world represent the most significant source, they are not the only places where systems exploiting the salinity gradient may be realized. There are particular sites characterized by high values of salinity, that is, the great salt lakes or seas that have a salt content of about one order of magnitude higher than the one characteristic of the oceans. As discussed in Section 9.2, higher salinity results in an increase of the work produced per mass unit fed to the system and this greatly helps the economic feasibility of the plant. The estimation of the potential power of some hypersaline lakes on a global scale is reported in Table 9.4. It is possible to realize that these sites could represent very interesting places to install facilities for a cheap production of electricity. Identifying a source of water with a lower salinity is important for such sites and even seawater would be fine, but environmental impact is the most significant limitation since brackish waters would significantly alter both the ecosystems.

9.7   Conclusions

This chapter presented the rationale behind the techniques for power generation by exploiting the salinity gradient. Technologies so far proposed to this purpose rely on different effects induced by the salinity gradient and face extremely different technical hurdles.

The most developed technologies are those based on osmosis and electrodialysis membranes, which can exploit the experience gained in preparation of commercial membranes

Table 9.4 Estimated osmotic power potential of some hypersaline lakes [36]

High-salinity water source	Salt (g/l)	Conjugate low-salinity water	Salt (g/l)	Lake size (km²)	Potential power (MW)
Australia					
Lake Torrens	Salt bed	Indian Ocean	35	5,698	2,000
Lake Eyre	330	Diamantina River / Indian Ocean	<1–35	9,690	3,300
Lake Gairdner	Salt bed	Indian Ocean	35	4,349	1,500
Central and West Africa					
Sebjet Tah, Western Sahara	Lowland	Atlantic Ocean	35	500	400
Lake Assal, Djibouti	348	Ghoubbet al-Kharab Hot Springs	35–45	54	200
Central Asia and Russia					
The Aral Sea, Kazakhstan	300	The Caspian Sea	1–1.2	68,000	16,000
Zaliv Kara-Bogaz-Gol	330	The Caspian Sea	10–15	18,200	4,500
Lake Baskunchak, Russia	300	The Volga River/The Caspian Sea	<1–1.2	115	40
North Africa and Middle East					
Chott el Jerid, Tunisia	Salt bed	The Mediterranean	35	5,360	2,000
Chott Melrhir, Algeria	Salt bed	The Mediterranean	35	6,800	3,000
Qattara depression, Egypt	Lowland	The Mediterranean	35	19,500	11,000
Lake Urmia, Iran	330	Zarrineh & Simineh / Caspian Sea	<1	5,700	800–1400
Lake Tuz, Turchia	330	Kizil Irmak River	<1	1,500	400
Arabian Peninsula	Lowland	Red Sea, Persian Gulf	4.5		varies
The Dead Sea, Israel/Jordan	330	The Mediterranean / The Red Sea	35–45	810	60–200
Western Hemisphere					
The Great Salt Lakes, USA	240	Bear, Weber Jordan Rivers	<1	440	400
Gran Bajo de San Julián	330	Atlantic Ocean	35	2,900	
Laguna Salada, Mexico	Salt bed	Gulf of California	35–45	1,000	500

for water desalinization and purification. Prototypal plants based on these membranes have effectively produced a net power output proving the technical feasibility of these concepts.

Nevertheless, there is a long way to go before these systems can reach a commercial application level. There are two main issues. Power density (specific to membrane area unit) attainable by the current membranes (approximately 1 W/m^2) is too low to make the technology cost-effective. However, the development of membranes for a specific purpose has just been started and significant improvements are expected in the next future in terms of performance, durability and cost. The second main issue is fouling caused by particles entrained by the streams contacted to membranes, which has to be controlled by expensive and possibly polluting water pretreatment processes. The latter problem is definitely avoided by the other two alternatives proposed, reverse vapor compression and hydrocratic generator, which on the other hand have not yet proved their technical feasibility.

In conclusion, although power generation from salinity gradient is still far to be used in commercial applications, its many prospective advantages in terms of source availability and supply dependability suggest that valuable benefits will be gained if the current efforts in R&D activities will pave the way to a future success.

References

1. Perrot, P. (1998) *A to Z of Thermodynamics*, Oxford University Press.
2. Greiner, W., Neise, L. and Stöcker, H. (1995) *Thermodynamics and Statistical Mechanics*, Springer-Verlag, p. 101.
3. Nijmeijer, K. and Metz, S. (2010) Chapter 5 Salinity gradient energy, in *Sustainability Science and Engineering* (eds I.C. Escobar and A.I. Schäfer), Elsevier, pp. 95–139.
4. Post, J.W., Veerman, J., Hamelers, H.V.M. *et al.* (2007) Salinity-gradient power: evaluation of pressure-retarded osmosis and reverse electrodialysis. *Journal of Membrane Science*, **288**, 218–230.
5. Kim, Y., Kang, M.G., Lee, S. *et al.* (2013) Reduction of energy consumption in seawater reverse osmosis desalination pilot plant by using energy recovery devices. *Desalination and Water Treatment*, **51** (4–6), 766–771. doi: 10.1080/19443994.2012.705549
6. McCutcheon, J.R. and Elimelech, M. (2006) Influence of concentrative and dilutive internal concentration polarization on flux behavior in forward osmosis. *Journal of Membrane Science*, **284**, 237–247.
7. Achilli, A., Cath, Y.T. and Childressa, E.A. (2009) Power generation with pressure retarded osmosis: an experimental and theoretical investigation. *Journal of Membrane Science*, **343** (1–2), 42–52.
8. Lee, K.L., Baker, R.W. and Lonsdale, H.K. (1981) Membranes for power generation by pressure-retarded osmosis. *Journal of Membrane Science*, **8**, 141–171.
9. Enomoto, H., Fujitsuka, M., Hasegawa, T. *et al.* (2010) A feasibility study of pressure-retarded osmosis power generation system based on measuring permeation volume using reverse osmosis membrane. *Electrical Engineering in Japan*, **173** (2), 8–20.
10. Thorsen, T. and Holt, T. (2009) The potential for power production from salinity gradients by pressure retarded osmosis. *Journal of Membrane Science*, **335**, 103–110.

11. Van Der Zwana, S., Pothofa, W.M.I., Blankertc, B. and Bara, I.J. (2012) Feasibility of osmotic power from a hydrodynamic analysis at module and plant scale. *Journal of Membrane Science*, **389**, 324–333.
12. Gerstandta, K., Peinemanna, K.V., Skilhagenb, S.E. *et al.* (2008) Membrane processes in energy supply for an osmotic power plant. *Desalination*, **224**, 64–70.
13. Gray, G.T., McCutcheon, J.R. and Elimelech, M. (2006) Internal concentration polarization in forward osmosis: role of membrane orientation. *Desalination*, **197** (1–3), 1–8.
14. Xu, Y., Peng, X., Tang, C.Y. *et al.* (2010) Effect of draw solution concentration and operating conditions on forward osmosis and pressure retarded osmosis performance in a spiral wound module. *Journal of Membrane Science*, **348** (1-2), 298–309.
15. Zhang, S., Wang, K.Y., Chung, T.S. *et al.* (2010) Well-constructed cellulose acetate membranes for forward osmosis: Minimized internal concentration polarization with an ultra-thin selective layer. *Journal of Membrane Science*, **360** (1-2), 522–535.
16. Tiraferri, A., Yip, N.Y., Phillip, W.A. *et al.* (2011) Relating performance of thin-film composite forward osmosis membranes to support layer formation and structure. *Journal of Membrane Science*, **367** (1–2), 340–352.
17. Ramon, G.Z., Feinberg, B.J. and Hoek, E.M.V. (2011) Membrane-based production of salinity-gradient power. *Energy and Environmental Science*, **4** (11), 4423–4434.
18. Skilhagen, S.E., Dugstad, J.E. and Aaberg, R.J. (2008) Osmotic power - power production based on the osmotic pressure difference between waters with varying salt gradients. *Desalination*, **220** (1–3), 476–482.
19. http://www.statkraft.com/energy-sources/osmotic-power/prototype/.
20. Achilli, A. and Childress, A.E. (2010) Pressure retarded osmosis: from the vision of Sidney Loeb to the first prototype installation – review. *Desalination*, **261** (3), 205–211.
21. Pattle, R.E. (1954) Production of electric power by mixing fresh and salt water in the hydroelectric pile. *Nature*, **174** (4431), 660–661.
22. Veerman, J., Saakes, M., Metz, S.J. and Harmsen, G.J. (2010) Electrical power from sea and river water by reverse electrodialysis: a first step from the laboratory to a real power plant. *Environmental Science and Technology*, **44** (23), 9207–9212.
23. Xu, T.W. (2005) Ion exchange membranes: State of their development and perspective. *Journal of Membrane Science*, **263** (1–2), 1–2.
24. Veerman, J., Saakes, M., Metz, S.J. and Harmsen, G.J. (2010) Reverse electrodialysis: evaluation of suitable electrode systems. *Journal of Applied Electrochemistry*, **40** (8), 1461–1474.
25. http://www.wetsus.nl/research/research-themes/blue-energy.
26. Veerman, J., de Jong, R.M., Saakes, M. *et al.* (2009) Reverse electrodialysis: comparison of six commercial membrane pairs on the thermodynamic efficiency and power density. *Journal of Membrane Science*, **343** (1–2), 7–15.
27. Vermaas, D.A., Saakes, M. and Nijmeijer, K. (2011) Doubled power density from salinity gradients at reduced intermembrane distances. *Environmental Science and Technology*, **45** (16), 7089–7095.
28. Olsson, M.S. (1982) Salinity-gradient vapor-pressure power conversion. *Energy*, **7** (3), 237–246.
29. http://www.waderllc.com.

30. Finley, W., Jones, A.T., Guay, C. (2004) Development of a salinity gradient power: upwelling tests using FLIP, Proceedings of MTTS/IEEE Techno-Ocean '04: Bridges across the Oceans, November 9–12, Kobe, Japan.
31. United States Patent no. 7,329,962 B2.
32. Ravilious, K. (February 2009) Salt solution. *New Scientist*, **28**, 40–43.
33. Stenzel, P. and Wagner, H.J. (2010) *Osmotic power plants: potential analysis and site criteria, Proceedings of 3rd International Conference on Ocean Energy*, Bilbao, Spain.
34. Wollast, R. and Duinker, J.C. (1982) General methodology and sampling strategy for studies on behaviour of chemicals in estuaries. *Thalassis Jugoslavia*, **18**, 471–491.
35. http://www.unep.org/dewa/giwa/areas/regions_and_network.asp.
36. Kelada, M. (2010) *Global Potential of Hypersalinity Osmotic Power*, MIK Technology.

10

Solar Process Heat and Process Intensification

Bettina Muster and Christoph Brunner
AEE - Institut für Nachhaltige Technologien, Gleisdorf, Austria

10.1 Solar Process Heat – A Short Technology Review

Beside the use of solar thermal energy for domestic purposes solar heat is also used for process heating from small systems of only a few kWth up to very large systems with thermal power of several MWth. There are applications for solar thermal systems in different temperature ranges: low temperatures from 30 °C up to 95 °C, for example, for washing processes; medium temperatures from 95 °C up to 250 °C, for example, for processes involving steam such as pasteurization or drying processes and high temperatures from 250 °C up to 400 °C. Typical industries with high heat demand are breweries, dairy, food processing, pharmaceutical, pulp and paper industries, as well as seawater desalination in some countries.

Up to now, there are only a few large-scale solar thermal applications for industrial processes, because it is often difficult to meet the expectations of the industry regarding pay-back time and flexibility of the solar thermal system. Often steam is used as transport medium in industrial distribution networks even if the temperature needed at the point of use in the process is considerably lower. In these cases, either steam must be produced with solar thermal energy or the heat distribution structure must be reorganized. Because it is still more costly to produce higher temperatures with solar energy, this affects the economy of the solar thermal system. Therefore, solar thermal energy is mainly used in such systems to preheat cold feed-in water for the steam process. Often, solar thermal heat would suit the heat demand of a process regarding temperature and load profile; however, feeding-in

Process Intensification for Sustainable Energy Conversion, First Edition.
Edited by Fausto Gallucci and Martin van Sint Annaland.
© 2015 John Wiley & Sons, Ltd. Published 2015 by John Wiley & Sons, Ltd.

solar heat into the process would need modifications of the machinery. Furthermore, industry often benefits from lower energy costs than private households, which makes it more difficult to achieve break-even points in a short period of time, even though heat generation with solar thermal fits in very well with the heat demand of industry in numerous applications.

Industrial sectors vary in structure and heat demand; the application of solar thermal systems in industry incorporates energy-efficiency optimization as a primary step. The minimization of the heat demand of an industry can be achieved by

- applying changes to the process (application of competitive energy technologies);
- applying changes to the energy distribution system (application of the heat integration systems); and
- applying changes to the energy supply system (application of heat pumps/co-generation systems and/or application of solar thermal systems).

Solar thermal systems vary in layout and design and this determines their suitability for application to the energy supply in production processes. The "integration point" specifies where solar heat is integrated within the thermal energy systems. Figure 10.1 gives an overview of different integration points in industry.

Basically, a differentiation can be made whether the integration of solar heat is done on the supply level or process level. Integrating solar heat in the supply level means to use solar heat within the central energy supply system of the production system, including heating of make-up water, heating of the distribution line or heating central energy storages. Bringing heat into the distribution line can be done in several ways: either by heating the return flow to the conventional heat source (e.g. increasing the condensate temperature) or by delivering heat to the supply line, which might be done indirectly via heat exchangers or directly (direct steam generation). When solar heat is not integrated centrally, but locally for

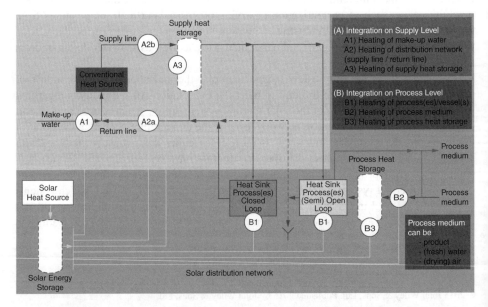

Figure 10.1 *Possible integration points for solar process heat applications [1]*

certain processes, we speak of process level integration. Again several options exist, which are basically depended on the existing process layout. In general, we can differ between heating the process vessel, heating the process medium and heating process energy storages. Detailed integration concepts may vary according to process specifications, existing heat transfer technologies and options to place additional heat exchangers (e.g. indirect heating vs. direct heating). A classification of integration concepts depending on the type of heat consumer (process) and the conventional way of heating has been developed in the frame work of IEA Task 49/IV and a dissertation by University of Kassel [2].

Two possible solar process heat concepts are described here:

10.1.1   Examples of solar process heat system concepts

For the metal surface treatment industry like the galvanic, solar process heat is a very suitable energy supply. The galvanic baths are mostly running at a temperature level of 60–70 °C. In the example shown in Figure 10.2, the solar thermal heat is stored in a storage tank. In countries with ambient temperatures below the freezing point, the fluid in the primary solar loop is a water/glycol mixture. The heat is transferred over a heat exchanger in the secondary solar loop and stored in the stratification tank. There are two inlets on different levels at the storage, which allows the storage of the energy also at lower temperatures. There are two outlets of the tank in order to provide energy on two different temperatures based on the process heat demand. In the case below, the galvanic fluid is heated up by an external heat exchanger and additional to that the backup system heats the galvanic bath by heating coils inside the bath. If possible, the galvanic bath can also function as storage, for example, during weekends when there is a production stop.

The solar support of process steam generation is usually only economical when a significant part of the steam is used in the processes directly (the steam network is an open or partly open system). Solar heating of the additional, demineralised make-up water can be attractive; heating of the condensate return flow or the feed water directly is more expensive because of the high temperatures. Additionally, at state-of-the-art installations, the feed water is usually pre-heated by an economizer.

In (partially) open steam networks, the demineralised make-up water is usually mixed with the returning condensate and has to be degassed before it can enter the steam boiler. This degasification is usually done thermally using process steam from the steam boiler.

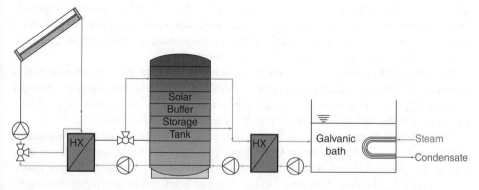

Figure 10.2   *Example of a solar process heat system concept heating a galvanic bath*

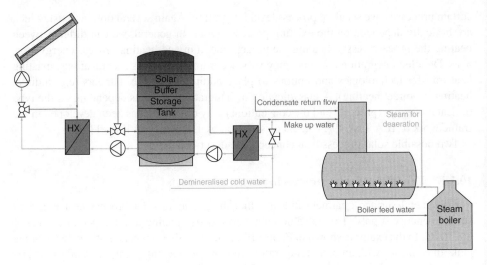

Figure 10.3　*Example of a solar process heat system concept for preheating make-up water*

With this steam, the feed water tank has to be heated up to 90 °C, often also up to slightly over 100 °C, when the feed water tank operates at an overpressure of 0.2 or 0.3 bar. It is therefore a good solution to pre-heat the decalcified, additional make-up water before it is mixed with the condensate and before the mixture has to be heated up. This way, less steam is consumed for degasification and since the steam supports many different processes in the factory, the solar thermal system can cover a significant fraction of the overall heat demand very elegantly just by adding one single heat exchanger (Figure 10.3).

10.1.2　Solar process heat collector development

For applications requiring temperatures up to 250 °C, there is limited experience and suitable collectors are still needed. Therefore, for these applications, the development of high-performance solar collectors and system components is necessary.

The aim of the solar thermal collector development is to improve and optimize them for the temperature level from 80 °C to 250 °C. There are different approaches like double-glazed flat plate collectors with anti-reflection coated glazing, stationary compound parabolic concentrator (CPC) collectors, evacuated tube collectors, vacuum flat plate collectors, small parabolic trough collectors, linear concentrating Fresnel collectors and a concentrating collector with a stationary reflector.

Beside the minimization of collector losses at higher temperatures, the investigation of materials suitable for medium temperature collectors is important, and therefore appropriate durability tests are applied to specific materials and components to allow the prediction of service lifetime and to generate proposals for international standards.

The collector efficiency curve is a good indication for the performance of a collector at different operating temperatures. Optical and thermal losses influence the efficiency of a collector. Thermal losses depend on the temperature at which the solar thermal collector is operating. Simple constructed collectors like an unglazed swimming pool collector have high losses already at relatively low temperatures. Process heat collectors have special

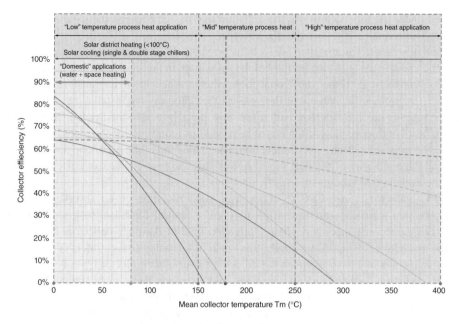

Figure 10.4 *Collector efficiency curve*

improvements in order to reduce these losses like convection or conduction losses and to have a higher efficiency at higher operating temperatures (Figure 10.4).

In the following, an overview of some collector types is given:

Flat plate collectors are the most common type of solar collector. They consist of: (1) a dark flat-plate absorber of solar energy; (2) a transparent cover that allows solar energy to pass through but reduces heat losses; (3) a heat-transport fluid (air, antifreeze or water) to remove heat from the absorber and (4) a heat insulating backing. In order to reach higher temperatures with a good level of efficiency, advanced flat plate collectors will be applied for providing process heat. These collectors are improved by the use of transparent covers to assure high transmittance and high durability, new absorber material, high-temperature residence insulation and casing which ensures stability and protects the absorber and the insulation against environmental impacts.

Evacuated tube collectors (ECTs) can be classified into two main groups:

Evacuated tube collectors have reduced conduction losses based on the insulation properties of the vacuum. With this collector technology, it is possible to reach good performances at relatively high temperatures (up to 120 °C).

Direct flow tubes: The fluid of the solar loop is also circulated through the piping of the absorber. If a single evacuated glass tube is used, the whole interior is evacuated. For this configuration, the flat or curved absorber as well as fluid inlet and fluid outlet pipes are inside of the vacuum; the absorber is coated with a selective surface. Currently the most common type is the Sydney tube collector, consisting of two glass tubes fused together; the vacuum is located between the two tubes. The outside of the inner tube is usually coated with a sputtered cylindrical selective absorber (Al-N/Al), while inside the heat is removed by copper u-tubes, which are embedded in a cylindrical (aluminum) heat transfer fin. Because the absorber is applied completely around the tube, a CPC reflector is often placed under the tube to also use the radiation that passes between the parallel mounted tubes. This radiation is reflected to the absorber.

Heat pipe tubes: The absorbed heat is transferred by using the heat pipe principle without direct contact to the heat transfer fluid of the solar loop. In this case, there are two different ways of connection: (1) the "dry" connection, where the heat has to be transferred from the condenser through the material of the header tube. This way the installation and removal of the tubes is much easier than with direct flow pipes brazed to the header. On the other hand, heat-conductive paste often has to be used, thus requiring that the pipes be installed professionally and (2) the "wet" connection, where the fluid of the solar loop directly flows around the condenser of the heat pipes. In this case, no heat-conductive paste is needed, but the exchange of tubes is more difficult.

CPC collectors use a CPC to concentrate solar radiation on an absorber. Because they are non-focussing (non-imaging), they are a natural candidate to bridge the gap between the lower temperature solar application field of flat-plate collectors ($T < 80\,^\circ C$) to the much higher temperature applications field of focussing concentrators ($T > 200\,^\circ C$). Flat-plate collectors have an enormous advantage over other collector types because they collect radiation coming from all directions, and therefore, they can be stationary on any given roof and all of the diffuse radiation is available to them. However, they also have the highest heat losses since they are proportional to the very large absorber area they possess. Because of these heat losses, the efficiency of flat-plate collectors at higher working temperatures of the solar loop is decreasing. Solar concentrators of the imaging focusing type have a small absorber area and therefore smaller heat losses. They provide high efficiency at high working temperatures. On the other hand, they have the disadvantage of having a smaller angle of view, and therefore, require a tracking system and cannot collect most of the diffuse radiation.

Collectors with CPCs can be designed so that they concentrate solar radiation by 1–2 factors and at the same time accept most of the diffuse radiation. Furthermore, these concentrators can be stationary or only need seasonal tilt adjustments. CPCs for process heat applications are line-concentrators with trough reflectors of different cross-sectional shapes.

Parabolic trough collectors concentrate the sunlight before it strikes the absorber. Mirrored surfaces curved in a parabolic shape linearly extended into a trough shape focus sunlight on an absorber tube running the length of the trough. A heat transfer fluid is pumped through the absorber tube of the collector where the solar flux is transformed to heat.

Parabolic troughs are collectors designed to reach temperatures over $100\,^\circ C$ and up to $450\,^\circ C$ and still maintain high collector efficiency by having a large solar energy collecting area (aperture area) but a small surface where the heat is lost to the environment (absorber surface). Although different definitions are used, in this chapter the concentration ratio refers to the ratio of the aperture area and the absorber surface (the surface that is hot and dissipates heat to the environment). The concentration ratio determines the temperature up to which the heat transfer fluid can be heated in the collector.

Linear concentrating **Fresnel collectors** use an array of uniaxially-tracked mirror strips to reflect the direct sunlight onto a stationary thermal receiver. The features of linear concentrating Fresnel collectors include their relatively simple construction, low wind loads, a stationary receiver and a high ground usage. Some applications allow for the use of the shaded area underneath the collector (e.g. as parking lots).

Almost flat primary mirrors are tracking the sun such that they reflect the direct sun rays to a stationary receiver. The primary mirrors can be made of flat glass because the slight curvature, which is needed for a focal length of several meters, can be applied by mechanical bending.

10.2   Potential of Solar Process Heat in Industry

Solar heat for industrial processes (SHIP) is currently at the early stages of development. Within IEA Task 49 so far (as of 2014) 164 operating solar thermal systems for process heat have been serveyed worldwide. 135 of these plants are reported in detail with an installed gross area of 141.355 m² (ship-plants.info). Most of these systems are of experimental nature and are relatively small scale. However, there is great potential for market and technological developments, as 28% of the overall energy demand in the EU27 countries originates in the industrial sector, majority of this is heat of below 250 °C.

According to a study [3] around 30% of the total industrial heat demand is required at temperatures below 100 °C and 55% of the total industrial heat demand is required at temperatures below 400 °C. The main part of the heat demand below 100 °C could theoretically be met with solar thermal systems using current technologies, if suitable integration of the solar thermal system can be identified. With technological development, more and more medium-temperature applications up to 400 °C will also become market feasible.

In several specific industry sectors, such as food, wine and beverages, transport equipment, machinery, textiles, pulp and paper, the share of heat demand at low and medium temperatures (below 250 °C) is around 60% (POSHIP, 2001). Tapping into this potential would provide a significant solar contribution to industrial energy requirements.

Solar heat has the technical potential to make a major contribution to the share of renewables in district heating and industrial low-temperature process heat. According to the Technology Roadmap of the International Energy Agency (IEA) envisions a contribution of solar heat in industrial applications of 7.2 EJ per year in 2050, corresponding to an installed capacity of over 3200 GWth [4].

10.3   Bottlenecks for Integration of Solar Process Heat in Industry

10.3.1   Introduction

In Europe, sectors with high potential for solar thermal integration commonly include production sites with a significant share of low-to-medium temperature requirement for their thermal energy demand (see Chapter 2), such as the food industry. These production sites currently face several challenges in energy supply of their processes. First, product quality has highest importance so that energy supply of the processes needs to be very secure and without failures. The large number of processes at one production site leads to discontinuous energy demand curves, as energy demand may vary considerably over time depending on the operational schedule of the processes. Additionally, many processes are operating in batch mode requiring high energy peaks within a short time for start-up or operation. Energy conversion plants have to be designed to cover these energy demand peaks.

However, as these peaks occur only seldomly compared to the overall yearly operation time, plants are operating in part load for the majority of the time.

Currently, there are many companies in the aforementioned sectors with high potential for solar heat which are eager to change their energy supply system to renewable energy conversion techniques. Depending on the current heat distribution medium (in most cases steam, hot water or thermo-oil), the available options for thermal energy generation requires more or less integration effort to supply the existing processes. For solar thermal integration, also the point of integration has an impact on the integration effort (see Chapter 1).

Challenges to integrate solar heat in industry will depend on the point of integration (including its specifications), the existing heat distribution medium and the type of solar collector chosen for heat supply. These challenges or bottlenecks can be further distinguished as bottlenecks of the industrial process system (impacts of the existing thermal energy system and the processes) or bottlenecks of the solar system (impacts of the solar collector and storage). This chapter focuses on bottlenecks of the industrial processes and deals with the question on how intensified process layout might affect solar integration possibilities.

10.3.2   Bottlenecks of the Industrial Process to Integrate Solar Heat Supply

The impact of the industrial processes for solar heat integration is an important aspect for the solar process heat system design. This impact is even more relevant for process-level integration, as here the solar heat supply directly interacts with the industrial process. However, obviously, there is also an influence of the industrial process to the overall thermal energy supply line and therefore relevant for supply line integration.

Basically we can define the following key bottlenecks of industrial processes to integrate solar heat:

- High process temperatures
- Low thermal transfer coefficients require large temperature gradients
- Fast heating rates
- Varying process loads (e.g. batch processes with high peak demands in energy supply)

Looking at the efficiency curve of solar collectors, we know that a decrease of the mean collector operating temperature leads to a higher efficiency of the solar collector. As outlined in Chapter 1, this mean collector operating temperature is influenced by the required supply temperature as well as by the return flow temperature of the solar collector. These temperatures are naturally influenced by the temperatures required on the process system side. High process temperatures are, therefore, one of the key bottlenecks for solar process heat integration. However, not only the supply temperature of the solar loop is relevant for the efficiency (directly linked to the process temperature) but the achievable return temperature has an impact on the mean collector operating temperature as well. This return temperature is again influenced by the heat transfer possibilities between process side and solar loop. The closer the return flow temperature of the solar loop can approach the process temperature, the larger will be the temperature difference between solar supply and return flow, thus lowering the mean collector operating temperature and increasing the collector efficiency. Especially in processes where thermal transfer coefficients are low (e.g. stirred tanks, see below), the return temperature, however, will have to increase to achieve

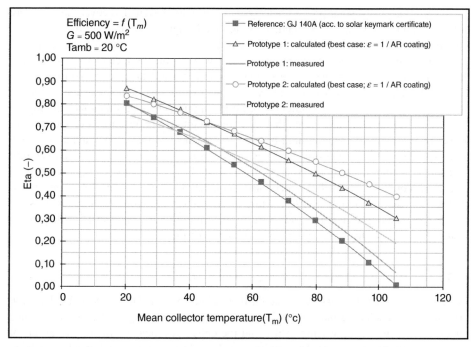

Figure 10.5 *Solar collector efficiency curve of an advanced flat plate collector (prototype 1 – single cover; prototype 2 – double cover)*

large temperature gradients between supply side and process side for an efficient heat transfer.

The collector efficiency curve shows very well the dependency of collector operating temperature and the efficiency of the collector. When the curve is very flat, the collector has still a good efficiency at higher temperature. Figure 10.5 gives an indication which technology improvement leads to a better performance (calculated with the Calculation Environment SOLAR, 1998–2013 by AEE INTEC).

For solar-assisted energy supply loops operating with hot water as energy medium, fast heating rates are a challenge. Either the temperature driving force (temperature difference between supply and process side) needs to increase lowering collector efficiencies or heat transfer areas become large and the required mass flow which needs to be pumped on the supply side increases. The latter two consequences will not enhance the competitiveness of the solar thermal process system design. Batch processes are an example of processes often requiring fast heating rates due to the high energy peaks (e.g. at start-up). Further the discontinuity of the load is a crucial bottleneck for solar thermal integration, as it adds to the necessary storage capacity.

As mentioned earlier, these bottlenecks are only partly applicable when we consider supply-level integration: naturally process temperatures will have an impact on the supply line temperatures and return flow temperatures will as well be influenced by the process design; however, this is most true for water-based distribution systems and less for steam systems. High-energy peaks and varying loads will pose no challenge for supply line integration, as the solar process heat system is usually designed to cover the base load of the overall thermal energy demand. This base load is quite constant over time.

Considering these facts, we can see a need for the development and/or application of new process systems in the manufacturing industry overcome these challenges. Changing the technology of a traditional process, however, requires a detailed consideration of the technological influence on product quality, or in other words on the chemical and biological processes that take place. The challenge for future engineering of production technologies, therefore, is to develop new plants that meet the qualitative requirements of the product in the best way, while enabling high process efficiencies and throughput and at the same time enable the use of renewable energies, which in many cases will be low-temperature heat supply media (district heat, solar thermal, waste heat). Here, it is highly important to take into account the effect of new technologies on the overall thermal energy management strategies of the production site. According to the authors' opinion, the final target is not only to reach an optimized process in terms of its driving forces, but a sustainable process system that shows a high exergy efficiency at the same time.

10.3.3   Bottlenecks of the Solar Process Heat System

Looking at the solar process heat system, there are also some key bottlenecks that limit the efficiency of solar thermal integration in industry. One challenge is the aforementioned relation between operating temperature and the collector efficiency (mainly relevant for non-concentrating collectors). This challenge is linked to one major bottleneck, which are thermal losses in heat transfer and storage. Even though new developments aim to reduce these thermal losses, heat losses in the collectors still amount to approximately 20% for advanced flat plate collectors at a mean collector temperature close to 100 °C (see figure 10.5). The thermal storages are another major source of heat losses, even though their rather large size in process heat plants will lead to quite small volume to surface ratio decreasing losses compared to conventional plants. In many production sites, beside good storage insulation, the heat management of the storage is crucial for realising minimal thermal heat losses. Heat storages are usually necessary in solar process heat plants because of the naturally occurring varying energy supply. The energy demand profile can act positive or negative on storage design; whether operation is largely in coincidence with sunshine hours or whether demand peaks are occurring during the night.

10.3.4   Engineering Intensified Process Systems for Renewable Energy Integration

The methodology how to select intensified technologies has been well defined by other authors. Figure 10.6 shows the basic steps according to Reay [6, 7]. To reach "intensified" processes, first bottlenecks of crucial processes need to be analysed and on this basis new solutions for overcoming these limitations need to be developed.

However, to reach the target of sustainable process systems, the engineer needs to consider not only the specific process that should be intensified, but also its effects on the overall energy efficiency and (thermal) energy management of the production site. Therefore, in the analysis of process intensification for new energy supply, we can actually expand the step "generating design concepts" with another step "analysing effects on overall energy

Figure 10.6   *The PI tower: Steps towards an intensified process [5]*

management". This analysis needs to take into account information of all other processes in the production systems and therefore adds to the complexity of the analysis. A basic representation of the methodology is presented in Figure 10.7.

To enable such an analysis process, modelling tools that allow the consideration of the complete production system will become important. Tools should enable modelling the impact of different technologies on the overall energy management considering potential heat recovery schemes and different renewable energy supply technologies. An example of this holistic approach is shown in the Greenfoods calculation, which is currently developed by a European project consortium (www.green-foods.eu).

10.4   PI – A Promising Approach to Increase the Solar Process Heat Potential?

When we speak of intensifying an industrial process, we usually aim at higher productivity with smaller reactor equipment. This increase in yield of products, while maintaining the raw materials and decreasing by-products and maintaining or decreasing energy input is in many cases achieved by an increase in mass and/or heat transfer. Often these phenomena are interlinked with each other, as in processes with heat transfer as limiting step, mass

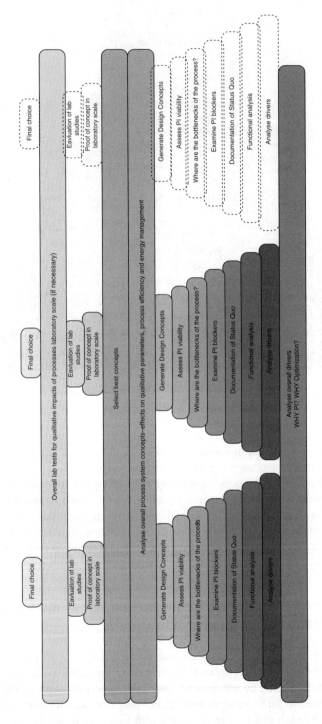

Figure 10.7   The PI tower for sustainable process systems

transfer can be enhanced as well once the heat transfer limitation is overcome. On the other hand, an increase of mass transfer rates induced by physical change of the process design (e.g. change from evaporation to a membrane based process) can allow processes to run at different temperatures and might change the possibility to integrate different energy supply. Naturally, all intensification strategies that affect heat transfer rates and/or change the heat supply to the processes might affect positively or negatively the solar process heat potential. Another important aspect for the integration of new energy supply is the continuity of process energy demand. In many cases, it is actually the change from a batch reactor to a continuous reactor, which will lead to mass and heat transfer improvements. A continuous energy demand profile may additionally be in better coincidence with the solar availability.

Some examples on the mentioned intensification strategies are discussed in what follows. Let us first focus on intensification by overcoming heat transfer limitations. Generally, there are several strategies to increase the heat transfer rate or overcome heat transfer limitations, such as the following:

- Increase of heat transfer area
- Increase of heat transfer coefficient
- Increase of temperature gradients
- Change to energy supply without thermal gradients.

These changes may or may not affect the basic reactor set-up of the existing plant, as they can – at least to some extent – usually be integrated into an existing plant by adaptions. Retrofit changes are naturally limited by existing space and existing construction and regulation. A change in technology will, of course, improve enhancements possibilities further.

Changing the physical process phenomenon on the other hand clearly requires a technology change and replacement of the existing equipment. Considering changes that affect the integration of energy supply possibilities, such as solar heat, we specifically need to look at

- increase in selectivity of separation processes (e.g. change from atmospheric evaporation to membrane assisted processes);
- electromagnetic action on molecules and microorganisms (e.g. change from thermal inactivation of microorganisms to non-thermal techniques, such as microwave or pulsed electric field).

Finally, we look at some specific examples for the change batch to continuous processes. Due to the elimination of peaks in heating/cooling demand, this has a large effect on energy supply and its design.

10.4.1   Intensifying the Industrial Process and Possible Effects on Solar Process Heat

10.4.1.1   Increase of Heat Transfer Area

While solar process heat integration often leads to lower temperature differences between heat supply and process temperatures, an increase in either heat transfer area or an increase

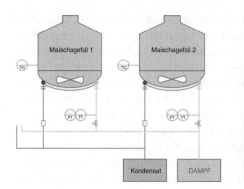

Figure 10.8   *Mash tuns (schematic figure and real plant) supplied by steam [8]*

in heat transfer coefficient is beneficial for the integration efficiency. This is more important for low-temperature solar heat production in flat plate or evacuated tube collectors than for heat produced at high temperatures in concentrating collectors (see Chapter 1). In many cases, as shown later, a combined enhancement of heat transfer coefficients and larger heat transfer area is applied for enabling low-temperature heat supply.

The increase of heat transfer area is a very basic enhancement strategy. By increasing the heat transfer area, the temperature gradient between energy supply and process can be lowered (when maintaining the same heat transfer coefficient). In this way, existing processes can be quite simply retrofitted for new low-temperature energy supply, such as district heat or low-temperature solar heat. It depends on the current process layout, however, if there are limits to adding new heat transfer area.

Basically, a "simple" addition of HX area without additional intensification would contradict the PI goal to reduce equipment sizes. It is, however, still a viable solution for integration of solar process heat. However, in most cases, the addition of new heat transfer area goes in hand with the enhancement of heat transfer coefficients. Figure 10.8 shows a design of a new mash tun retrofitted with additional heat exchange area (including improvement of heat transfer) for enabling the heating of the mash tun with solar process heat. This system has been recently introduced in a Heineken brewing plant in the European funded project SOLARBREW, designed by AEE INTEC and GEA Brewing Systems. The new heating plates have a special surface layout for enhancing heat transfer coefficients and the application of heat supply with lower temperature should also lead to less fouling. The plates had to be designed in a specific manner to enable integration to the existing vessel with positive effects on mixing and cleaning possibilities (Figure 10.9).

Speaking of heat transfer increase in process intensification, one immediately thinks of compact heat exchangers. Compact heat exchangers integrate maximum heat transfer areas in small and compact equipment. The degree of compactness is described by the term HX compactness factor that shows the ratio of heat transfer surface over the equipment volume $[m^2/m^3]$. The large heat transfer areas and usually high heat transfer coefficients would enable integration of solar heat and any low-temperature heat supply very well. Examples of compact heat exchangers with extended surfaces are, for example, plate-fin heat exchangers with a specific area of $800\,m^2/m^3$ [9].

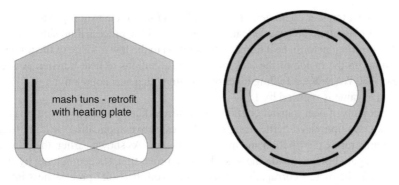

Figure 10.9 *Heating plates on the inside of the mash tuns for low-temperature heat supply [8]*

10.4.1.2 Increase of Heat Transfer Coefficient

There are several means to increase heat transfer coefficients. In many processes in industry heat transfer coefficients are limited by the heat transfer coefficient (α) on the process fluid side. Intensification strategies therefore mainly aim to overcome this limitation.

Heat transfer coefficient limitation is especially pronounced in stirred tanks, where the velocity inside the tank is comparable small and α-values on the process fluid's side are therefore low. Figure 10.10 shows an example of the overall heat transfer coefficients in a stirred tank with a high-viscosity process medium with stirrer tip speeds of 1–7 m/s. When

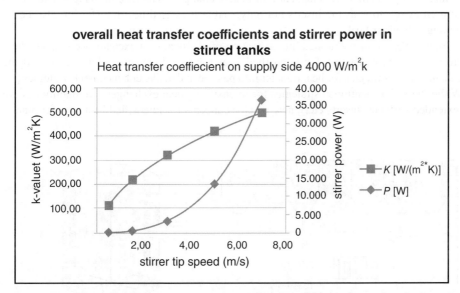

Figure 10.10 *Exemplary data for heat transfer coefficients and stirrer power for a high-viscosity process medium in a batch stirred tank with heating jackets [5]*

assuming a heat transfer coefficient α-value 4,000 W/m^{2}K on the supply side, k-values are one order of magnitude below, so clearly the α-value of the process fluid side in the stirred tank is the limiting factor in heat transfer. In addition to the low α-value of the process fluid side, fouling on the inside of the tank may pose limitations to heat transfer. Assuming a k-value of 350 W/(m^{2}K), a fouling layer of 0.5 mm with a heat conductivity of 1 W/(m K), the overall k-value is reduced by 15%.

Enhancements of heat transfer coefficients in stirred tanks can be achieved by improved mixing (e.g. via optimized baffle design for the respective application) leading to a more uniform temperature distribution and higher turbulence. As shown earlier, the introduction of new corrugated plates inside the vessel can improve transfer coefficients by inducing turbulences at the plate's corrugation. This shows well that solar process heat integration is also possible as retrofit to existing stirred tanks.

Some processes in which pumping of process fluid is not critical will allow placing heat exchangers outside the process bath, thus increasing the design flexibility and enabling the integration of low-temperature heat supply more easily. In external heat exchangers, the possibilities for increasing convective flow (and therefore Nu values) on the process and on the supply side are usually much larger than in internal heat exchangers placed in process vessels. Limitations may arise due to limits in process fluid velocity to reduce shear stress on the product or due to limits in temperature increase of the process fluid in processes where temperature must be kept in very precise ranges (e.g. galvanic baths) (Figure 10.11).

Increasing heat transfer coefficients in heat exchangers has been researched quite extensively, recently reviewed by Anxionnaz *et al.* [10]. Plate heat exchangers performance can be improved by different surface effects (corrugation), extended surfaces (e.g. plate-fin HX) or inserts (e.g. foams). This last class of heat exchangers with inserts (plate reactors) can be used as reactors, as the inserts not only enable increased heat transfer but also better mixing behaviour [10].

Next to the so far mentioned intensification of classical heat exchanger concepts, there are also different reactors for intensified heat transfer. Reay *et al.* have classified the enhancement strategies for heat transfer into passive and active enhancement strategies [6]. While the so far mentioned strategies for intensifying heat exchangers are mainly passive (extended surfaces, inserts etc.), new reactor concepts for intensified heat transfer mainly

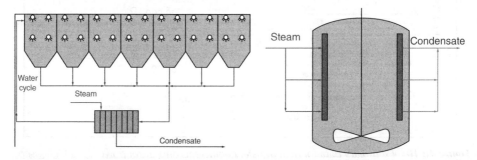

Figure 10.11 *Schematic presentation of vessel heated by external (left) and internal (right) heat exchanger*

Heat exchanger /reactor	spinning disc	Compact multifunctional heat exchanger (offset strip fins)	Plate heat exchanger	Oscillatory flow reactor	Tubular exchanger reactor	Batch reactor with external HX	Batch reactor with double jacket
Scheme	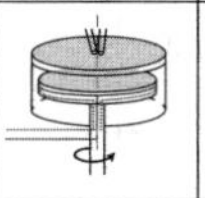					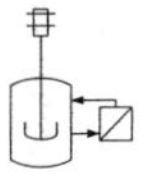	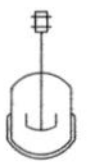
Heat transfer coefficient (W/m²K)	15000 (1)	5000 (2)	2000-4000 (3)	1500-2500 (4a)	500 (5)	1000 (5)	400-800 (5;6)
Specific area (m2/m³)		800 (2)		400 (4b)	400 (5)	10 (5)	2,5 (5)
Maginuted of residence times	seconds	seconds - minutes	seconds - minutes	minutes - hours	seconds - minutes	minutes - hours	minutes - hours

Figure 10.12 *Increase in heat transfer efficiency from batch stirred tanks to intensified heat exchangers (1) [6], (2) [9], (3) own data, (4a) own calculations based on comparison with batch stirred tank (base data [11], (4b) estimated based on tubular heat exchanger, (5) [12], (6) [13])*

rely on active enhancement strategies. Examples are rotational reactors (e.g. spinning discs), fluidized bed (membrane) reactors or oscillatory flow reactors. Another interesting approach is the active intensification of heat transfer rates in heat exchangers over ultrasound [6]. Figure 10.12 shows the comparison of various heat transfer data and the magnitude of residence time in a table-based form based on the work of Ferrouillat *et al.* [9].

The effects of enhanced heat transfer in two reactor types, spinning discs and oscillatory flow reactors, are discussed based on two application examples.

Application Example Spinning Disc Operated at Uniform Temperature. In spinning discs, heat and mass transfer coefficients are significantly enhanced via the rotational speed of the reactor [6]. In a spinning disc reactor, the liquid flows on top of a rotating plate, which is heated/cooled via a heat transfer fluid (typically glycol or water) from below (see Figure 10.7). The reason behind the significant higher heat transfer coefficient realised in spinning disc reactors is the formation of a characteristic film flow of the process medium on the spinning disc. Process film heat transfer coefficients up to $50\,kW/m^2K$ can be realised, shifting the limitation of heat transfer towards the heat supply side or heat conductivity of the plate. Overall film coefficients, however, still reach approximately $10\,kW/m^2K$, allowing much lower temperature difference driving forces [6]. In this sense, they could increase the solar thermal potential by allowing for lower temperature heat transfer. However, assuming a disk operated at a uniform temperature in this example, runback temperatures will be very close to supply temperatures, which is again important in the solar process heat design.

Figure 10.13 shows a basic scheme of a spinning disc supplied by solar heat. Assuming a dT of 2–3 K between supply and return temperature of the heat transfer medium heating the disc, the design of the process side HX will be essential to define the temperatures of the storage and consequently the temperatures of the solar plant. A process side HX with a very small temperature driving force ($dT_{log,process}$) would allow the solar plant to operate at temperatures very close to the required process temperature, however large mass flows on

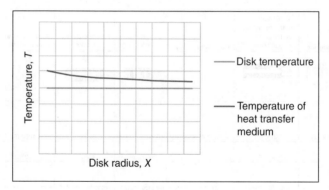

Figure 10.13 *Exemplary illustration of temperature evolution of a heat transfer medium in a spinning disk operated at uniform temperature*

the solar process heat system side will be necessary to enable a small temperature difference between T_{solar} and $T_{solar, return}$.

$$Q_{process} = k_{HX,process} * A_{HX,process} * dT_{log,process}$$

With $dT_{log,process}$ being defined via the temperatures from the solar storage and the process (assuming a countercurrent HX):

$$dT_{log,process} = \frac{(T_{storage,unload} - T_{process,supply}) - (T_{storage,unload,return} - T_{process,return})}{\ln \frac{(T_{storage,unload} - T_{process,supply})}{(T_{storage,unload,return} - T_{process,return})}}$$

Considering a certain temperature loss over the storage, the temperature profile of unloading the solar storage on the process side will naturally be similar to the temperature profile of loading the storage on the solar side. If $dT_{log,process}$ is chosen very small (e.g. 1 K), the temperature difference between the temperature on top of the solar storage and at the bottom will be as well very small. This can lead to problems in heat management and will not enable a good stratification of the energy storage. In case the solar side HX is again designed with a very small temperature driving force, we will end up with a solar supply temperature very close to the process supply temperature (e.g. $T_{solar} = T_{process,supply} + 5\,K$), which would seem positive for operating the solar plant. However, also the solar return flow will be at a similar temperature. This requires large mass flows in the collectors, leading to large pumps and piping dimensions, which will substantially increase the costs of the solar thermal process heat system and the operation costs (Figure 10.14).

$$Q_{solar} = m_{solar} * cp_{solar} * (T_{solar} - T_{solar,return})$$

This effect will be even more pronounced when the spinning disc would be directly connected to the solar buffer tank (Figure 10.15a). One alternative would be to design the process side HX and solar side HX with larger temperature driving forces, which would increase the required temperature produced by the solar collector, but decrease the mass flow (Figure 10.15b). Another alternative is depicted in Figure 10.15c where the mass flow of the heat transfer medium is only partly heated by the solar system to higher temperatures and later remixed. In both of these alternatives, the temperature difference of the

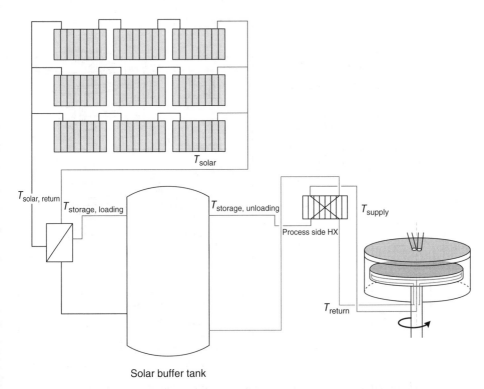

T_solar

T_solar, return

T_storage, loading

T_storage, unloading

Process side HX

T_supply

T_return

Solar buffer tank

Figure 10.14 *Basic scheme of a spinning disk heated by solar thermal energy over an energy storage tank*

heat delivered by the solar collectors and of the storage return will increase, thus leading to a lower mass flow and smaller dimensioning of pipes and pumps. However, this will be realised at the expense of the collector's efficiency, which will for non-concentrating collectors decrease due to the higher mean collector operating temperature. This trade-off has to be analysed economically. For concentrating collectors, the relation between collector operating temperature and collector efficiency at temperatures up to 300 °C is less pronounced.

Basically the same conclusions are true for any intensified reactor. In case the intensified reactor allows a process to be heated with a small dT over the heat exchanger, but at the same time realises a substantial difference between solar supply and return temperature, there will be a high potential for heating the reactor with solar thermal energy with low-temperature collectors. This is further depicted in the next example, where an oscillatory flow reactor is analysed with gradual heating of the process medium.

Application Example: Gradual Heating Within an OBR. Oscillatory flow reactors have the very special feature of decoupling flow velocity from mixing and heat transfer performance [14], thus having large potential in enabling low-temperature heat transfer also for processes with slow intrinsic kinetics. While the process medium flows at low Re numbers, high heat transfer coefficients are enabled over the oscillatory motion and therewith induced turbulences between the baffles. Heat transfer enhancements are reported in the

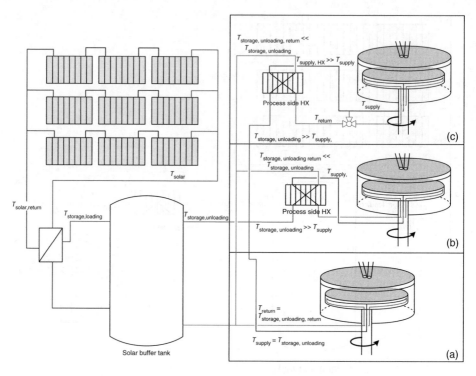

Figure 10.15 *Alternative schemes for solar heating of a spinning disk*

order of magnitude from 20% to 300% compared with stirred tanks and/or tubular reactors without oscillations [11, 15].

With oscillatory reactors being very promising candidates for intensifying slow processes with gradual heat requirement, a possible temperature distribution of heating medium and process medium could be very close to the temperature profiles of a counter-current tubular heat exchanger (see Figure 10.16).

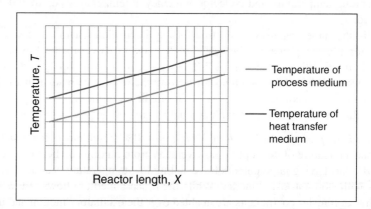

Figure 10.16 *Exemplary illustration of temperature evolution of a heat transfer medium in an OBR gradually heating a process medium*

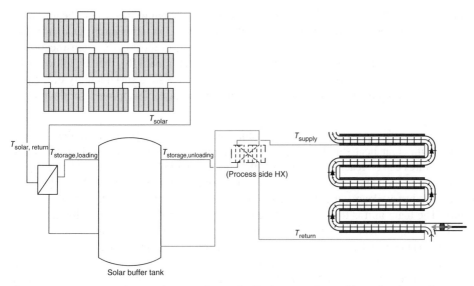

T_{solar}

$T_{solar, return}$ $T_{storage,loading}$ $T_{storage,unloading}$ T_{supply}

(Process side HX)

T_{return}

Solar buffer tank

Figure 10.17 *Basic scheme of an oscillatory baffled reactor heated by solar thermal energy over an energy storage tank*

Figure 10.17 shows a very basic scheme of a solar heated oscillatory baffled reactor, which is gradually heating a process medium. In contrast to the solar heated spinning disk example above, where return temperatures were very close to supply temperatures, here the temperature difference between supply and return temperatures will be much larger, due to the heating requirement of the process itself. This situation will allow the solar buffer storage operating at different temperatures between top and bottom, leading to a reasonable mass flow on the solar side.

In conclusion, it becomes clear that the process heating requirement (temperature profile) will still remain the most important issue in determining the viability of solar thermal energy supply. Any intensified reactor with high heat transfer coefficients will be a promising candidate for integrating solar heat with the *process temperature profile* (rather than the process temperature) mainly determining the technical and economical sensibility.

10.4.1.3   Increase of Temperature Gradients

For the increase of temperature gradients, obviously two approaches exist for processes that are indirectly heated: either increasing the temperature of the energy supply medium or lowering the process temperature. For the integration of solar process heat, the aim will rather be to lower the process temperature as an increase in supply temperature will not favour the efficiency of the solar system. In regions with low direct radiations, the designer might rather aim to lower the *supply temperature* to design a more efficient solar thermal plant. For concentrating systems that produce steam, the supply temperature is not so critical but nevertheless the solar thermal plant will not increase in efficiency if the pressure and temperature of the steam has to be increased. Generally, we can therefore conclude that in many plant designs of solar thermal process heat systems where the process itself is not changed the integration of solar heat will actually lead to a decrease in temperature gradients. This decrease has to be balanced by a corresponding increase in heat transfer surface and/or coefficient. We could argue that a decrease in temperature gradients does not fall under the basic idea of intensification; however, in most cases intensification of heat exchangers aims at an increase in heat transfer coefficient [10], thereby allowing for low-temperature gradients. Additionally, if a measure serves the sustainability of the process while maintaining its efficiency, it is a valid and sustainable approach for the process system design.

Lower Process Temperatures. When it comes to *increasing* temperature gradients in processes with solar heat supply, the process temperature must be lowered. Lower process temperatures can be reached by changing the chemical pathway of a process: This happens in the textile industry with the use of low-temperature detergents. Similarly, in the metal surface industry, degreasing and washing baths are aimed to be operated at low temperature. The main aim is to reduce heat transfer losses from these large, usually uncovered baths. Such approaches enhance the potential for low-temperature heat supply such as solar process heat especially in countries with low to medium solar radiation.

Runback temperatures can obviously be lowered when process temperatures are lower and are largely influenced by the heat exchanger design, which has been shown in Figures 10.13 and 10.16. In a counter-current plate heat exchanger, the runback temperature approaches the temperature at the inlet of the process medium. In such designs, the heat transfer coefficient is so high that temperature gradients can be low. When a vessel is heated at similar temperatures, the runback temperature approaches the vessel temperature in each time step; therefore, we are facing a co-current heat exchanger with much larger temperature gradients. For the design of a solar thermal plant, the first approach is much better which is in line with the common intensification strategy to change from vessel operation to more structured plug flow reactors [16].

Technologies that lead to lower process temperatures are also many membrane based processes. This is further discussed in Section 4.1.5.

Another important approach which is related to this discussion is the decrease of process heating rates for solar heat supply. In this case, however, strictly speaking, the temperature gradients are not lowered, but the heating profile of the process is changed. Obvious examples of this approach can be found when changing from a batch process to a continuous process, eliminating peaks in heating demand.

Direct Steam Injection. Processes with large thermal gradients that have integrating potential for solar heat are direct steam injection processes. One example is the PDX reactor which was developed and marketed by Pursuit Dynamics, UK. In PDX reactors, mixing and heating is performed in a supersonic region leading to intensified mass and heat transfer.

Steam injectors are heated directly with steam; therefore, concentrating collectors would be a feasible option for solar heat integration. A basic scheme of such a PDX reactor for wort heating, heated with concentrating collectors is shown in Figure 10.18. The heat generated by the concentrating collectors is stored in the solar buffer tank, typically with thermo-oil, high pressurized water or steam as storage medium. The heat is then transferred to a steam drum where conditioned water is evaporated to conditioned steam, which consequently is transferred to the reactor.

In concentrating collectors, the effect of the mean collector operating temperature on the collector efficiency is much less pronounced as in non-concentrating systems. However, it

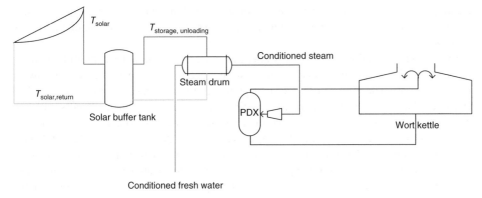

Figure 10.18   *Basic scheme of a steam injector by solar thermal energy over an energy storage tank*

will still be important to reach a distinct temperature gradient between the supply temperature of the solar collectors and the return temperature to enable small piping dimensions and pumps on the solar side. In a direct steam heated system this will be well possible with a well-designed steam drum. The substantial temperature increase from conditioned freshwater to steam would theoretically allow a very low runback temperature to the solar buffer tank. Here, the steam drum design is decisive.

10.4.1.4   Gradientless Process Energy Supply

An important aspect in eliminating heat transfer limitations and intensifying processes is the shift towards gradientless processes, in which driving forces do no longer depend on temperatures of product/process medium and supply medium. Gradientless processes are no longer thermally driven, but rather by electrical energy. There are a number of very promising processes in which this intensification approach has been realised.

Microwave heating is a very promising example as the heat is generated exactly where it is needed and there is no heat for convective/conductive heat transfer to the particles. A typical thermal process in which microwave drying has been widely studied is drying; however, there are also a number of other application examples in which often an intensification of mass transfer is realised together with heat transfer enhancement. Sturm *et al.* recently published a novel continuous flow reactor concept for which temperature distribution analysis has been performed with thermal images (see Figure 10.12).

Microwave heating as well as other electromagnetic-induced heating such as IR or RF have no potential for being heated with solar thermal energy. However, in some cases, the combination of an electromagnetic intensified process with low-temperature heat could be envisioned.

10.4.1.5   Increase in Selectivity of Separation Processes

Membrane processes increase selectivity and efficiency in transporting specific components and can thus improve the performance of reaction (e.g. by shifting the equilibrium of a certain reaction). Due to these facts, membrane processes may intensify energy-intensive techniques [17]. This intensification of transport and reaction efficiency may enable lower process temperatures, which can be exemplary shown for membrane distillation or pervaporation.

Membrane distillation is a promising alternative over conventional evaporation. In membrane distillation, the targeted evaporation runs at much lower temperature than in conventional evaporation due to the difference in vapour pressure over the membrane. Figure 10.19 shows the basic process in a membrane distillation process. An aqueous feed solution is heated on one side of a hydrophobic microporous membrane, which does not allow aqueous molecules to penetrate. Only volatile substances will pass the membrane and are collected on the permeate side. The low operating temperature, due to the fact that the liquid does not need to be heated above its boiling point, is only one of the advantages of membrane distillation next to reduced fouling and low operating pressure. Condensation of the volatile substances on the permeate side can be achieved via several means (aqueous solution; cooled air gap, seep gas transportation to a condenser or vacuum). For further details, the reader is referred to specific literature [17, 19].

Solar thermal energy could be used in membrane distillation to preheat the feed solution prior to entering the membrane module. Figure 10.20 shows a possible process scheme.

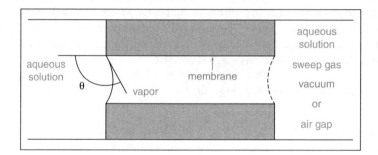

Figure 10.19   *Process schematic of membrane distillation [18]*

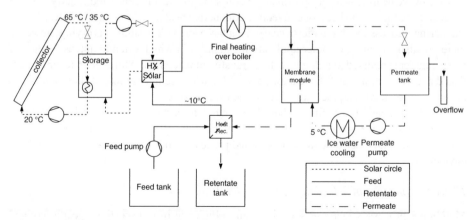

Figure 10.20   *Basic scheme of a membrane distillation unit heated by solar thermal energy over an energy storage tank (AEE INTEC, [20])*

The feed solution is pumped through a first heat exchanger in which the waste heat of the hot retentate preheats the incoming feed stream. The feed is then further heated with solar energy and (if necessary) heated to its target temperature in a final heating unit. When the feed is only once circulated over the membrane, the temperature difference between incoming feed stream and target temperature will be sufficient for efficient operation of the solar system. In such plant designs, the collectors can operate at target temperatures close to (or even below in case of preheating only) the process temperatures.

A similar example of a promising application of solar heat for intensified process systems is pervaporation. In pervaporation, a selective membrane is used as barrier between two phases, the liquid feed and the vapour permeate. The process depends on the sorption equilibrium and the mobility of the components through the membrane and is rather independent of the vapour liquid equilibrium. The desired component, which is in liquid form in the feed, permeates through the membrane and evaporates while passing the membrane, because the partial pressure of the permeating component is kept lower than the equilibrium vapour pressure [21]. Permeabilities depend on the solubility and diffusion rates through the membrane.

Pervaporation units have promising potential to be used in combination with conventional distillation columns to overcome azeotropes in distillation [17]. Feed temperatures are often in the range between 20 °C and 90 °C, showing a potential for solar heat integration even with non-concentrating collectors. A process scheme of a pervaporation/distillation unit supplied by solar thermal energy would be basically similar to solar heat integration for membrane distillation, as also in pervaporation a suitable place for integrating solar heat is when preheating the feed stream prior to entering the membrane module.

10.4.1.6   *Electromagnetic Action on Microorganisms*

When speaking of intensifying thermal processes, the electromagnetic action on microorganisms is an important field of research. In the food industry it is studied as an alternative to thermal pasteurization and sterilization techniques. Table 10.1 shows a short summary of different electric processes currently studied as alternatives to thermal processes in the food industry. These technologies clearly will not lead to an increase of solar process heat potential; however, rapid pasteurization and sterilization processes have also until now not been highly important applications for solar process heat due to the fast heating rates and high potential for heat recovery.

10.4.1.7   *Changing from Batch to Continuous Processes*

In earlier sections, it has become clear that a major impact on the solar heat potential is the required temperature profile of the process itself. The intensification strategy from batch processes to continuous processes is a perfect example of changing the temperature profile to more continuous gradual heating rates, which positively enhances the performance of the solar loop.

The general advantages of a continuous process can be summarized as follows:

- High process efficiency, small residence time distribution, structured processes
- Good process controllability

Table 10.1 Electromagnetic techniques as alternatives to thermal processes in the food industry [22–36]

Technique	Application	Intensification	Challenges
Direct electroheating/ ohmic heating	Sterilization, pasteurization of pumpable foods; blanching of vegetable purees;	Heat transfer decoupled from heat transfer coefficients; uniform heating based on electrical conductivity; very fast change in temperature; microbial inactivation due to thermal effects and electric field	Faster inactivation of enzymes; fouling (proteins) and corrosion (mainly at low frequencies); electrical conductivity temperature dependent (temperature runaway possible); material with non-conductive parts (particles/(fat)globules) & food with high consistency may not be heated uniformly; precise temperature/mass flow control required;
Indirect electroheating/ microwave heating	Drying (including MW-freeze drying and MW vacuum drying), cooking, blanching, concentration, pasteurization, thawing, tempering, roasting and baking; products: meat, fish, dairy products, potatoes, pasta, grains, fruits, vegetable and juices etc.	Thermal intensification, reduction in pasteurization and cooking time; rapid heating; high potential for hybrid processing (with hot air; vacuum, osmotic, freezing, IR/RF)	Non-uniform heating (design of oven); dielectric properties of material change during treatment – effects must be well understood;
Indirect electroheating/ IR heating	Pasteurization	Bacterial deactivation possible at low temperature (e.g. 40 °C)	

Indirect electroheating/ RF heating	Meat cooking, packaged ham pasteurization; disinfestation of fruits; liquid food heating (starch solutions, guar solutions)and pasteurization (milk, fruit juices)	Thermal intensification, reduction in pasteurization and cooking time; rapid heating	Uniformity of temperature in solid food (outside vs. inside); liquid food: temperature distribution and penetration depth needs further analysis on larger scale; generally still lack of information regarding impact on quality; missing dielectric property data; scale-up; lack of mathematical modelling tools
UV irradiation/ pulsed light	Pasteurization/sterilization (e.g. sterilizing films ofpackaging material)	Non-thermal technique; inactivation of microorganisms without chemicals	Poor penetrating power of light
Pulsed electric field	Preservation of pumpable fluid or semi-fluid foods (e.g. milk, liquid whole egg; vegetable soup, fruit juice)	non thermal technique; inactivation of microorganisms without chemicals and heat but via high voltage pulses; combination with mild heat treatment potentially interesting	Research need (e.g. efficiency of PEF treatment; equipment design; effect on food properties); non-uniformity of electric field strength may lead to over-processing and deteriorate quality

- Low energy intensity (no peaks in heating/cooling demand)
- Low cleaning requirements
- Flexible processes
- Decreased energy distribution losses due to continuous heat demand

Figure 10.21 shows the flow characteristics in a typical stirred tank versus a plug flow reactor. While the motion of an incoming particle is hardly defined in a stirred tank, the flow is much better predictable in structured plug flow reactor [16]. This fact shows that reactions in plug flow reactors are much better controllable due to the small residence time distribution, which is again linked to advantages in terms of energy efficiency and supply.

For the performance of the solar process heat system, the continuous heat load has a major impact on the solar buffer storage. Additionally, novel continuous reactors will allow much better (than stirred tanks) for small temperature driving forces, thus bringing the solar supply temperature close to the process temperatures. As discussed earlier, it will be further decisive which temperature difference between solar supply and solar return temperature can be achieved, which basically relates to the process heating rates.

In the following, an example of a storage tank is shown that is loaded with solar heat at 80–90 °C. Two process profiles for unloading the storage tank are compared: a discontinuous process profile with varying heat load and (to show a drastic example) a completely smooth process profile over time. Both processes require the same amount of energy and in total much higher energy than delivered from the solar system, the solar fraction being around 30%. For delivering 90% of energy from the solar collectors over the storage to the process, the required storage tank size is $250\,m^3$ for the discontinuous process profile. In contrast, the continuous process profile would only require a storage tank size of $170\,m^3$ for the same energy supply from solar heat. Figure 10.22 shows the power requirement of the processes and the solar gain as well as storage temperatures (at the top of the storage) and the power of the heat exchangers unloading the solar storage for the process.

The difference between the two process profiles is most obvious when looking at the weekends: While the continuous process can still unload the storage tank on weekends, the discontinuous process has no energy requirement on Saturday and Sunday. This leads to the fact that the storage is fully loaded on these days, which leads to higher storage size requirement and also higher energy losses of the storage tank. The increase in energy

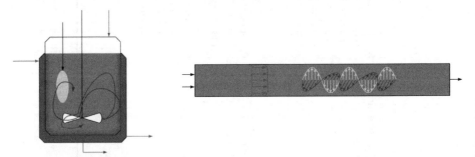

Figure 10.21 *Flow characteristics in stirred tanks versus plug flow reactors (Reprinted (adapted) with permission from Ref. [16]). Copyright (2009) american chemical society*

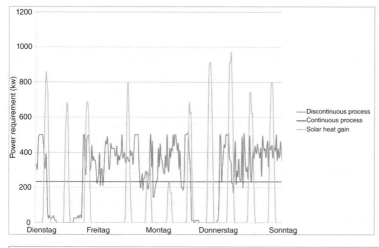

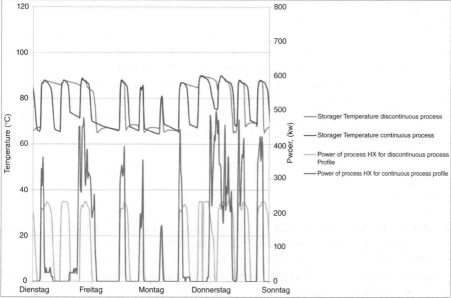

Figure 10.22 *Storage and heat exchange simulation example of a storage tank loaded with solar energy and unloaded with a discontinuous process and a continuous process (simulated in SOCO developed by AEE INTEC and TU graz 2013)*

losses of the storage tank unloaded with the discontinuous process profile is 2% based on this simulation example. On Mondays, the storage tank can now be unloaded and the heat load withdrawn from the heat exchanger on the process side peaks to almost 500 kW. In the continuous process profile, heat is withdrawn whenever available at a rather constant rate of around 220 kW. Thus, obviously heat exchanger sizes can be reduced when a smooth heating profile is possible.

10.5   Conclusion

This chapter has discussed whether intensified process layout will have effects on the potential of solar process heat and its efficiency. Basically, it can be concluded that the most important factor for low-temperature energy supply is the temperature profile of the process itself. The lower the process temperature and the larger the temperature difference of the process medium during the heating process, the easier the integration of low-temperature heat transfer media. However, it depends also on the chosen process technology, how closely the process temperature profile is actually represented in the energy supply temperature profile. In process technologies with low heat transfer coefficients, temperature gradients often need to be high to enable the required heat transfer with the existing heat exchange area. In such cases, any intensified technology with high heat transfer will enhance the efficiency of solar heat. Additionally, any measure which lowers heat demand peaks will be beneficial. An important aspect is the change from batch to continuous processing. The sun will always remain a batch process itself, and the combination with a continuous load profile instead of another contradicting batch profile will increase the efficiency of the solar process heat system.

Finally, we can conclude that process modelling will remain an important aspect in the future designing of sustainable production sites, where the effects of new process technologies on the overall energy management of the site can be analysed.

Concluding it can be forecasted a reasonable potential for the combination of solar thermal energy and process intensification in various industry sectors. New research developments and realized applications will lead hopefully to follow-ups and to a significant reduction of energy consumption and CO_2 emissions.

References

1. Muster-Slawitsch, B., Schmitt, B., Krummenacher, P. *et al.* (2014) in *Integration Guideline* (ed B. Muster-Slawitsch), IEA Task 49/IV.
2. Schmitt, B. (2014) Integration of solar heating plants for supply of process heat in industrial companies (in German language), Dissertation University of Kassel, Shaker Verlag, Aachen, Germany.
3. Pardo, N., Vatopoulos, K., Krook-Riekkola, A., Moya, J.A., Perez, A. (2012) Heat and cooling demand and market perspective, Joint Research Centre - Institute for Energy and Transport. European Comission, EUR -Scientific and Technical Research series, ISBN 978-92-79-25310-2, doi: 10.2790/56532
4. International Energy Agency (2012) Technology Roadmap Solar Heating and Cooling, available from: http://www.iea.org/publications/freepublications/publication/Solar_Heating_Cooling_Roadmap_2012_WEB.pdf, last accessed February, 2015
5. Muster-Slawitsch, B. (2014) Thermal energy efficiency and process intensification for the food industry - methodology development for Breweries, Dissertation. Technical University Graz, Graz.
6. Reay, D., Ramshaw, C. and Harvey, A. (2013) *Process Intensification: Engineering for Efficiency, Sustainability and Flexibility*, Butterworth-Heinemann.

7. Reay, D. (2008) The role of process intensification in cutting greenhouse gas emissions. *Applied Thermal Engineering*, **28**, 2011–2019.
8. Brunner, C., Mauthner, F., Hubmann, M., Muster-Slawitsch, B. (2013) Process integration design as basis for the call for tender for brewery production sites, Deliverable 2.1 within FP7 project Solarbrew, Project Number: 295660. Gleisdorf.
9. Ferrouillat, S., Tochon, P., Valle, D.D. and Peerhossaini, H. (2006) Open loop thermal control of exothermal chemical reactions in multifunctional heat exchangers. *International Journal of Heat and Mass Transfer*, **49**, 2479–2490.
10. Anxionnaz, Z., Cabassud, M., Gourdon, C. and Tochon, P. (2008) Heat exchanger/ reactors (HEX reactors): concepts, technologies: state-of-the-art. *Chemical Engineering and Processing: Process Intensification*, **47**, 2029–2050.
11. Stephens, G.G. and Mackley, M.R. (2002) Heat transfer performance for batch oscillatory flow mixing. *Experimental Thermal and Fluid Science*, **25**, 583–594.
12. Trambouze, P. and Euzen, J.P. (2002) *Les réacteurs chimiques de la conception à la mise en oeuvre*, Editions Technip, Paris.
13. Nassar, N.N. and Mehrotra, A.K. (2011) Design of a laboratory experiment on heat transfer in an agitated vessel. *Education for Chemical Engineers*, **6**, e83–e89.
14. Ni, X., Mackley, M.R., Harvey, A.P. *et al.* (2003) Mixing through oscillations and pulsations—a guide to achieving process enhancements in the chemical and process industries. *Chemical Engineering Research and Design*, **81**, 373–383.
15. Solano, J.P., Herrero, R., Espín, S. *et al.* (2012) Numerical study of the flow pattern and heat transfer enhancement in oscillatory baffled reactors with helical coil inserts. *Chemical Engineering Research and Design*, **90**, 732–742.
16. Van Gerven, T. and Stankiewicz, A. (2009) Structure, energy, synergy, times - the fundamentals of process intensification. *Industrial & Engineering Chemistry Research*, **48**, 2465–2474.
17. Drioli, E., Stankiewicz, A.I. and Macedonio, F. (2011) Membrane engineering in process intensification—an overview. *Journal of Membrane Science*, **380**, 1–8.
18. Wang, Z., Gu, Z., Feng, S. and Li, Y. (2009) Applications of membrane distillation technology in Energy transformation process-basis and prospect. *Chinese Science Bulletin*, **54**, 2766–2780.
19. El-Bourawi, M.S., Ding, Z., Ma, R. and Khayet, M. (2006) A framework for better understanding membrane distillation separation process. *Journal of Membrane Science*, **285**, 4–29.
20. Brunner, C., Herzog, U., Muster, B. (2012) Darstellung des Potentials zur Effizienzsteigerung und zum erhöhten Einsatz Solarer Prozesswärme durch den Einsatz neuer Technologien für die betrachteten Prozess, Projekt Solar Foods im Rahmen des Programms "Energie der Zukunft".
21. Wee, S.-L., Tye, c.-T. and Bhatia, S. (2008) Membrane separation process— pervaporation through zeolite membrane. *Separation and Purification Technology*, **63**, 500–516.
22. Jakób, A., Bryjak, J., Wójtowicz, H. *et al.* (2010) Inactivation kinetics of food enzymes during ohmic heating. *Food Chemistry*, **123**, 369–376.
23. Stancl, J. and Zitny, R. (2010) Milk fouling at direct ohmic heating. *Journal of Food Engineering*, **99**, 437–444.

24. Icier, F. (2012). Ohmic heating of fluid foods. In P. J. Cullen, B. K. Tiwari, & V. P. Valdramidis (Eds.), *Novel Thermal and Non-Thermal Technologies for Fluid Foods*, Academic Press, Elsevier Inc., USA, pp. 305–368. doi:10.1016/B978-0-12-381470 -8.00017-7.

25. Umesh Hebbar, H. and Rastogi, N.K. (2012) Microwave heating of fluid foods, Academic Press, Elsevier Inc., USA, pp. 369–409.

26. Mao, W., Oshima, Y., Yamanaka, Y. *et al.* (2011) Mathematical simulation of liquid food pasteurization using far infrared radiation heating equipment. *Journal of Food Engineering*, **107**, 127–133.

27. Marra, F., Zhang, L. and Lyng, J.G. (2009) Radio frequency treatment of foods: review of recent advances. *Journal of Food Engineering*, **91**, 497–508.

28. Pereira, R.N. and Vicente, A.A. (2010) Environmental impact of novel thermal and non-thermal technologies in food processing. *Food Research International*, **43**, 1936–1943.

29. Laycock, L., Piyasena, P. and Mittal, G.S. (2003) Radio frequency cooking of ground, comminuted and muscle meat products. *Meat Science*, **65** (3), 959–965.

30. Orsat, V., Bai, L., Raghavan, G.S.V. and Smith, J.P. (2004) Radio-frequency heating of ham to enhance shelf-life in vacuum packaging. *Journal of Food Process Engineering*, **27**, 267–283.

31. Awuah, G.B., Ramaswamy, H.S. and Piyasena, P. (2002) Radio frequency (RF) heating of starch solutions under continuous flow conditions: effect of system and product parameters on temperature change across the applicator tube. *Journal of Food Process Engineering*, **25** (3), 201–223.

32. Awuah, G.B., Ramaswamy, H.S., Economides, A. and Mallikarjuanan, K. (2005) Inactivation of Escherichia coli K-12 and Listeria innocua in milk using radio frequency (RF) heating. *Innovative Food Science and Emerging Technologies*, **6** (4), 396–402.

33. Piyasena, P. and Dussault, C. (2003) Continuous radio-frequency heating of a model viscous solution: influence of active current, flow rate, and salt content on temperature rise. *Canadian Biosystem Engineering*, **45**, 327–334.

34. Geveke, D.J. and Brunkhorst, C. (2004) Inactivation of Escherichia coli in apple juice by radio frequency electric fields. *Journal of Food Science*, **69** (3), FEP 134–FEP 138.

35. Geveke, D.J. and Brunkhorst, C. (2008) Radio frequency electric fields inactivation of Escherichia coli in apple cider. *Journal of Food Engineering*, **85** (2), 215–221.

36. Toepfl, S., Mathys, A., Heinz, V. and Knorr, D. (2006) Review: potential of high hydrostatic pressure and pulsed electric fields for energy efficient and environmentally friendly food processing. *Food Reviews International*, **22**, 405–423.

11

Bioenergy – Intensified Biomass Utilization

Katia Gallucci and Pier Ugo Foscolo
Department of Industrial Engineering, University of L'Aquila, 67100, L'Aquila, Italy

11.1 Introduction

A major objective of European energy policy is to move towards more sustainable development based on a diverse mix of resources, in particular, renewable resources including biomass. Programmes devoted to this end address the pressing challenges of security of supply and climate change imponderables thereby supporting the future competitiveness of European industry.

In contrast to centralized power generation systems based on fossil energy, bioenergy is amenable to distributed (local) production and deployment in areas close to those of biomass availability; it is therefore in good accord with current policies of power market deregulation and power generation decentralization.

This chapter addresses the broad field of power and combined heat and power (CHP) generation from biomass: more specifically, advances in biomass gasification technology aimed at increasing the overall conversion and efficiency – and hence in a decreased cost of electricity.

Poly-generation strategies (for combined heat, power and chemical production applications) are also considered, with particular reference to recent technological innovations in hot gas cleaning and conditioning; these have been developed to achieve the required improvements in syngas quality and have been validated under industrially relevant conditions.

Process Intensification for Sustainable Energy Conversion, First Edition.
Edited by Fausto Gallucci and Martin van Sint Annaland.
© 2015 John Wiley & Sons, Ltd. Published 2015 by John Wiley & Sons, Ltd.

11.2   Biomass Gasification: State-of-the-Art Overview

Biomass gasification is a thermo-chemical conversion process for the production of a fuel gas. Various combinations of air, oxygen and steam may be used as the gasification agent. Because gasification is an endothermic process, it necessitates the provision of thermal energy to the reactor, which can be achieved in various ways that are described in the following text. Most practical applications involve the combustion of part of the gasification product and sometimes also of auxiliary fuels produced during the downstream purification process. Hence, the requirement for an oxygen/air supply to the reactor system, which may be expressed as an *equivalence ratio* (ER), the ratio of the oxygen requirement to what would be needed for complete biomass combustion – values typically of about 0.3 being found in practice. The inclusion of steam in the reactor feed has the effect of improving the performance of hydrocarbon and char reforming processes and enhancing the water-gas shift towards higher hydrogen yield.

This chapter focuses on steam gasification processes carried out in fluidized bed reactors for the production of H_2-rich syngas. Fixed bed down-draft and up-draft gasifiers have also been used, but these are restricted to small-scale applications of biomass gasification with air and are not amenable to scale-up or to *in-situ* gas purification; they are also difficult to operate under steady-state and intrinsically safe conditions.

Steam gasification produces a fuel gas rich in hydrogen and carbon monoxide with a significant methane content. Carbon dioxide, steam and nitrogen are also present in the product gas, in addition to organic (tar) and inorganic (H_2S, HCl, NH_3, alkali metals) impurities and particulates. Tar – high-molecular-weight hydrocarbons – is an undesirable and noxious by-product [1, 2] that may cause major clogging, fouling, efficiency loss and unscheduled plant stoppages [3]. Tar yields can be reduced by careful control of the operating conditions (temperature, biomass heating rate, etc.), appropriate reactor design and a suitable gas conditioning system [4–6].

Biomass gasification plants, integrated with gas cleaning and heat and power generation facilities, have been built and are in operation in many European countries. These demonstrate convincingly the potential of such technologies to sustainable energy production in both developed and developing regions of the world (Table 11.1).

Among these, the well-known combined heat and power plant in Güssing, Austria, is particularly noteworthy not only as a successful industrial reality, but also in view of the light it sheds on the feasibility of the technical concepts involved and the economic benefits derived from the diversification of biomass syngas applications in energy and chemical production systems.

The capacity of the Güssing plant is about 8 MW (electrical output 2 MWe, district heating output approximately 4.5 MWth). Gasification is carried out in a dual fluidized bed arrangement (the fast internally circulating fluidized bed – FICFB – gasification system) developed at the Institute of Chemical Engineering of Vienna University of Technology (TUV) in cooperation with AE Energietechnik [9, 10].

The basic idea of this gasification system is separation of the gasification and the combustion reactions in order to obtain a largely nitrogen-free product gas. The endothermic gasification of the biomass fuel takes place in a stationary bed fluidized with steam. This is connected via a chute to the combustion section that is operated as a circulating bed fluidized with air. Here, any non-gasified fuel particles transported along with the bed

Table 11.1   *Overview of worldwide biomass gasification plants and systems [7,8]*

Country	Capacity (MW$_{th}$)	Technology	Location
Austria	8	TUV FICFB CHP	Güssing
	8.5		Oberwart
	15		Villach
	2	down-draft CHP demonstration	Wiener Neustadt (not in operation any more)
Denmark	5	VØlund up-draft CHP demonstration	HarbØre
	0.7	Viking two-stage gasification and power generation	Lyngby
	3.125 and 0.833 (Japan)	TK Energi three-stage, gasification process demonstration	Gjøl
	30	Carbona Renugas fluidized bed CHP demonstration	Skive
Finland	4–5	Bioneer up-draft gasifiers	8 in Finland and one in Sweden
	60 (50–86)	Foster Wheeler Energy CFB co-firing plant	Lahti (Ruien, Belgium)
	40	Foster Wheeler Energy fluidized bed metal recovery gasifier	Varkaus
	7	NOVEL up-draft demonstration	Kokemäki
Germany	130	Commercial waste to methanol plant (fixed bed + pressurized entrained flow)	Schwarze Pumpe
	100	Lurgi CFB gasifier firing cement kiln	Rüdersdorf
	0.5	Fraunhofer Umsicht CFB pilot plant	Oberhausen
	1	CHOREN Carbo-V 2-stage entrained pilot plant	Freiberg
	3–5	Future energy pyrolysis/entrained flow GSP gasifier	Freiberg
	11.5	TUV FICFB CHP	Ulm (2011)
Italy	1	ENEA CFBG pilot plant	Trisaia (similar plant operated in China)
Netherlands	85	AMER/Essent/Lurgi CFB gasification co-firing plant	Geertruidenberg
	250 (35 MWe from biomass)	Biomass co-gasification shell entrained coal gasification plat	Willem-Alexander Centrale
	3	CFBG Plan	Tzum
		Several pilot plants at ECN	Petten

(*continued overleaf*)

Table 11.1   (continued)

Country	Capacity (MW$_{th}$)	Technology	Location
New Zealand	2	Page Macrae up-draft BMG plant	Tauranga
Sweden	30	Foster wheeler energy CFBG	Karlsborg paper mill
	20	Foster wheeler energy CFBG	Norrsundet paper mill
	30	Gotaverken CFBG	Södra Cell paper mill
	18	Bioflow/Sydkraft/Foster wheeler energy CHP demonstration at	Värnamo
	32	TUV FICFB CHP	Göteborg
Switzerland	0.2	Pyroforce down-draft BMG system	Spiez (scale-up to 1 MWe plant in Austria)
United Kingdom	100 KWe	Rural generation down-draft BMG system	Northern Ireland
	Up to 250 KWe	Biomass Engineering Ltd., down-draft BMG CHP systems	Northern Ireland
	Up to 300 KWe	Exus energy down-draft BMG CHP systems	Northern Ireland
	7 MWe	Charlton energy rotary kiln waste gasification	Gloucestershire
	2 MWe	Compact power two-stage waste gasification plant	Bristol
USA	Up to 120	Primenergy gasification/combustion systems	6 in USA
	Up to 22 KWe	Community Power Corporation small modular down-draft gasification systems FERCo SilvaGas dual CFBG process RENUGAS fluidized bed BMG process FERCo SilvaGas dual CFBG process RENUGAS fluidized bed BMG process	

material are fully burnt, together with additional fuel fed directly to the bed, to provide the heat required for the gasification reactions. The heated bed material is then separated in a cyclone and fed back into the gasification section. Literature references on such dual fluidized bed systems for biomass gasification can be found in Koppatz *et al.* [11].

The gas cleaning process was developed by RENET, Austria. It operates in three stages. Fine hydrated lime is first injected into the system to reduce the tar content to less than 1 g/Nm^3, accompanied by enhanced recovery of sensible heat from the product gas stream;

particulate filtration is then carried out at about 150 °C; finally, the fuel gas is contacted with biodiesel (RME) in a scrubbing column to reduce the tar content to about 20 mg/Nm^3 at a temperature of about 50 °C. The calorific value of the product gas is 12–14 MJ/Nm^3; it can be either used in a gas engine or upgraded to synthesis gas.

Construction of the Güssing plant commenced in September 2000 and electricity was first generated in April 2002; since then the plant has been operating regularly and is presently continuously on stream for about 80% of the year. Similar facilities now in operation have been designed and constructed in accord with this technology (Table 11.1).

The dual fluidized bed gasification system combines the advantages of steam as the gasification agent (lower tar) and a nitrogen-free product gas (higher calorific value). This is made possible by integration of the separate gasification and combustion zones.

An alternative and simpler approach that results in a product gas of comparable quality is biomass gasification with oxygen and steam in a single reactor with a single gaseous output stream [12]: a well-known application being the revamping of the pressurized Värnamo plant, Sweden [13]. In this case, the advantages of a gasification system less complex to design, construct and operate are counterbalanced by the need to use oxygen in place of air. The 18 MWth Värnamo plant operated at 18 bar pressure. The raw gases are cleaned without condensation by using candle filters and combusted in a closely integrated Typhoon gas turbine to generate 6 MWe and 9 MWth heat for district heating. The cost of oxygen has been falling steadily over the years – today the vast majority of coal gasification processes use oxygen blown gasifiers [14] – so that this option will certainly become increasingly attractive in the medium term and for smaller scale applications.

An innovative steam/oxygen gasifier (1 MW load) has been commissioned at the Trisaia research centre of the Italian National Agency for Renewable Energy and the Environment (ENEA) and adapted to test the integrated gasification and hot gas cleaning *UNIQUE* concept to be described in some detail later in this chapter and in Foscolo *et al.* [15]. The fluidized bed is fitted with a vertical partition plate separating two interconnected bubbling sections. The bed inventory, which contains primary catalyst and sorbents, is made to circulate around these two sections by maintaining different fluidization fluxes on either side of the partition, thereby promoting good mixing of the light fuel particles within the much denser granular bed of mineral material.

A further key feature of the design provides for self-regulation of the overall amount of solid material in the bed by means of a weir over which the overload flows to an adjacent chamber from which it may be withdrawn. The reactor product gas passes through a candle filter installed in the bed freeboard, and so arranged that the fine particle agglomerates formed at its surface fall to the top of the bed and so over the weir.

This gasifier configuration was developed with the aid of a cold model, tested and verified by reference to a similar set-up for air gasification of sawdust and rice husks – at the Liaoning Institute for Energy Resources in China – which exhibited good temperature homogeneity throughout the system [16].

The International Energy Agency (http://www.ieatask33.org/content/thermal_gasification_facilities) provides regular updates of the above information.

11.2.1   Cold Gas Cleaning and Conditioning: Current Systems

Gas cleaning is normally carried out by filtration and scrubbing of the product gas, a procedure that can drastically reduce the particulate and tar contaminants.

Table 11.2 *Classification of gas cleaning processes*

Temperature range	Basic type	Effectiveness	Gas cleaning system
20–60 °C	Wet	Tar, particles, metal compounds, permanent pollutant gas fraction	Packed column scrubber, quench column, venturi scrubber, (wet) electro-static precipitator (ESP), etc.
140–300 °C	Dry	Particles, metal compounds, permanent pollutant gas fraction	Dust ESP, filtration deduster, etc.
300–800 °C	Dry	Particles, permanent pollutant gas fraction, (tar)	Filtration deduster, dust ESP, etc.

Source: Reproduced from Ref. [17].

Such mechanical/physical methods, summarized in Table 11.2, fall into two major categories: dry and wet gas cleaning [18]. Dry gas cleaning – using cyclones, hot gas filters, sand bed filters, adsorbers, etc.– is carried out prior to gas cooling at temperatures typically above 500 °C. Dry gas cleaning using fabric filters takes place typically below 200 °C after gas cooling. Wet gas cleaning (in scrubbers, wet electrostatic precipitators, wet cyclones, etc.) is usually performed at 20–60 °C.

Wet cleaning processes operate of necessity at close to ambient temperatures. Under such conditions, the most immediate option for the cleaned gas is power generation in a gas engine, for which, however, electric conversion efficiencies are low: reported values close to 25–30%, no more than can be achieved from combustion plants coupled to steam turbines. This penalizes notably the overall economic performance, which would benefit from a higher ratio of electricity over heat production – even more so in consideration of the incentives for green electricity offered in most countries. In addition, the tar removal process is generally less effective, results in a reduction of gas yield and gives rise to waste streams, which are difficult to process, dispose of or recycle. Further drawbacks of cold gas cleaning result from the corrosive nature of wet gases and condensates.

A series system of tar removal technologies is often required to achieve the required threshold limit for downstream applications: for gas engines the requirements are less than 5 mg/Nm^3 particulates and less than 25 mg/Nm^3 tar.

Typical separation efficiencies for mechanical/physical gas cleaning systems are shown in Figure 11.1 for various particle and droplet sizes [20, 19].

The tar-removing capabilities of rotating particle separators (RPS) and fabric filters have been reported to be 30–70% and 0–50%, respectively [19]. Tar deposited on fabric filters cannot be easily cleaned, and tar accumulation on the filter surfaces leads to eventual plugging [21].

Additional methods for tar removal are fixed bed adsorbers made of lignite coke – providing good adsorption at low cost – activated carbon and sand bed filters. Activated carbon filters are usually installed upstream a fabric filter and downstream the RPS and are able to remove at least 70% tar. Sand bed filters are also suitable for removing tar from biomass gasification products, achievable reduction levels ranging from 50 to 97% [19], Pathak *et al.* [22] reported them achieving a 90% reduction of particulates.

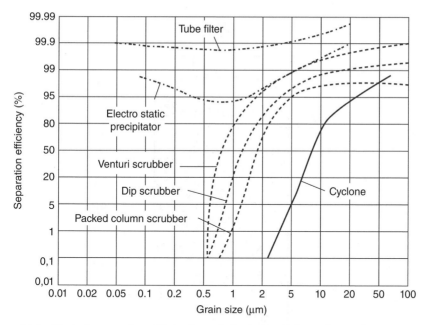

Figure 11.1 *Typical separation efficiencies of mechanical/physical gas cleaning systems.* (Source: Reproduced from Ref. [11], with permission from Elseiver)

The most common types of wet scrubbers are spray towers, venturi scrubbers, packed bed scrubbers and impingement scrubbers. Spray towers are the simplest, least expensive, however not very efficient, tar separation devices [23].

Bhave *et al.* [24] investigated the performance of a wet packed bed scrubber (Figure 11.2) in product gas cooling/cleaning applications. This unit is able to reduce tar and dust content by 75% of its inlet gas concentration levels. For impingement scrubbers, the reported overall efficiency is also about 70%. Higher tar removal efficiencies can be achieved by connecting wet impingement scrubbers in series, three being required to obtain efficiencies of above 95%. This gas cleaning arrangement has the additional desirable feature of being of simple construction [25].

Venturi scrubbers (Figure 11.3) can also achieve high removal capabilities, ranging from 50 to 90%, for tar and particulates.

Swirl cyclone-scrubbers [26, 27] (Figure 11.4) exhibit high and stable particle collection efficiencies, negligible pressure drop (110–120 mm H_2O) and low construction, operation and maintenance costs. In addition, this system successfully resolves the problem of clogging by salt formation and/or sticky particulates within collection devices.

The scrubbing process is a unit operation in which contaminants in the gaseous stream are selectively absorbed into a liquid phase: the gas phase contaminant must therefore be soluble to some extent in the scrubbing liquid [28].

Water as a washing medium is often unsatisfactory regarding regeneration efficiency and continuous operation: severe problems can result from saponification, low solubility of hydrocarbon compounds, surface tension effects and clogging [17]. Lee *et al.* [26] also

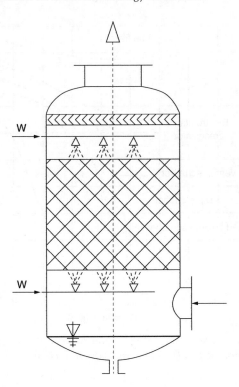

Figure 11.2   *Wet packed bed scrubber. (Source: Reproduced from Ref. [17])*

reported significant clogging and fouling problems in conventional wet scrubbers, resulting from salt formation within and around nozzles and at the tube walls.

Further disadvantages of conventional wet scrubbing are high cost of treatment, disposal of waste liquid and sludge streams and high operational costs incurred in maintaining separation efficiency. In addition, wet cleaning processes lead to a lowering of the product gas heating value and of the net energy efficiency of the process.

To mitigate the disadvantages of wet scrubbing, various washing agents can be included in the liquid feed – which can then range in composition from pure water, through water–oil emulsions, to water–oil mixtures. With the water medium, removal of tar content can be attributed mainly to condensation, the water temperature being lower than that of the entering gas stream: tar condenses as an oily liquid on the water surface [29].

Higher tar removal efficiencies have been found with the use of vegetable and engine oils – but the tar content at the exit of a diesel and biodiesel oil scrubber has even been reported to increase, due probably to solvent evaporation [29].

The sorption efficiency of light aromatic hydrocarbon tars has been found to depend on whether they are hydrophilic or hydrophobic. On scrubbing with pure water, it has been observed that only phenol is removed – phenol being polar, for example hydrophilic. Whereas the other light aromatic hydrocarbon tars – benzene, toluene, xylene, styrene and indene – are non-polar hydrophobic compounds.

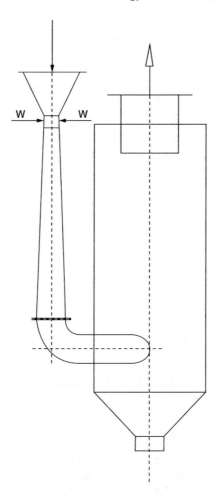

Figure 11.3 *Venturi scrubber Ref. [17]*

On this basis, it is to be expected that the absorption efficiency for polar phenol should be lower in non-polar oil scrubbers than in water scrubbers. However, this has not been found to be the case: no significant differences in the absorption efficiency of phenol in other scrubbing mediums, other than diesel fuel, have been observed. A possible explanation for this could be that phenol molecules are removed by the oily material as a result of van der Waals forces, thereby rendering oils a more effective scrubbing medium than water.

Sorbent viscosity has also been found to play an important role, lower viscosities in general enhancing sorption efficiency [30]. Other factors, however, may play a more decisive role. It has been found, for instance, that although diesel has lower viscosity than biodiesel oil, the sorption efficiency of diesel for light aromatic hydrocarbon tars is lower than that of biodiesel oil. This could be due to evaporation of some of the additives present in the diesel oil, for example, xylene and phenol.

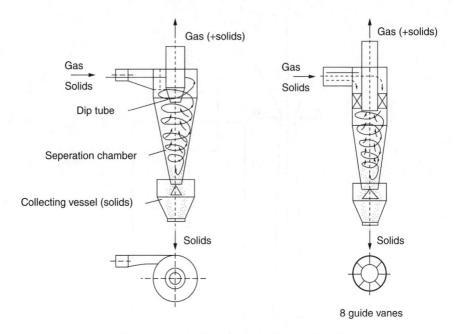

Figure 11.4   *Swirl cyclone scrubbers Ref. [17]*

Table 11.3   *Absorption efficiencies of tar components by various sorbents (%)*

	Water	Diesel fuel	Biodiesel fuel	Vegetable oil	Engine oil
Benzene	24.1	77	86.1	77.6	61.7
Toluene	22.5	63.2	94.7	91.1	82.3
Xylene	22.1	−73.1	97.8	96.4	90.7
Styrene	23.5	57.7	98.1	97.1	91.1
Phenol	92.8	−111.1	99.9	99.7	97.7
Indene	28.2	97.9	97.2	97.6	88.7
Naphthalene	38.9	97.4	90.3	93.5	76.2

Source: Adapted from Ref. [29].

Naphthalene is a key component of tars present in biomass gas. Its absorption efficiency in oily liquids can be ranked as follows: diesel fuel > vegetable oil > biodiesel fuel > engine oil (Table 11.3). Diesel fuel however is expensive: from an economic viewpoint, vegetable oil becomes the best option for biomass tar removal.

Absorbed light aromatic hydrocarbon tars (especially benzene, toluene and xylene) can be recovered from the exhausted oily scrubbing liquid by desorption (stripping); this liquid is frequently used in the gasification process as an auxiliary fuel source – the chosen option in the Güssing plant.

To reduce costs, waste vegetable oil – resulting from food frying, for example – can be used as a sorption medium. Gravimetric tar removal efficiencies of fresh vegetable and waste cooking oils have been compared for scrubbers operating under otherwise

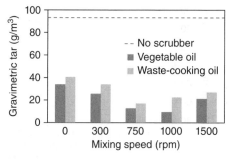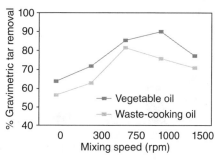

Figure 11.5 *Turbulent mixing effect in vegetable oil and waste-cooking oil scrubbers on tar removal. (Source: Reproduced from Ref. [31], with permission from Elsevier)*

identical conditions. Somewhat better removal efficiencies were found with the pure vegetable oil. This is to be expected given the oxidative, hydrolytic and thermolytic reactions that take place in the food frying process, which give rise to contaminants that are difficult to remove by pretreatment and which may contribute to a decrease in tar removal efficiency by blocking sorption sites of the waste cooking oil molecules.

Turbulent mixing effects in oil scrubbers have also been investigated by installing a stirrer in the liquid pool at the bottom of a laboratory-scale scrubber [31]. It was found that the best conditions corresponded to rotating speeds within the range of 750–1000 rpm (Figure 11.5). This result can be explained qualitatively in terms of the factors influencing the scrubbing performance: an increase in mixing speed giving rise to an increase in gas/liquid contact area and hence an increase in the van der Waals interactions between oil and tar molecules. On the other hand, the contact time of tar with the scrubbing medium decreases with stirrer speed, this effect coming to dominate at stirring speeds above a certain value.

Another physical tar removal technique is dry adsorption on a solid sorbent [31]. Such systems are simple to design and operate. Activated carbon is an effective sorbent due to its high porosity; it is readily available from carbon-containing materials such as coal and biomass. Char generated from rice husk gasification, for example, is very effective as a tar removal medium [32]. For industrial applications, hybrid systems, including wet scrubbing followed by dry sorption on activated char, provide high performance and low-temperature tar removal solutions (Figure 11.6).

The tar sorption effectiveness of rice husk char in sorbent beds has been characterized in terms of breakthrough curves for a variety of tar components. Retention effects were not found to change abruptly, sorbent saturation being reached after long periods of operation. Exhaust char can be regenerated using simple thermal, extractive or chemical procedures. Alternatively, it may be used as a fuel in the gasifier – during the initial heating-up period, for example, by mixing it in with the bed inventory (silica/olivine sand).

Another example of a commercial-scale hybrid cold gas cleaning application is provided by the Harbøøre up-draft gasification plant in Denmark [33]. Here, the gas is first processed in a water scrubber, with further water/tar aerosol and dust removal being carried out in a wet ESP (Figure 11.7). This combined operation results in the contents of both tar and dust in the cleaned gas being brought to below 25 mg/Nm^3, at 40 °C.

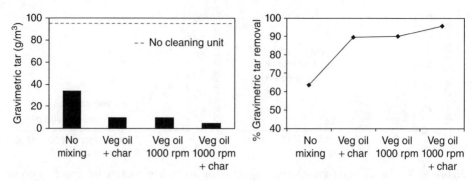

Figure 11.6   *Tar removal performance for the combination of a vegetable oil scrubber (operated under various conditions) followed by a rice husk char sorption bed. (Source: Reproduced from Ref. [31], with permission from Elsevier)*

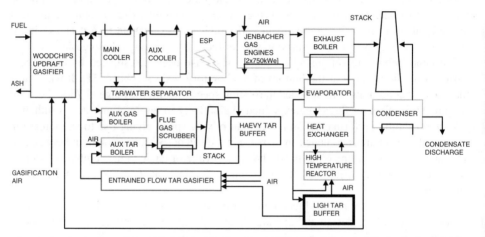

Figure 11.7   *Process scheme for the harbøøre gas cleaning system with water-based tar scrubber, wet ESP and TARWATC system. (Source: Reproduced from Ref. [33])*

Water-based scrubbing technology results in huge amount of tar-contaminated water. The Harbøøre plant produces 0.6 kg of waste water per kg of biomass [34]; the heavy tar in this case is recovered by coalescence and used to augment district heating capacity during peak consumption periods.

Tar-contaminated water cannot be discharged in the sewage system because of its contaminants: phenol ($\approx$10 g/l), total organic carbon (TOC $\approx$ 45 g/l) and acids (mainly acetic and formic acids, pH $\approx$ 2). Cleaning is necessary in a TARWATC system that uses heat from the engine exhaust boilers to distil off the tars (low heating value (LHV) about 14 MJ/kg). The resulting mildly contaminated steam is stripped in counter-flow with the clean flue gas/steam stream from the high-temperature reactor before being passed to the reactor itself, the temperature of which is further increased by the burning of the light tars; the cleaned steam is then condensed in a district heating exchanger. This water-based scrubbing procedure for cleaning the gas sufficiently for certain applications and in compliance with environmental regulations is however costly [34] – regardless of the fact that scrubber oil emulsions act as solvents, supporting the cleaning procedure and protecting the system from clogging [17].

Fuel gas cleaning with oil scrubbers has been applied successfully downstream the fast internally circulating fluidized bed (FICFB) gasifier in Güssing [9, 33, 35] where waste streams are appropriately recycled to minimize disposal problems (Figure 11.8).

Here the product gas exiting the gasifier at about 850 °C is cooled to about 150 °C in a pressurized water exchanger, the recovered heat being used for district heating purposes. The fuel gas is then de-dusted in a fabric filter. The removed dust, which contains fine char particles, is then returned to the combustion chamber of the gasifier. The downstream scrubber, which employs as solvent rapeseed oil methyl ester (RME), significantly reduces the concentration of tar, ammonia and acidic components in the gas; when loaded with tar, the solvent is returned as auxiliary fuel to the combustion section. Such strict integration makes it possible to recycle all waste streams back into the gasification reactor [17].

For economic reasons, the use of RME as scrubbing liquid requires initial tar concentrations in the product gas to be relatively low. At the Güssing plant, tar concentrations downstream the fabric filter are reduced to approximately 2.5 g/Nm^3 by the preliminary dry de-tarring treatment referred to earlier: injection of fine limestone particles into the gasifier freeboard. Benzene and remaining tar compounds are then almost completely removed in the RME scrubber, operating at 5 °C [33].

An alternative to RME scrubbing for gas cleaning is provided by the oil-based OLGA tar removal technology that reduces the scrubbing liquid consumption (Figure 11.9).

OLGA oil gas washing technology was developed by ECN and Dahlman [35, 37] and involves multi-stage scrubber processing. In the first stage (the collector), where the gas is gently cooled by the scrubbing oil, more than 99% of the heavy tars are condensed and, after separation, recycled to the gasifier. In the second stage (the absorber/stripper), lighter gaseous tars are absorbed by the scrubbing oil, which is then regenerated in the stripper. Air loaded with light tars is fed as gasifying medium to the gasifier. The net effect is that heavy and light tars are recycled to the gasifier where they are consumed, contributing to an increase in energy conversion efficiency [38].

OLGA installation costs are high however, and energy is lost in cooling down the RME scrubber – factors that influence the economic performance in similar ways to secondary waste water treatments in less sophisticated gas cleaning installations [31].

An overview of concepts implemented for gas cleaning in various plant configurations is reported in Table 11.4 [17].

Although recycling of residues in gasification installations has many very clear advantages, care is needed in taking into account its effect on operating conditions and performance, such as temperature levels and additional air requirements to ensure complete combustion, also with regard to the control of pollutants from organic and inorganic matters.

11.3 Hot Gas Cleaning

11.3.1 Contaminant Problems Addressed

It is now universally recognized that for optimal energy efficiency, the cleaning and conditioning of gasification product gas should be carried out at high temperature, which is close

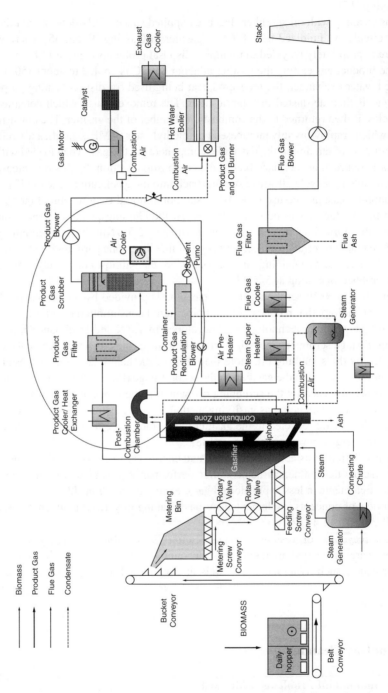

Figure 11.8   *Fuel gas cleaning with dry dust precipitation and wet RME scrubbing in güssing. (Source: Reproduced from Ref. [36])*

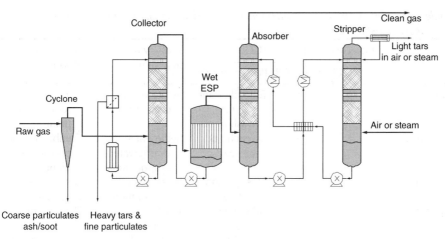

Figure 11.9 *Process scheme of the OLGA tar removal technology. (Source: Reproduced from Ref. [33])*

to that of the gasifier. This criterion points naturally to the development of systems for tar conversion and the abatement of particulates in the gas that may be integrated directly in the gasifier itself in a simple compact design solution. Such considerations are of even more relevance of the specific case of steam gasification, coupled perhaps with a high-temperature fuel cell or a downstream catalytic processing facility. The significant amount of water vapour contained in the gas stream, which in the case of cold gas cleaning would be lost by condensation, can now usefully promote the reformation of CH_4, the water-gas shift reaction towards H_2 and the elimination of carbon deposits on the catalytic surfaces.

The fact that hot gas cleaning is a focal point of applied research in the gasification field is testified by its extensive treatment in the research literature and, perhaps more significantly, by the large number of European projects funded under the FP6 and FP7 European framework programs: Clean and Efficient Energy from Waste and Biomass, COPOWER, CHRISGAS, BIGPOWER, AER-GAS II, GREEN FUEL CELLS, BIOCELLUS, UNIQUE, CLEANEX, UNIfHY, MCFC-CONTEX, THERCHEM, GREENSYNGAS, ENERMIN – all dealing with this issue in one way or another.

It is worth pointing out at this stage that modifications and additions to gasification units brought about by the need to clean and condition the product gas – such as those considered above – contribute substantially to both the overall investment and operating costs. This can have serious bearing on decisions of whether or not to adopt biomass gasification-based solutions to problems of energy provision. In many regions, such options, involving small-to-medium-scale gasification facilities, could be ideally suited for a variety of reasons – not least the desire to establish secure, independent distributed power generation facilities, using low-cost, locally produced biomass, in the face of scarce social acceptability of large thermal conversion installations.

As discussed earlier, syngas produced by biomass gasification suffers from the presence of a range of contaminants: particulates, tar as well as various trace components. These can harm the catalytic media that in the advanced conditioning treatments, to be discussed later in this chapter, are necessary for the efficient conversion of tars and the adjustment

Table 11.4 *Overview of gas cleaning methods and secondary treatment of waste water in various plant installations*

	Gas cleaning process (GCP)			Waste water treatment	Waste water recycling
	Dry GCP	Wet GCP	Process detail		
Güssing	X	X	Tube filter and wet tar washing	Waste water evaporation	Combustion of residues in the plant
Harboøre		X	Quench and wet ESP	Sedimentation Waste water evaporation	Combustion of residues in the plant
Wiener Neustadt		X	Quench and wet ESP	Waste water evaporation	Disposal of residues
Pyroforce	X	X	Tube filter and wet tar washing	Waste water storage and disposal	Disposal of residues or utilization in the process
IWT test facility /shaft gasifier	X	X	Tube filter and wet tar washing	Staged waste water treatment, evaporation, vapour residue recycling	Recycling in the process, discharge of waste water into the sewer system possible
DTU test facility /2-stage gasifier	X		Dry gas dedusting with tube filter	Treatment unnecessary	Recycling in the process
IWT test facility /multi-stage gasifier	X		Dry gas dedusting with tube filter	Treatment unnecessary	Recycling in the process

Source: Reproduced from Ref. [17].

of H_2/CO ratios. After tar, the most important contaminants are alkalis, ammonia, sulphur compounds, hydrogen chloride and particulates – all calling for advances in the cleanup technologies. Alkali salts can cause filter plugging as well as fouling; H_2S and HCl can cause corrosion as well as poisoning of catalysts in tar reforming and the final energy and chemical conversion processes.

11.3.1.1   Solid Particulates

These contaminants originate during the gasification process and are entrained by the syngas. They can block gas passages in the downstream equipment and are one of the major causes of local air pollution. In innovative, highly efficient power generation systems, including solid oxide fuel cells (SOFC) and micro turbines (mGT) [39], particles ranging from the sub-micron to a few microns in size match the pore sizes on the anode surface of the solid oxide fuel cell, so causing blockages and negatively affecting the performance.

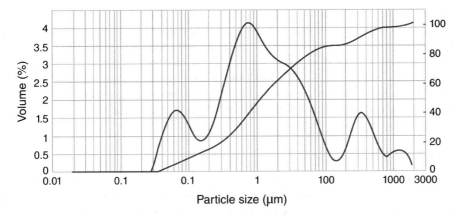

Figure 11.10 *Mastersizer 2000 granulometric analysis of ash, char and bed particles collected on a filter candle placed in the freeboard of a laboratory-scale fluidized bed gasifier*

In the case of fluidized bed gasifiers, fine powders are formed by erosion of the mineral bed inventory due to limitations on its resistance to attrition and ash and char formation in the biomass conversion process; they are elutriated by the gas leaving the reactor.

Figure 11.10 shows the size distribution of particles collected on the filter surface following an almond shell gasification test of a few hours in a 100 mm ID laboratory-scale gasifier; the bed inventory was of olivine particles characterized by good resistance to fluidized bed abrasion. Bed particles surface weighted mean diameter $(d_{3,2})$ was about 400 μm, the corresponding value in the dust cake being 3.5 μm. This means that a high fraction of elutriated particles are of a size below the standard capture range of cyclones and would therefore require a barrier filter for their capture. Typical dust loads in the raw product gas from fluidized bed gasifiers are of the order of 10 g/Nm^3, while the tolerance limit for solid particulates of most power conversion devices, including high-temperature fuel cells, is of the order of 1 mg/Nm^3.

11.3.1.2   Tar (Heavy Hydrocarbons)

These contaminants originate from the gasification process and are contained in the product gas. The vapour-phase organic compounds produced during devolatilization [40] may either undergo further decomposition to yield smaller molecules or permanent gases or else polymerize further, leading to the formation of tars that condense at temperatures below 400 °C, blocking gas passages and fouling downstream equipment. Primary oxygenated pyrolysis products are mainly transformed into high-molecular-weight hydrocarbons that include single-ring to five-ring aromatic compounds, along with various oxygen-containing and complex polycyclic aromatic hydrocarbons [41]. Thermodynamic equilibrium calculations [42] do not predict the relatively high tar concentrations found experimentally however; this is because of the limited residence times and low conversion reaction rates under normal gasifier operating conditions with temperatures in the 800–900 °C range.

As discussed earlier, the presence of tar among the products of gasification reduces gas yield and conversion efficiency. In addition, these contaminants are responsible for carbon deposition that can block the porous media of a fuel cell anode. The tolerance limit of high-temperature fuel cells to tars has yet to be fully defined in the literature due to the novelty of such systems and shortage of data on their long-run performance; a value of the order of 100 ppmw has however been proposed by Aravind and de Jong [43].

Tar sampling of the product gas stream is usually carried out in accordance with technical specification CEN/TS 15439 [44]. In addition to gas chromatographic analysis, innovative methods have been proposed for monitoring and characterizing tar produced in gasification processes: the methods include HPLC-UV and fluorescence spectroscopy, which provide rapid and relatively cost-effective procedures [45].

11.3.1.3   Sulphur Compounds

These contaminants are produced in the gasification process as a result of the presence of small amounts of sulphur in the biomass feed. The most important of these is H_2S, followed by COS. H_2S is chemisorbed on catalyst surfaces, thereby blocking active sites in the catalytic gas conditioning systems and limiting fuel cell performance in power generation applications. The loss of catalytic activity resulting from sulphur contamination is usually reversible in the systems dealt with in this chapter: removal of H_2S from the fuel gas results in the restoration of catalytic activity to the original level.

Biomass has a comparatively low sulphur content with respect to coal: less than 0.1% by weight for lignocellulosic biomass, less than 1% for refuse-derived fuels. A large part of the sulphur is removed with the ashes and as a consequence the sulphur compound content of raw syngas rarely exceeds the order of 100 ppmv. Nevertheless, the tolerance limit may be of the order of 1 ppmv, or even lower in downstream gas catalytic conversion processes and for high-temperature fuel cell operations below 1000 °C [43].

Metal oxides are frequently used to absorb H_2S from high-temperature gas streams:

$$MeO + H_2S \leftrightarrow MeS + H_2O$$

However, a major limitation to this procedure follows from the reductive properties of syngas, which can lead to the formation of metals, with consequential loss of sorption capacity and metal evaporation; further problems can arise from the substantial presence of H_2O in the cases of steam gasification, resulting in thermodynamic equilibrium constraints to the reduction of hydrogen sulphide in the gaseous phase below the threshold limits for utilities and downstream process applications.

11.3.1.4   Alkali Metals

Alkali metal contaminants are released as halides (e.g. chlorides) during the gasification process. Of the total alkali content of biomass (mainly comprising sodium and potassium salts, potassium salts usually being the major contributor), only a minor fraction finds its way into the gas product phase, at concentrations usually of the order of 1 ppmv (slightly above this limit in the case of straw gasification); however, purification down to the order of 10 ppbv is necessary for gas turbine applications in order to avoid corrosion and fouling problems. Alkali metals can also lead to ash softening and melting and thus to filter blockages. They are present in the vapour phase at temperatures above 600 °C and condense at lower temperatures, so that a further gas cleaning step becomes necessary when a hot gas filtration system functioning above this temperature is in place and sorbents for alkali metal salt capture are not included in the gasifier bed inventory. If they are allowed to condense in cooler parts of the plant, they can block gas passages and anode surfaces of fuel cells. They are thus one of the major influencing factors for fouling.

Several investigations on chemical hot gas cleaning have been carried out that demonstrate the suitability of aluminosilicates for alkali sorption [46]. Further investigations on the chemisorption mechanism illuminate the fundamental role of steam, as shown by the following stoichiometric relation:

$$Al_2O_3 \cdot xSiO_{2,(s)} + 2AlkCl_{(g)} + H_2O_{(g)} \leftrightarrow 2AlkAlO_2 \cdot xSiO_{2,(s)} + 2HCl_{(g)}$$

This expression illustrates the increase of hydrogen chloride gas content brought about by the alkali removal process.

11.3.1.5 Heavy Metals

These trace elements are released during gasification from pollutants in the biomass. Some of them can poison Ni catalysts and the anodes of fuel cells. How important this could be for power generation applications and catalytic system performance has yet to be subjected to detailed analysis and experimental investigation.

11.3.1.6 Hydrogen Chloride

Chlorides in the biomass may be released as HCl during gasification, in addition to being retained in the ashes. It is highly corrosive, especially with respect to the interconnecting material of downstream equipment, including SOFC, causing degradation of various system components. Current recommendations advise reducing HCl levels in the gas to no more than a few ppmv.

Sorbents based on alkali or alkaline earth metal compounds are recommended for high-temperature capture of hydrogen chloride (and halides in general). However, metal vapours can form from these metallic compounds at high temperatures, with danger of condensation during fuel gas utilization.

11.3.1.7 Ammonia Compounds

Fuel-bound nitrogen in the biomass is released into the product gas during gasification – primarily as ammonia with only small amounts of HCN. On combustion, NH_3 in the fuel gas has a tendency to form nitrogen oxides (NO_x), pollutants that are difficult to remove and precursors of *acid rain*. On the other hand, NH_3 has no significant effect on fuel cell operation for anode feed gas concentrations lower than about 1% by volume – which is more than that usually to be found in the gasifier product. Ammonia, as it happens, can be a fuel for SOFCs: it dissociates at the anode to form H_2, which is then oxidized, and N_2, the overall reaction being endothermic and increasing the efficiency of the cell operation [43].

11.3.2 Dust Filtration

Gas filtration at high temperatures [47] provides protection for downstream equipment (heat exchangers, catalyst units, turbines, scrubbers, etc.) against erosion and fouling, facilitates process intensification and simplification and avoids the blocking of filter elements by the condensation that would occur at lower temperatures. It may be carried out at temperatures ranging from above 260 °C up to some 900 °C, under variable pressure conditions, in both oxidizing and reducing atmospheres and in the presence of chemically aggressive compounds. It is therefore necessary that the filter media and the containment vessel be mechanically and thermally stable under the operating conditions of temperature and pressure, chemically stable with respect to the chemical composition of the gas and erosion resistant with respect to the dust. For these reasons, only rigid self-supporting ceramic or metallic filter elements may be employed; flexible filter media, suitable for low-temperature operation, would be destroyed at high temperatures by the mechanical stresses brought about by the instantaneous reverse gas flow back-pulse procedure used to remove the dust cake build-up on the filter surfaces and thereby control the pressure drop.

From an industrial viewpoint, hot gas cleanup of particulates using ceramic or metallic barrier, or granular moving bed, filters [48, 49] has proved a commercial success [50]. The only disadvantage is the higher investment cost, due to construction material requirements and the higher gas volume to be dealt with as a result of high-temperature operation. However, in most cases, this is more than compensated for by the reduction in the overall process costs.

Additional extensive reviews of high-temperature gas cleanup by means of, among other devices, candle filters have been summarized by Di Carlo and Foscolo [51]. Smith and Ahmadi [52] review the field of hot gas filtration in relation to pressurized fluidized bed combustion (PFBC) and integrated combined cycle gasification (IGCC).

Candle filters have in general proved highly efficient, even for very fine particles of sizes down to the order of 1 µm. However, a number of problems remain to be solved, such as the build-up of dust cake on the filters leading to occasional cake bridging between candles, filter failure and breakage. A number of hot gas filtration systems have been developed and tested under industrially relevant conditions: an advanced particle filter (APF) containing 284 candles was installed at the 70 MW (electric) Tidd PFBC demonstration plant at Brilliant, Ohio; gas flow and particle deposition have been modelled by Ahmadi and Smith [53] for the Tidd filter vessel and by Mazaheri and Ahmadi [54] for the Siemens–Westinghouse particulate control device (PCD).

Considerable research effort has been put in the studies of the mechanisms controlling the dispersion and deposition of fine particles: under turbulent flow conditions, the particles are conveyed by the mean motion and dispersed by the fluctuating velocities; numerous reviews on such turbulent transport and deposition processes are available [55]. In order to improve the effectiveness and reliability of the hot gas filtration process, further work directed towards a more complete understanding of the nature of the particle transport and deposition processes in the filter vessel is required.

The most commonly adopted filtering media is a high-density ceramic characterized by an asymmetric structure, a support material covered with a thin membrane layer containing very small pores. The membrane collects the fine particles and by making it thin the differential pressure of the filter element is kept low. The ideal solution is to have a very thin layer without defects that just covers the support material, so that purely surface filtration takes place. Experimental evidence at industrial scale shows that this ideal solution can be achieved in practice, penetration of particles into the support structure of the filter element being prevented, and that the element can be effectively regenerated by the clean gas back-pulsing procedure described by Cocco *et al.* [56]. It has also been confirmed at laboratory scale that, after removal of the dust cake accumulated on the filtration surface, the pressure drop across the cleaned candle returns to its original value [51]. The filtration efficiency of high-temperature ceramic filters reaches 99.99% and can be even higher, resulting in particulates concentration in the filtered gas of 5–10 mg/Nm^3 and thus fully adequate to meet the requirements of almost all energy generation systems.

At high temperatures, dust softening or sintering can occur, resulting in a sticky dust layer on the filter element surface. The softening temperature depends on the chemical composition of the dust. Chlorides, such as NaCl, KCl or CaCl$_2$, decrease the softening temperature, and, in particular, for eutectic mixtures such a decrease can become significant. Ceramic filters are also susceptible to reactions with alkali vapours, which can lead to filter plugging and even structural failure.

However, particle stickiness problems do not impose additional requirements on the aspect of high-temperature filtration with which we are primarily concerned here – its coupling with fluidized bed gasifiers. This is because fluidized bed behaviour is also affected by inter-particle sticking phenomena that should be carefully avoided to assure a stable reactor operation. The evolution with temperature of instantaneous pressure fluctuations in a fluidized bed provides an effective measure of the approach to the minimum sintering temperature of the bed inventory [57]: particle adhesion and aggregation into clusters brings about changes in the quality of fluidization that affect the dominant frequency and RMS of pressure fluctuations (fewer and bigger bubbles, leading to bed de-fluidization in extreme cases). De-fluidization is accompanied by a large and rapid increase in bed pressure drop and the development of gas preferential channels through the granular material. This phenomenon occurs at temperatures slightly higher than the minimum sintering temperature and can be predicted by dilatometric methods [58]. A strong correlation exists between temperature excess beyond the minimum sintering point and excess gas fluidization velocity needed to avoid undesired cohesion phenomena [59].

If the reaction products include cohesive ashes, it is possible to maintain the fluidization regime by improving segregation among easy-to-sinter particles, adding the bed inert particles also promotes the removal of the cake from the surface of filter candles.

Gasification and pyrolysis of biomass provide an interesting field of application for hot gas filtration. Many small hot gas filter installations, containing 1–100 candles, have been in operation in laboratory and demonstration units around the world since the early 1990s; such applications merge well with current plans to rely extensively on biomass gasification in the perspective of a more sustainable energy market based on renewable sources. Gas filtration has so far been usually carried out in the temperature range of 500–600 °C [60, 61]. Under these conditions, condensation of tars does not take place and the tendency of the dust, which typically has a high alkaline content, to become soft and sticky is low. New advanced concepts in biomass gasification, however, favour hot gas filtration to be carried out at the gasification temperature of 800–900 °C [62]. In this way, the filtered gas is ready for downstream catalytic processing to effectively promote the gas and tar reforming reactions at the same temperature at which it leaves the gasifier. Gas cool down and subsequent heat up are thus avoided; the catalyst unit is protected from particle deposition and fouling, and the process is much simplified and rendered highly energy efficient. The concept is shown schematically in Figure 11.11.

Experimental tests on the filtration of syngas from biomass gasification at temperatures close to 800 °C have shown promising results [63]. However, additional long-term trials are still required to confirm stable filtration under industrial conditions of operation.

Figure 11.11   *General steps for an energy efficient biomass gasification process*

In parallel to experimental research, Eulerian-Lagrangian CFD models can provide insight into the mechanisms governing dust cake formation on the filter media and particle trajectories in the immediate surroundings. A particle elutriation model (PEM) was proposed to account for elutriation and attrition in a fluidized bed. The model is based on the assumption that the generation of fines by attrition, a nonlinear function of time, depends on the percentage of agglomerated fines. Elutriation rate constants and attrition rates are evaluated for various particle diameters by using experimental data and the assumption that the entrainment rate at low air velocities is affected by inter-particle adhesion forces. The PEM is then interfaced with a CFD code in order to simulate transport of particles in the filter vessel, the deposition of fine particles on the filter candle and thus the pressure drop due to cake formation. A k-ε model is used for turbulence and a discrete random walk (DRW) model for the turbulent dispersion of particles. In order to simulate the filter candle, the filtration surface is considered as a trap for solid particles and as a porous medium for the gas phase. The position of the trapped particles and their diameters enable the overall resistance to permeability for the gas phase to be computed and expressed as a combination of that related to the original permeability of the clean candle and that of the particle layer permeability, computed using the Ergun law applied to the cake. Figure 11.12 illustrates typical results obtained using this modelling approach [51]; it shows a 2D simulation of particle velocity vectors, streamlines and distribution around the candle for specific operating conditions. Figure 11.13 relates to the same conditions as Figure 11.12. It shows, as a function of time, a comparison of the experimentally determined and calculated pressure drop profiles due to the formation of the cake of fines elutriated and deposited on the candle surface.

11.3.3   Catalytic Conditioning

The main problems influencing the industrial viability of biomass gasification technology in general relate to the presence of tars in the product gas. Milne *et al.* [64] provide the following collective definition of these compounds: "The organics, produced under thermal or partial-oxidation regimes (gasification) of any organic material, are called tars and are generally assumed to be largely aromatic".

Hot gas conditioning, with its potential for thermal integration within the gasification process itself, can provide the most complete solution to this problem: tars are eliminated by converting them into desired product gas components, thus retaining their chemical energy in the product gas, contributing to the gas yield and increasing the thermal efficiency of the process as a whole.

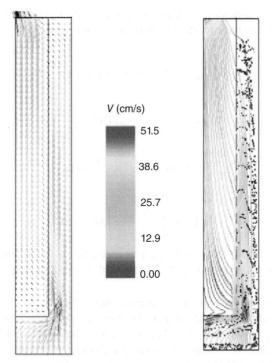

Figure 11.12   *(left) velocity vectors, (right) streamlines and particles distribution around the candle, obtained by simulation for the case 89 m/h filtration velocity and 20 cm of static bed at t = 50 sec. (Source: Reproduced from Ref. [51], with permission from Elsevier)*

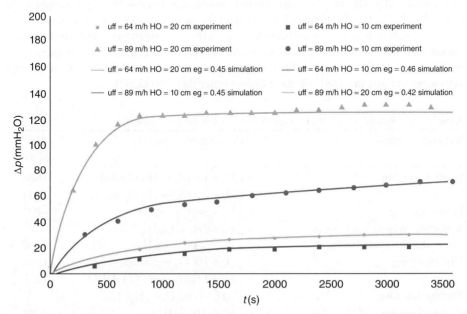

Figure 11.13   *Pressure drop calculated by the model, compared with experimental results. (Source: Reproduced from Ref. [51], with permission from Elsevier)*

The main reactions whereby the high molecular weight tars formed during gasification may be broken down in the conditioning process are as follows [65]:

Steam reforming:

$$C_nH_m + nH_2O \leftrightarrow nCO + \left(n + \frac{m}{2}\right) H_2$$

Dry reforming:

$$C_nH_m + nCO_2 \leftrightarrow 2nCO + \left(\frac{m}{2}\right) H_2$$

Cracking:

$$C_nH_m \leftrightarrow nC + \left(\frac{m}{2}\right) H_2$$

Hydro-cracking:

$$C_nH_m + \left[\frac{(4n - m)}{2}\right] H_2 \leftrightarrow nCH_4$$

A more detailed picture of possible reaction mechanisms for hydrocarbons undergoing hot gas conditioning is provided by Table 11.5 with reference to naphthalene as a model tar compound.

Thermal cracking of the tars takes place at considerably higher temperatures (above 1100 °C) than that of the operating level of most biomass gasifiers (800–900 °C). It would thus require a secondary reactor unit where the fuel gas is partially oxidized to increase its sensible energy content; the hydrocarbons become thermodynamically unstable under such conditions, thereby resulting in a clean syngas with a very low methane content [42]. However, the energy requirement for increasing the temperature of the gasifier product gas to that required for the thermal cracking reactions has a negative effect on the overall process efficiency (CGE). This may be expressed in terms of an increase in the oxygen equivalence ratio (ER). Table 11.6 [33] quantifies this effect for various cracking temperatures.

Steam reforming of the tars has been considered the most appropriate route to their elimination [64]. Once again, however, the chemical kinetics of the gaseous-phase homogeneous reactions are satisfactory only at high temperatures, that is, above 1200 °C, due to the high activation energies involved – in most cases greater than 350 kJ/mol.

Table 11.5 *Possible reactions of naphthalene, a key tar component*

Name of reaction	Stoichiometric equation	ΔH^0 kJ/mol[a]
Steam reforming	$C_{10}H_8 + 10H_2O \rightarrow 10CO + 14H_2$	1676
	$C_{10}H_8 + 20H_2O \rightarrow 10CO_2 + 24H_2$	1704
Hydrocracking	$C_{10}H_8 + 36H_2 \rightarrow 10CH_4$	−825
Reduction to benzocyclohexane	$C_{10}H_8 + 2H_2 \rightarrow C_{10}H_{12}$	−47
and decahydronaphthalene	$C_{10}H_{12} + 3H_2 \rightarrow C_{10}H_{18}$	−260
Dry reforming	$C_{10}H_8 + 10CO_2 \rightarrow 20CO + 4H_2$	1648
	$C_{10}H_8 + 14CO_2 \rightarrow 24CO + 4H_2O$	1637
Thermal cracking	$n\, C_{10}H_8 \rightarrow mC_xH_y + pH_2$	
Carbon formation	$C_{10}H_8 \rightarrow 10C + 4H_2$	−77

[a]Standard values of enthalpy of formation.

Table 11.6 *Effect of thermal tar cracking on the product gas and the process efficiency*

Tgasifier (°C)	Tcracker (°C)	Required ER (–)	HHV product gas (MJ/Nm^3)	CGE (%)
850	–	0.21	7.3	82
850	1100	0.28	6.0	76
850	1200	0.31	5.3	72
850	1300	0.34	4.8	69

Source: Adapted and reproduced from Ref. [33].

Catalytic conditioning. With the use of catalysts, steam and dry reforming reactions become an effective way to convert the tar components in the fuel gas at lower temperatures, compatible with those of the gasification processes.

11.3.3.1 *Primary Catalytic Treatment: Within the Gasifier*

A major advantage of fluidized bed gasifiers lies in the possibility of utilizing a low-cost mineral bed material that is catalytically active for the tar conditioning reactions in the presence of steam. Various catalytic systems are able to decrease the activation energies for the reforming reactions of high-molecular-weight organic compounds to such levels that reasonable reaction rates and conversions become possible at temperatures compatible with those of typical fluidized bed gasifier operations (800–900 °C – the upper value being limited by ash sintering and consequent de-fluidization limitations). The integration of gasification and catalytic gas conditioning in a single reactor vessel is referred to as *primary* tar conversion and the catalysts in the gasifier bed inventory as *primary* catalysts [66, 67]. Calcined dolomite, limestone and magnesite have all been found to lead to increases in the gas hydrogen yield [68–72].

Much research into biomass gasification in fluidized beds has concerned the use of dolomite [$(Ca,Mg)CO_3$] and olivine [$(Mg,Fe)_2SiO_4$] as bed inventory. Of the two, olivine shows a slightly lower activity in biomass gasification and tar reforming, but has a higher attrition resistance than dolomite [73–75].

Olivine activity, or more specifically olivine activation, depends on its iron oxide content [76, 77]. Iron can be present either in the olivine crystalline structure or free at different oxidation states depending on levels of high-temperature pretreatment and exposure to the reducing/oxidizing conditions in the gasifier. How this is able to account for catalytic activity in biomass gasification has been well documented [78].

It has been found that in industrial-scale steam gasification plants a gradual modification of the bed material occurs due to the interaction of olivine with biomass ash components and additives [79]. Experimental tests into this phenomenon have shown that with prolonged use the catalytic activity of olivine increases, leading to an increase of hydrogen content in the product gas, enhancement of the exothermic water-gas shift reaction and substantial decrease of tar content. This, at first sight rather surprising, result has been found to be ascribable to the formation of a calcium-rich layer on the surface of olivine particles, as clearly shown in Figure 11.14 and quantified by the results of energy-dispersive X-ray (EDX) spectroscopy reported in Table 11.7 [80].

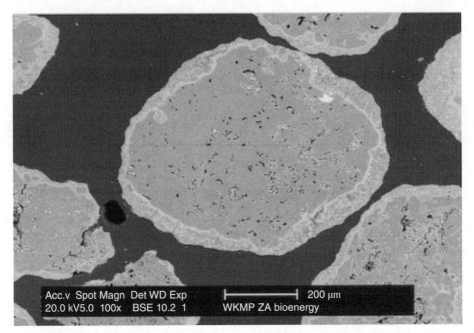

Figure 11.14 *SEM image of a calcium-rich layer on the surface of the used olivine. (Source: Reproduced with permission from Ref. [80]. Copyright © 2011, American Chemical Society)*

Table 11.7 *Results of energy-dispersive X-ray (EDX) spectroscopy*

| Elements | Unused olivine | Used olivine | | |
	Particle inside (wt%)	Particle inside (wt%)	Inner layer (wt%)	Outer layer (wt%)
C	n/a	n/a	5.3	9.8
O	15.2	14.2	12.9	13.7
Mg	37.4	30.3	8.6	18.5
Al	0.1	0.7	0.1	0.1
Si	35.5	32.2	17.3	15.4
P	n/a	0.0	n/a	0.3
K	n/a	8.5	2.6	3.2
Ca	n/a	1.8	46.5	33.5
Cr	n/a	n/a	2.7	0.4
Mn	n/a	0.0	0.9	1.2
Fe	11.8	12.5	3.0	3.7

Catalysts for fluidized bed gasification of biomass should be efficient for the reforming of hydrocarbons and have high selectivity for syngas and high resistance to attrition and carbon deposition. They should also be relatively low cost, because the formation of ash and char necessitates the continual removal, and replenishment with fresh or regenerated material, of the bed inventory. In dual fluidized bed gasifiers, the inventory is also exposed to the additional stress of alternate operations under reducing and oxidizing conditions. Most reforming catalysts, developed for other industrial applications, fail to meet these varied requirements. As a result, a considerable amount of research has gone into the development of dolomite- and olivine-derived catalysts to improve both activity with respect to the various reactions they are there to promote and stability under the harsh environmental conditions to which they are to be subjected.

Commercial Ni-based reforming catalysts exhibit high activity and selectivity for tar conversion into hydrogen-rich gas, but suffer from a number of severe limitations regarding their other properties such as mechanical fragility; rapid deactivation, mostly due to sulphur, chlorine, alkali metals and coke formation; metal sintering. The overall effect is a limited active lifetime [2].

Very positive results with respect to all these drawbacks have been obtained by impregnating olivine with nickel [81]. The mechanism of active phase formation in Ni-olivine under biomass gasification conditions is well understood [82]. The positive features of natural olivine – good mechanical resistance and activity in tar reformation [73] – combine with those of nickel catalysts – high activity in reforming of hydrocarbons – without the disadvantages encountered with commercial products. A study of steam reforming of toluene as a tar model compound demonstrated clearly the much higher activity and selectivity of Ni-olivine towards syngas than that of olivine alone. A kinetic model – zero order for water, first order for toluene – was proposed and the parameters fitted from experimental data from a fixed bed micro-reactor. When used as a primary catalyst for tar destruction and methane reforming in a pilot-scale (100 kW thermal) dual fluidized bed gasifier, the Ni-olivine catalyst proved well its resistance to attrition and coke formation and gave rise to an order of magnitude reduction in the tar content of the product fuel gas [67].

A major disincentive to the extensive application of Ni-olivine catalysts for fluidized bed gasification concerns the biomass ash waste product: contamination with Ni, a heavy metal, would greatly inflate the cost of its disposal. For this reason, research is being carried out into the improvement of catalytic activity of dolomite and olivine by impregnation with inexpensive and non-toxic metallic elements. The same synthesis methods developed for Ni can be applied to other transition metals such as Co, Cu or Fe – the last being the most promising for both economic and environmental reasons; moreover, as with nickel, iron can form solid solutions with magnesium (FeO-MgO) over a wide range of composition. It is also to be expected that such catalytic systems should perform well in relation to the water-gas shift reaction, iron oxide being well known to be active in that respect under high-temperature conditions.

A comparative study on the catalytic activities of nickel- and iron-doped alkaline earth oxides (including calcined dolomite) showed that the presence of iron leads to significant improvements in the catalytic activity of CaO and MgO substrates for toluene steam reforming, rendering them more active than $(CaMg)O$ and hence worthy of consideration for scale-up applications [83–85]. Carbon deposition resistance during toluene steam reforming was also found to improve markedly on the addition of iron. Ni/$(CaMg)O$ catalyst was also found to exhibit a stronger interaction between metal and substrate (the NiO-MgO solid solution), indicating superior attrition resistance – an important consideration in large-scale fluidized bed applications.

An attrition-resistant Fe-doped olivine catalyst has been successfully tested under industrially relevant operating conditions at ICB-CSIC (Zaragoza, Spain). By using an

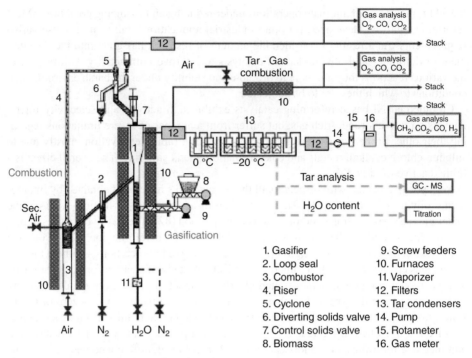

Figure 11.15 *Scheme of the bench-scale catalytic biomass gasification unit at ICB-CSIC (zaragoza, spain). (Source: Reproduced from Ref. [86], with permission from Elsevier)*

optimized impregnation technique, Fe content of olivine was enriched with an additional 10 wt% – resulting in a total iron content of about 16 wt%. As mentioned earlier, olivine activation depends on its iron oxide content; on thermal treatment, iron is partially expelled from the crystalline structure $[MgO_xFeO_{(1-x)}SiO_2]$ to form free oxides, which are responsible for its catalytic activity. Catalytic biomass gasification experiments were performed at a biomass feedstock rate of 250 g/h in the bench unit illustrated in Figure 11.15 – which was designed to simulate an industrial dual fluidized bed gasifier/combustor system [86]. The abatement of tar in the fuel gas is shown as a function of gasification temperature, both qualitatively (by visual comparison of liquid samples collected in the tar condensers) and quantitatively, in Figure 11.16, where the results are also compared with those of similar tests conducted with (catalytically neutral) sand and un-doped olivine. The Fe-olivine material has a double effect on tar destruction: it acts both as a catalyst for tar and hydrocarbon reforming; while transferring oxygen from the combustor to the gasifier, part of the oxygen is used to burn volatile compounds. Characterization tests performed after the runs showed the catalyst structure to have been maintained, despite the large number of redox cycles to which it had been subjected.

Similar positive results for the inclusion of iron were obtained for steam gasification in a stationary bubbling fluidized bed operating at slightly higher biomass throughput rates: the inclusion of 10 wt% Fe in the olivine gives rise to about 40% increase in the gas yield and 88% increase in the hydrogen yield compared to those obtained using olivine alone.

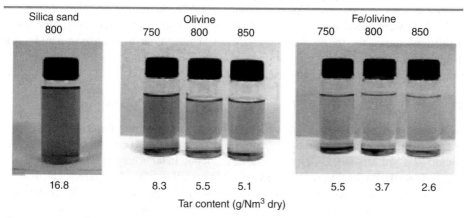

Figure 11.16 *Effect of bed material and gasification temperature on tar content. (Source: Reproduced from Ref. [86], with permission from Elsevier)*

At the same time, methane in the syngas was reduced by 16% and tar production per kg of dry ash free biomass by 46% [87].

Complete characterization and micro-reactor reactivity data are also available for Fe-olivine materials containing up to 20 wt% iron [88]. With 10 wt% Fe-olivine, the results for toluene conversion (91%) and hydrogen production (0.066 molH$_2$/h/g$_{cat}$) turned out to be three times greater than those obtained with olivine alone. Similar positive results with this catalyst are reflected by the kinetic constants obtained for methylnaphthalene steam reforming. Figure 11.17 compares these results with those obtained using olivine

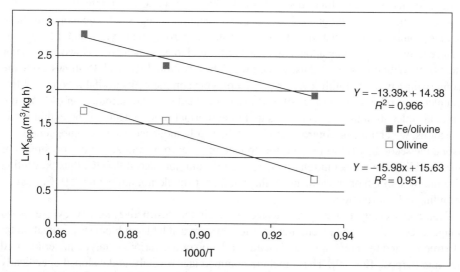

Figure 11.17 *Temperature dependency of the apparent reaction rate constant according to arrhenius' law for 10% fe/olivine and olivine Ref. [89]*

alone under the same conditions of operation: concentration of 1-methylnaphthalene 5 g/Nm^3, water in excess, 1-MN/steam ratio of 1/127. Activation energies of 111 and 133 kJ/mol were found for 10 wt% Fe/olivine and olivine, respectively [89] – the lower activation energy for the iron-containing catalyst indicating its increased activity for 1-methylnaphthalene steam reforming and providing further confirmation of the benefits of iron in promoting catalytic efficiency in general.

11.3.3.2   Secondary Catalytic Treatment: Downstream of the Gasifier

As a result of the improvements to catalyst efficiency documented above, primary *in-bed* hot gas cleaning can perform the crucial role of decreasing the tar content of the fuel gas to some 1–2 g/Nm^3 dry gas. Despite this drastic reduction in tar achieved within the fluidized bed gasifiers themselves, the raw syngas has yet to meet the stringent requirements imposed by the utilities for power generation and chemical syntheses applications. Secondary gas treatment is therefore required, either to remove the tars at low temperature, as previously described or, preferably, to convert them catalytically in a downstream process [90]. A major problem with such secondary hot gas conditioning concerns the optimization of the process layout to conserve as far as possible the sensible energy content of the fuel gas, so that it may be processed catalytically at high temperature with minimal loss of thermal efficiency. This is not an easy task because of the particulate content of the fuel gas leaving the gasifier, as discussed in an earlier section on hot gas filtration. When a filter is followed by a catalytic fixed bed reactor, the inevitable loss of thermal energy by the gas during the filtration process needs to be compensated for by re-heating the gaseous stream, which involves additional partial oxidation of the fuel gas at the expense of its chemical energy and of the energy efficiency of the process as a whole. The alternative approach, a secondary catalytic reactor followed by a filter, dictates the use of monolith modules so as to avoid the risk of catalyst clogging by solid particles; these may be placed directly at the gasifier outlet [91, 92]. Monoliths are ceramics blocks of parallel, straight channels on the walls of which a thin film of catalytically active material is deposited. The monoliths have a honeycomb structure that tolerates the presence of particles in the gas, and their performance in this gas cleaning application has been found to be satisfactory. However, a major drawback to their commercial use is their rapid deactivation. Figure 11.18 shows a scheme comprising a two–monolith-layer reactor, employing commercially available Ni monoliths, coupled to a pilot-scale gasifier [92]. Provision is made for the introduction of four separate air inlet streams to effectively supply the thermal requirements of the endothermic reactions, which may be controlled to maintain near homogeneous temperature conditions along the secondary reforming reactor. In fact, in order to maximize the life-on-stream of the monolith, the temperature at its front face should not exceed 900 °C, thus preventing fouling by sticky biomass ash, and at the exit face it should not fall below 750 °C, thereby avoiding coke formation.

Numerous catalysts have been tested for their tar destructive activity over a broad range of conditions, and recent literature reviews are available on biomass gasification tar destruction and the preparation and characterization of the various catalytic materials used for this purpose [28, 50]. These include commercial Ni catalysts developed specifically for steam reforming, which have found application in biomass gasification tar conversion and as secondary catalysts in separate fixed and fluidized bed reactors [2].

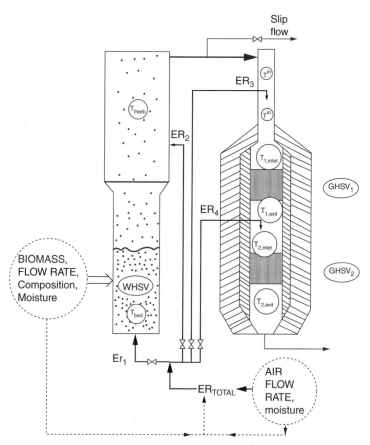

Figure 11.18 *Scheme of the gasification-gas cleaning process: Two monolith layers reactor utilizing commercially produced ni monoliths, coupled with a pilot-scale gasifier. (Source: Reproduced with permission from Ref. [92]. Copyright © 2006, American Chemical Society)*

Figure 11.19 shows the process flow sheet for a pilot-scale fluidized bed gasifier, capable of processing some 20 kg/h of biomass feed, coupled with a thermal cracker and reformer reactor. The reformer is loaded with fluidizable nickel-based reforming catalyst and fitted with gas analysis ports at its inlet and outlet. The system has been used to evaluate catalyst activity and the decay of hydrocarbon conversion with time from a slip stream sample of the raw fuel gas. In this way, it is possible to quantify the frequently reported phenomenon of commercial catalyst deactivation, sometimes quite rapid, from high activity of fresh samples to lower residual activity brought about by various factors, including the presence of poisons (sulphur, chlorine) and coke formation.

The catalyst screenings also examine the effect of the support material, for example zirconia and alumina, over a wide range of operating conditions (in particular the secondary reactor temperature), on the reforming capacity and the deactivation kinetics of active metals and promoters, such as Co, Ni, Fe, Cr, Ce and Pt [93] – although the prospect for commercial application of metals other than Ni and Fe currently appears remote.

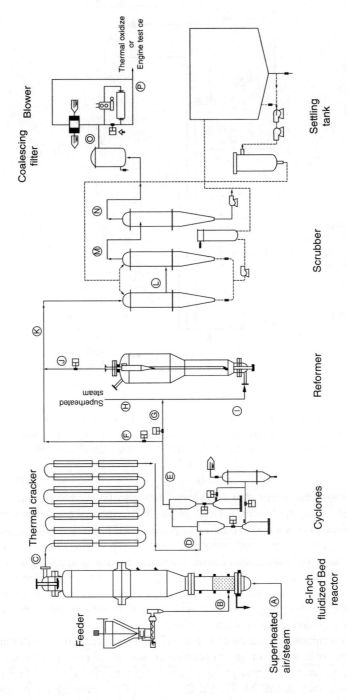

Figure 11.19 *Process flow sheet for a pilot-scale (~20 kg/h of biomass feedstock) fluidized bed system to evaluate catalyst activity. . (Source: Adapted with permission from Ref. [2]. Copyright © 2005, American Chemical Society)*

11.3.3.3 Simultaneous Particulate Cleaning and Tar Conditioning

A promising innovation has recently been proposed to deal with the dual problems of particulates and tar in the product gas: *catalytic filters*, coupled to the gasification unit, to remove both of these contaminants simultaneously. Convincing proofs of this concept have been provided at laboratory- and pilot-scale levels, as well as on a slip stream from an industrial installation.

A catalytic filter, rather than a filter followed by a catalytic bed, or a monolith followed by a filter, simplifies the process flow diagram, reduces equipment requirements, leads to higher thermal efficiency and reduces costs. Ceramic hot gas filters lend themselves naturally to adaptation for this dual purpose.

Various procedures have been proposed and investigated for carrying out the required adaptation. That developed by Pall Filtersystems GmbH calls for a design modification of the ceramic hot gas filter candle involving a porous inner tube to accommodate the catalyst particles (Figure 11.20a, [94]). The advantage of this approach is the high flexibility it provides with regard to potential applications: the catalyst phase is easily integrated within the filter structure.

An alternative approach is illustrated in Figure 11.20b and involves a catalytic coating of the porous inner structure of the filter elements. In addition, there is potential here for increasing the catalyst capacity by filling the free hollow cylindrical volume of the filter element with catalytic particles or foam of high active surface area.

Filtration, together with tar reforming, has been carried out in these units on synthetic gases at temperatures ranging from 800 to 850 °C. Various tar reforming catalyst systems have been considered, with varying NiO loadings and diverse catalyst support materials. Complete conversion of naphthalene at 800 °C has been achieved, also in the presence of 100 ppmv of H_2S [95].

The technical and economic optimization of such nickel catalytic filters has been the subject of various experimental investigations, which include studies of the effect of doping the catalyst with a second low-cost metal in attempts to improve its activity. In fact, it is well known that the formation of mixed Ni/Fe alloys can provide good resistance to carbon fouling [96]. Other transition metal catalysts, such as Co, are also known to be effective for hydrocarbon and bio-ethanol reforming [97].

Much work is also underway on manufacturing procedures for the cost-effective production of industrial-scale catalytic filter candles and on means of enhancing their chemical resistance and active working life under harsh process conditions to which they are subjected.

11.3.4 The *UNIQUE* Concept for Gasification and Hot Gas Cleaning and Conditioning

A new concept for cleaning and conditioning of biomass syngas has recently been developed [62, 98, 99]. It involves positioning catalytic filter elements for particle and tar removal directly in the freeboard of a fluidized bed steam gasifier. This results in a compact unit providing a good example of process intensification. It is also cost-effective because the investment costs of gas cleaning equipment are reduced as well as the space required for its installation. A further advantage is that the gas temperature is in the correct range for the catalytic tar reforming reactions, so that no reheating of the gas or auxiliary electric heating of pipes and filter vessel is necessary. As was reported above,

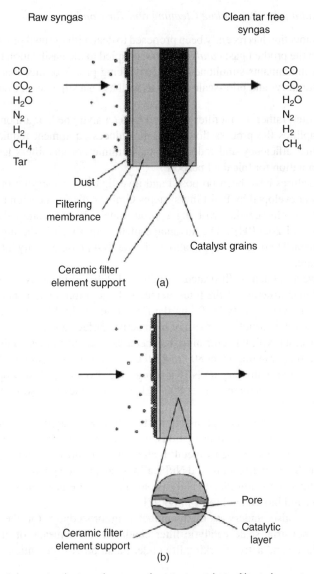

Figure 11.20 *Schematic designs for tar reforming catalytic filter elements with integrated fixed bed (a) or catalytic layer (b). (Source: Reproduced from Ref. [94], with permission from Elsevier)*

the temperature for fluidized bed biomass gasification is typically in the range of 800 to 900 °C, which is also the required range for catalytic tar reforming.

Figure 11.21 illustrates the principle of this new concept. Its major positive features include the realization of notable system simplification and process intensification; thermal losses significantly reduced and cooling and heating steps avoided; increased activity of catalysts and sorbents; abatement of particulates and tar in the gasifier outlet. In all, a combination of the benefits of traditional primary and secondary hot gas treatments without

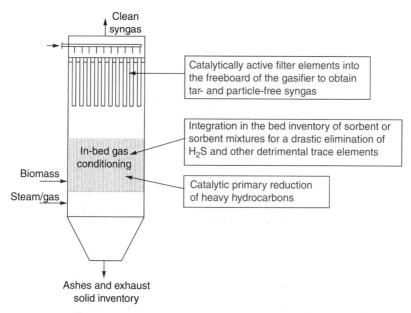

Clean syngas

Catalytically active filter elements into the freeboard of the gasifier to obtain tar- and particle-free syngas

Integration in the bed inventory of sorbent or sorbent mixtures for a drastic elimination of H_2S and other detrimental trace elements

In-bed gas conditioning

Biomass

Steam/gas

Catalytic primary reduction of heavy hydrocarbons

Ashes and exhaust solid inventory

Figure 11.21 *The UNIQUE concept for integration of gasification with hot gas cleaning and conditioning in one reactor vessel [100]. (Source: Reproduced with permission from Ref. [100]. Copyright © 2009, American Chemical Society)*

their well-established drawbacks – catalyst clogging by particulates, loss of gas chemical and thermal energy, etc. – all achieved in a compact, cost-effective and energy-efficient single reactor vessel, capable of delivering high-purity syngas for potential use in diverse power generation applications, including those of small-to-medium-scale CHP and power plants, thereby contributing effectively to the overall economic revenue. The *UNIQUE* concept for hot gas cleaning and conditioning provides a concrete contribution to the target of reducing the cost of electricity obtained by means of advanced biomass energy systems.

Tests carried out under both simulated and real process conditions confirmed that nickel-based catalytic filters of varying design can be successfully applied to high-temperature reforming of tars and removal of particulates from biomass gasification product gas. The laboratory tests involved positioning a 40 cm filter candle segment in the 10 cm diameter freeboard of an electrically heated bench-scale gasifier; and the Güssing dual fluidized bed gasification plant – operated by Biomasse Kraftwerk Güssing GmbH & Co KG – was used for the tests on a fully functioning commercial installation: a prototype candle being inserted in the gasifier freeboard for processing a slip stream of raw syngas.

Figure 11.22 shows the layout of a bench-scale facility used for the laboratory tests [100], and Figure 11.23 provides a sketch and the P&I flow sheet of the test module realized in Güssing [101]. In the latter case, the filter candle is cleaned periodically to remove dust cake by the conventional back-pulsing procedure used for rigid hot gas filter elements, the blowback gas being nitrogen preheated to avoid condensation problems.

In both bench- and industrial-scale tests, it has been verified that dust and char are absent in the gas leaving the candle; in the case of the Güssing plant, the concentrations of these particulates in the gasifier freeboard for olivine fluidization without the filter candle are

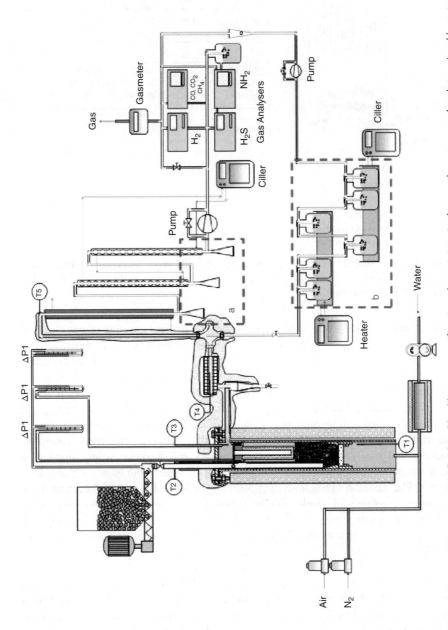

Figure 11.22 *Scheme of the continuous fluidized bed biomass gasification plant: (a) tar in the condensate samples determined by total organic carbon (TOC) analysis; (b) tar fraction sampled in 2-propanol, according to the technical specification CEN/TS 15439 analysed by GCMS or HPLC/UV. (Source: Adapted with permission from Ref. [100]. Copyright © 2009, American Chemical Society)*

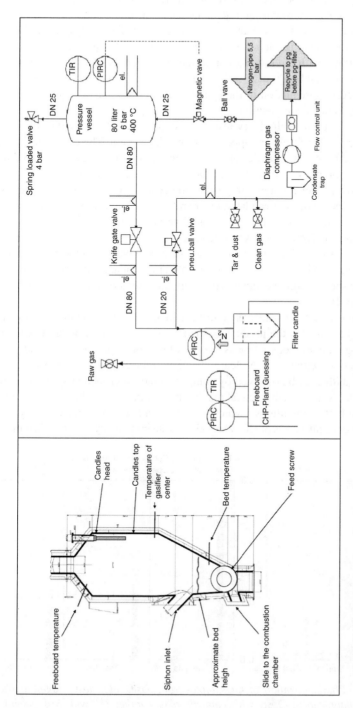

Figure 11.23 *Sketch and P&I flow sheet of the test module realized in güssing for gas filtration within the gasifier freeboard [101]*

56.3 and 23.6 g/Nm³, respectively. Tar reduction is also high, about 95% in both cases. The bench-scale tests also showed a significant methane conversion, up to a maximum of 40%, whereas at plant scale only half that value was achieved. The dry gas yield improvement obtained with the catalytic filter candle and quantified in the bench-scale tests is in the range of 70–100%, with the total carbon of the permanent gaseous phase increasing as a result of tar conversion by about 30%. This substantial increase in the gas yield is partially counterbalanced by a reduction in the heating value of the fuel gas by 13–16%, brought about by changes in its composition.

On the basis of these preliminary tests, the adoption of a filter candle solution for gas cleaning in the gasifier freeboard appears a promising option. The overall technical feasibility of long-term operation under industrial conditions is yet to be established however; innovative ceramic porous structures resistant to the harsh environment and the temperature peaks frequently encountered in gasifier freeboards have recently been developed by Pall Filtersystems GmbH and validated in laboratory tests; and experiments are in progress at the Güssing plant to verify the long-term reliability of such devices – a prerequisite for a breakthrough from R&D to market potential.

Figure 11.24 reports characterization tests on tar samples from the bench-scale gasification runs: results with a catalytic filter candle are compared to that without it; among the various tar species, toluene appears the more prevailing compound following catalytic reformation [102].

It is noteworthy that in the reference test less than 60% of the hydrogen content of the biomass is to be found in the product gas as H_2, whereas in tests with the catalytic candle this value greatly exceeds 100% as a result of hydrogen production from hydrocarbon reforming reactions with steam. A clear indication of the outstanding reforming activity of the catalytic candle may be obtained by comparing the experimental values of steam conversion in each test with its theoretical maximum, obtained on the basis of thermodynamic equilibrium under test conditions – which implies complete conversion of the hydrocarbons and char, with concentrations of CO, CO_2, H_2 and steam fixed by the equilibrium condition

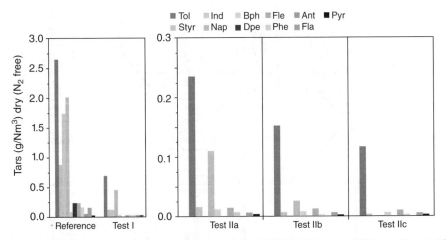

Figure 11.24 *Characterization of tar samples from gasification tests. Reference test without catalytic filter candle. Tests with catalytic filter candle: (1) with S/B (steam/biomass ratio) = 0.84; (II) with S/B = 1.1. (Source: Reproduced with permission from Ref. [102]. Copyright © 2012, American Chemical Society)*

for the water-gas shift reaction.

$$C_{21}H_{30}O_{14} + 7H_2O \rightarrow 21CO + 22H_2$$

$$CO + H_2O \leftrightarrow CO_2 + H_2,$$

where $C_{21}H_{30}O_{14}$, the chosen formula for biomass, corresponds to the composition of the almond shells used in the gasification tests.

Figure 11.25 shows water conversions obtained experimentally with the catalytic candle to be usually quite close to this limiting thermodynamic reference level. This is a noteworthy result, as low water conversion is often considered a drawback for steam gasification [103].

The net energy demand for the reactions that take place within the catalytic candle – steam reforming of methane and tar and the water-gas shift – amounts to some 0.5 MJ/kg of biomass (estimated on the basis of experimental and thermodynamic data). With secondary enhancement of the reforming reactions taking place within the gasifier vessel, the supply of this thermal load can be provided in a more energy-efficient manner than is possible with alternative process layouts involving downstream gas conditioning, where energy from additional sources is often required [104].

Pressure drop measurements across the catalytic filter candle have been performed under varying temperature conditions, both in the unloaded state and as a function of time during gasification with consequential build-up of dust cake on the filtration surface. These results are reported in Figure 11.26 [103]. The impregnation of the porous filter structure with catalyst – thereby decreasing its porosity – and, where relevant, the inclusion of a catalyst fixed bed both result in increase in pressure drop with respect to that of a non-catalytic filter of the same structural characteristics (Figure 11.13).

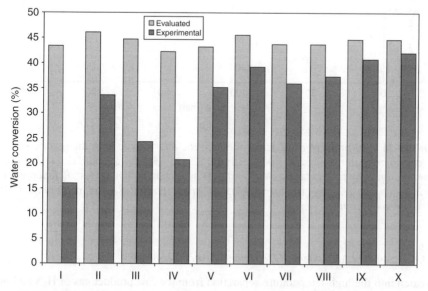

Figure 11.25 *Experimental and equilibrium steam conversion values [103]. Test I: without catalytic filter candle. Test II–X: with catalytic filter candle. (Source: Reproduced from Ref. [103], with permission from Elsevier)*

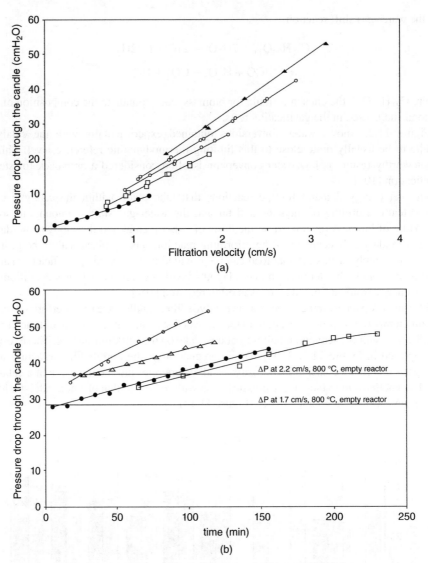

Figure 11.26 *(a) pressure drop across the catalytic filter candle with the reactor empty of particles; temperature levels:(●) 20°C, (□) 300°C, (■) 450°C, (○) 600°C, (▲) 800°C. (b) pressure drop across the catalytic filter candle, as a function of time on test; filtration velocity: (○) 2.3 cm/s, (△) 2.2 cm/s, (●) 1.8 cm/s, (□) 1.7 cm/s. (Source: Reproduced from Ref. [103], with permission from Elsevier)*

11.3.4.1   In-Bed Abatement of Detrimental Trace Elements

Research into the high-temperature separation from the raw product gas of H_2S and other detrimental trace elements – such as alkalis, HCl and heavy metals – has been carried out within the framework of the European Union research project *UNIQUE*.

Gibbs free energy minimization provides a practical basis for investigating the fate and removal of alkali species and sour gas components from biomass gasification product gas as a function of temperature and composition of inlet streams. The methane content of the product gas – a result of non-equilibrium conversion – is accounted for in the procedure by setting as inert compounds the corresponding fractions of carbon and hydrogen in the biomass feedstock [42].

Following theoretical investigations [105–107], experimental tests have been designed and performed on both conventional and innovative synthetic solid sorbents and sorbent mixtures deemed suitable for use under gasifier operating conditions. The idea being that these could then be added to the reactor granular bed inventory and so ensure very low concentration levels of the relevant contaminants in the product gas – comparable with threshold values recommended for SOFC and the production of biofuels by chemical syntheses.

The integration of biomass steam gasifiers with high-temperature fuel cells (specifically, solid oxide fuel cells) appears a likely candidate for the realization of efficient bioenergy utilization at relatively small scale. It is increasingly becoming recognized that power generation by means of high-temperature fuel cells can compete in terms of efficiency with large IGCC installations, such units being able to make use of the major products – H_2, CO, CH_4 – of the gasification process. Although more resistant to contaminants than low-temperature cells, provision should nevertheless be made to drastically reduce the fuel gas content of alkali and sour gas compounds. These classes of contaminants are typically removed during downstream processing, attention being generally focused on the extent of sorbent capacity rather than the resultant gas purity – the latter being more often than not back-calculated from the sorbent conversion. The alternative of including suitable additional sorbents in the fluidized bed inventory of the gasifier of the type illustrated in Figure 11.21 has the further beneficial effect of enhancing the catalytic activity for tar reforming in the freeboard candle processers. So that inorganic contaminants along with tar and particulates are subjected to high-temperature elimination in a single efficient and compact processing unit.

Experimental evidence suggests that hydrogen sulphide, which is present in the product gas at concentrations of some 100 ppmv, does not cause irreversible deactivation to the catalyst particle layer in the filter candle, although it certainly reduces its activity [108]. However, at such levels, its presence can be expected to cause more serious problems for the anodes of fuel cells, for which well-defined threshold limits have yet to be specified because of the novelty of such applications; concentrations not exceeding a few ppmv have nevertheless been recommended for operation at temperatures below 1000 °C [109].

An additional issue is separation of the sorbents from the remaining constituents of the gasifier bed inventory for their regeneration and cyclic reuse; separation based on particle size differences appears the most appropriate means of achieving this.

Calcium-based sorbents have long been recognized as effective for high-temperature H_2S capture, the sulphation of both calcined and uncalcined limestone having been studied extensively for this purpose [110, 111]. Thermodynamic limitations however, especially in the presence of large amount of steam, make it unlikely that H_2S concentrations as low as those required by SOFC can be achieved.

Alternative systems all suffer from one drawback or another: reduction of sorption capacity with temperature for iron oxide; metal vapourization for zinc oxide; oxide reduction by H_2 and CO for copper oxide, etc. Combinations of different metal oxides and high dispersion of the active phase on a support are strategies commonly explored to overcome these problems.

Experimental work on H_2S removal, both under actual operating conditions and with simulated coal gas, has led to the conclusion that fuel gas composition has little influence on the desulphurization process except at minimum residual concentration levels where the thermodynamic equilibrium is worsened by the presence of CO, CO_2, H_2O [112]. With CeO_2, indications are that the presence of H_2O has no negative impact [113], and this seems to be also true with CuO-Al_2O_3 sorbents [114].

Alkalis and heavy metals cause fouling and corrosion on condensation. Aluminosilicates have been shown to reduce the alkali concentrations to the ppb-level under gasification conditions, and when rich in alumina to also remove chlorine and heavy metals such as zinc [115, 116].

The *UNIQUE* sorption experiments reported here [107] were performed under *packed bed* conditions at atmospheric pressure. The experimental setup, illustrated in Figure 11.27, consisted of a tube furnace with five independent heating zones encasing the sorbent sample bed through which the gas in all sorption tests flowed at a temperature of 700–900 °C progressively saturating the sorbent; the cleaned gas composition being determined by a molecular beam mass spectrometer (MBMS).

The gas stream for the alkali sorption experiments consisted of 94 vol% He, 4 vol% H_2 and 2 vol% H_2O; it was passed over a KCl source to saturate it with KCl (about 20 ppmv). Helium was used as carrier gas, resulting in high signal intensities in the MBMS, extending the detection limit down to some 100 ppbv. For the H_2S and HCl sorption experiments, a synthetic biomass syngas was used as their sorption characteristics can be influenced by a number of the syngas components.

The results of the KCl sorption tests confirm aluminosilicates as suitable sorbents for KCl removal below 100 ppbv at 800–900 °C (Figure 11.28).

The results of H_2S sorption experiments conducted at between 700 and 900 °C and using a variety of sorbents are reported in Figure 11.29. These show Ca- and Cu-based sorbents to be unsuitable for reducing H_2S concentrations below 1 ppmv at these temperatures. The sorbent composition is important – slag lime, which contains several oxides beside calcium oxide, achieving the best H_2S reduction (down to 50 ppmv) of all the Ca-based sorbents.

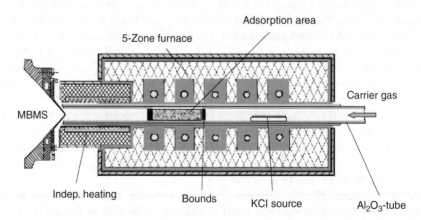

Figure 11.27 *Experimental setup for sorption experiments [107]. (Source: Reproduced from Ref. [107], with permission from Elsevier)*

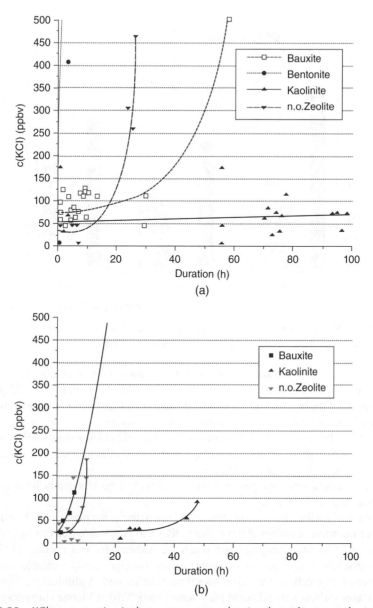

Figure 11.28 *KCl concentration in the gaseous stream leaving the sorbent sample: (a) 800°C; (b) 900°C [107]. (Source: Reproduced from Ref. [107], with permission from Elsevier)*

In order to meet the requirement of 1 ppmv H_2S, a new sorbent had to be developed. Thermodynamic considerations suggest that stabilized Ba-based sorbents should be effective for the purpose. As a result of this, sorbent oxide mixtures were considered, the "CaBa" sorbent being prepared from a mixture of 10 mol% $BaCO_3$ and 90 mol% $CaCO_3$.

As hoped for, the CaBa sorbent achieved H_2S concentrations of below 0.5 ppmv – the detection limit in these experiments – for temperatures above 800°C. The stabilization effect was confirmed by XRD analysis: Figure 11.30 shows the occurrence of a BaS phase

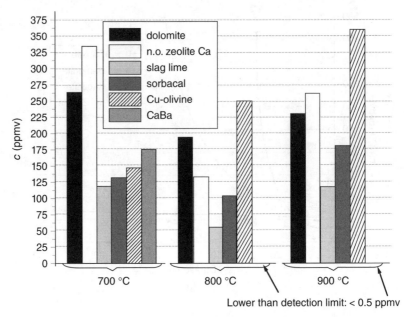

Figure 11.29 *H_2S concentration in the gaseous stream leaving the sorbent sample [107]. (Source: Reproduced from Ref. [107], with permission from Elsevier)*

in the CaBa sorbent, which should normally be unstable under these conditions. At temperatures below 760 °C, the H_2S concentration in the effluent stream rises to 175 ppmv due to carbonation of the sorbent, which is an indication that CaBa sorbent regeneration could be possible by cooling the saturated sorbent under a non-oxidizing atmosphere.

The sorption data also showed the CaBa sorbent capable of reducing the HCl gas content to below 1 ppmv over the temperature range 800–900 °C, the formation of $BaCl_2$ phases being confirmed by XRD spectra (Figure 11.30).

Neither the alkali nor sour gas removal was found to be kinetically limited at space velocities as high as 9800 h^{-1}.

Fluidized bed gasification tests at atmospheric pressure in the presence of bauxite for the removal of the most abundant alkali halides, KCl and NaCl, released during gasification have confirmed the suitability of aluminosilicate sorbents for this purpose [117]. These materials (bauxite, bentonite and kaolinite) are of low cost, easily available and without environmental implications for their disposal when exhausted. Additional gasification tests with the CaBa sorbent, carried out at pilot scale (100 kWth) at Vienna University of Technology, confirmed the tendency towards lowered H_2S content in the product gas, but only in terms of qualitative trends: more work needs to be done on the optimization of the operating conditions, such as sorbent particle size and amount charged to the gasifier.

11.3.4.2   Ideas for Further Developments of the UNIQUE Concept

As discussed earlier in this chapter, an important issue governing the configuration of biomass gasification installations is that of thermal energy transfer to this intrinsically endothermic operation.

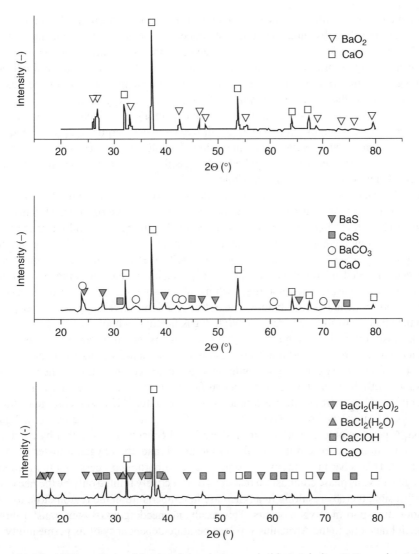

Figure 11.30 *XRD analysis of fresh sulphur-laden and chlorine-laden CaBa sorbent [107].*
(Source: Reproduced from Ref. [107], with permission from Elsevier)

In Güssing-type dual fluidization systems, this task is accomplished by circulating the
bed material (olivine sand, char and ash) to a separate combustion chamber where char
together with additional fuel is burnt with air to supply heat to the granular bed; so that by
careful adjustment of the solid circulation rate and the temperature gap between the gasification and combustion zones, the necessary thermal energy is provided for the endothermic gasification reactions. Though an elegant solution, and indeed one that has displayed
impressive results at fully operative level, this plant configuration is by no means a low-cost
gasification system.

An alternative solution to avoid contamination with nitrogen of the product gas, which results in a substantial reduction of its heating value, is at present under test at 1 MWth pilot plant level by ENEA Trisaia, again proposed within the framework of the European *UNIQUE* programme. This involves a single bubbling bed reactor fluidized by an oxygen-steam mixture, an altogether more simple concept than that of the dual bed approach but one that assumes the availability of oxygen or oxygen-enriched air.

With small-to-medium-scale gasification units operating at ambient pressure, air separation could be accomplished by selective nitrogen–oxygen sorption, a process requiring feed gas compression: specifically, compression of a volumetric gaseous stream five times that of the oxygen stream to be supplied to the gasifier – with substantial penalty for the energy efficiency of the process as a whole.

A real breakthrough could come about with the development of ionic transport membranes to separate oxygen from air without the need for pressurized gas streams. R&D progress along these lines has recently been reported with regard to endothermic reacting systems – among them auto-thermal methane steam reforming – in which oxygen transfer is achieved by means of ionic transport membranes. The feasibility of such an approach has been demonstrated at laboratory scale [118, 119], and, in a recent development, a 10 cm ID laboratory-scale gasifier incorporating such membranes within the fluidized bed has been studied and modelled [120], thereby integrating this innovative oxygen transfer process with the other innovative features of *UNIQUE* gasification technology.

The feasibility of including a cluster of cylindrical oxygen transport membranes in the gasifier fluidized bed, in addition to the catalytic filter candles in its freeboard, is being evaluated. Preliminary calculations confirm the geometric compatibility of the reactor configuration with the overall oxygen transfer surface required; these were based on literature data on membrane permeability as a function of temperature [121] and common values of the oxygen equivalence ratios (ER) imposed by the requirements for auto-thermal gasification. A positive outcome arising from these initial considerations is that by conducting the combustion reactions in the vicinity of the membrane, the oxygen transfer process is enhanced. This is because the *fast* heterogeneous char burning process (not to mention the homogeneous gas phase combustion reactions) is effectively limited only by the rate of oxygen diffusion, thereby resulting in an increase in temperature in the region close to where the ionic transport of oxygen occurs. This results in speeding up oxygen transfer through the membrane due to the Arrhenius-type temperature dependency of its permeability.

11.4   Conclusions

The application of process simplification and intensification concepts to biomass gasification not only leads to more simple, efficient and cost-effective plant installations but also provides the opportunity for overcoming some of the major obstacles inhibiting the large market acceptance of such technologies – particularly at the small to medium scale of operations (<10 MWth) that could prove economically attractive both in developed and developing countries. In contrast to fossil fuels – which are concentrated in a few, at times politically unstable, geographic areas – biomass is widely available; and it possesses a comparatively *dilute* heating content that favours local consumption. The combination of biomass feedstock with renewable sources that are not amenable to accumulation and storage – such as solar and wind – conforms well with plans to put into effect a distributed and more sustainable energy policy that could be expected to receive widespread public approval.

The unearthing of new advanced catalysts, sorbents and high-temperature filtration media are all contributing to the development of innovative processes for the production of clean fuel gas from biomass, an alternative energy source directly suited to efficient conversion into electrical power by means of high-temperature fuel cells, micro gas turbines and advanced integrated heat and power plant installations; it is also amenable to further catalytic processing for the production of second-generation biofuels (liquid fuels, hydrogen) and chemicals and to the implementation of poly-generational strategies adaptable to varying market demand.

High-temperature gas cleaning and catalytic conditioning remain the key issues to be resolved in the promotion of industrial applications of biomass for energy and chemicals. To avoid unacceptable increases in plant and operating costs, gas treatment should be strictly integrated with biomass conversion and carried out under closely similar temperature conditions. These objectives, together with optimal energy efficiency, are achieved by a combination of the gasification and gas cleaning operations in a single reactor vessel.

The *UNIQUE* gasification technology, the fruit of an extensive R&D programme with the participation of ten research organizations and private companies throughout the European Union, has provided the opportunity for the development of technical innovations addressing the essential issues involved and their implications for existing and new industrial installations. They have been tested under real and simulated industrial conditions over a wide range of scales, from bench-scale laboratory units to full industrial operating installations.

References

1. Bridgwater, A.V. (2003) Renewable fuels and chemicals by thermal processing of biomass. *Chemical Engineering Journal*, **91**, 87–102.
2. Bain, R.L., Dayton, D.C., Carpenter, D.L. *et al.* (2005) Evaluation of catalyst deactivation during catalytic steam reforming of biomass-derived syn gas. *Industrial and Engineering Chemistry Research*, **44**, 7945–7956.
3. Bergman, P.C.A., van Paasen, S.V.B. and Boerrigter, H. (2002) *The novel "OLGA" technology for complete tar removal from biomass producer gas, Pyrolysis and gasification of biomass and waste, expert meeting*, Strasbourg, France.
4. Simell, P.A., Kurkela, E., Stahlberg, P. and Hepola, J. (1996) Catalytic hot gas cleaning of gasification gas. *Catalysis Today*, **27**, 55–62.
5. Caballero, M.A., Corella, J., Aznar, M.P. and Gil, J. (2000) Biomass gasification with air in fluidized bed. Hot gas clean-up with selected commercial and full-size nickel-based catalysts. *Industrial and Engineering Chemistry Research*, **39**, 1143–1154.
6. Van Paasen, S.V.B., Kiel, J.H.A. (2004) Tar formation in a fluidized bed gasifier: impact of fuel properties and operating conditions, ECN-C-04-013 Report, 1-58.
7. Babu, S.P. (2006) Work Shop No.1: perspectives on biomass gasification, in *Task 33: Thermal Gasification of Biomass of IEA Bionergy Agreement*, Prentice Hall.

8. Massi, E., Devianto, H. and Gallucci, K. (2012ISBN: 978-1-4471-2368-2) Digesters, gasifiers and biorefineries: plants and field demonstration, in *Fuel Cells in the Waste-to-Energy Chain*, Springer, Berlin, pp. 81–94.

9. Hofbauer, H., Rauch, R., Loeffler, G., Kaiser, S., Fercher, E., Tremmel, H. (2002) Six years experience with the FICFB-gasification process, 12th European Conference on Biomass for Energy, Amsterdam, The Netherlands, 17–21 June 2002.

10. Hofbauer, H. and Knoef, H. (2005) Success stories in biomass gasification, in *Handbook Biomass Gasification*, BTG Biomass Technology Group BV, Enschede, Netherlands, pp. 115–161.

11. Koppatz, S., Schmid, J.C., Pfeifer, C. and Hofbauer, H. (2012) The effect of bed particle inventories with different particle sizes in a dual fluidized bed pilot plant for biomass steam gasification. *Industrial & Engineering Chemistry Research*, **51**, 10492–10502.

12. Gil, J., Aznar, M.P., Caballero, M.A. *et al.* (1997) Biomass gasification in fluidized bed at pilot scale with steam-oxygen mixtures. *Energy and Fuels*, **11**, 1109–1118.

13. Albertazzi, S., Basile, F., Brandin, J. *et al.* (2005) The technical feasibility of biomass gasification for hydrogen production. *Catalysis Today*, **106**, 297–300.

14. Shelley, S. (2006) Coal gasification comes of age. *Chemical Engineering Progress*, **102** (6), 6–10.

15. Foscolo, P.U., Germanà, A., Jand, N. and Rapagnà, S. (2007) Design and cold model testing of a biomass gasifier consisting of two interconnected fluidized beds. *Powder Technology*, **173**, 179–188.

16. Xiao-hua, Y., Ma, L., Chen, P. *et al.* (2005) Experimental research on biomass gasification in clapboard-type inner circulation fluidized bed. *Taiyangneng Xuebao*, **26**, 743–746.

17. Lettner, F., Timmerer, H. and Haselbacher, P. (2007) Biomass gasification – state of the art description, in *Guideline for safe and eco-friendly biomass gasification Intelligent Energy–Europe (IEE)*, Graz University of Technology-Institute of Thermal Engineering, Austria.

18. Anis, S. and Zainal, Z.A. (2011) Tar reduction in biomass producer gas via mechanical, catalytic and thermal methods: a review. *Renewable and Sustainable Energy Reviews*, **15**, 2355–2377.

19. Hasler, P. and Nussbaumer, T. (1999) Gas cleaning for IC engine applications from fixed bed biomass gasification. *Biomass Bioenergy*, **16**, 385–95.

20. Thambimuthu, K.V. (1993) *Gas Cleaning for Advanced Coal-Based Power Generation*, IEA Coal Research.

21. Han, J. and Kim, H. (2008) The reduction and control technology of tar during biomass gasification/pyrolysis: an overview. *Renewable and Sustainable Energy Reviews*, **12**, 397–416.

22. Pathak, B.S., Kapatel, D.V., Bhoi, P.R. *et al.* (2007) Design and development of sand bed filter for upgrading producer gas to IC engine quality fuel. *International Energy Journal*, **8**, 15–20.

23. Battelle Memorial Institute (1986) Engineering analysis of biomass gasifier product gas cleaning technology, in *Biofuels and Municipal Waste Technology Division*, PNL-5534, Richland, Washington.

24. Bhave, A.G., Vyas, D.K. and Patel, J.B. (2008) A wet packed bed scrubber-based producer gas cooling–cleaning system. *Renew Energy*, **33**, 1716–1720.
25. Khummongkol, D. and Tangsathitkulchai, C. (1989) A model for tar-removal efficiency from biomass-produced gas impinging on a water surface. *Energy*, **14**, 113–121.
26. Lee, B.K., Jung, K.R. and Park, S.H. (2008) Development and application of a novel swirl cyclone scrubber – (1) experimental. *Journal of Aerosol Science*, **39**, 1079–1088.
27. Park, S.H. and Lee, B.K. (2009) Development and application of a novel swirl cyclone scrubber (2) Theoretical. *Journal of Hazardous Materials*, **164**, 315–321.
28. Shen, Y. and Yoshikawa, K. (2013) Recent progresses in catalytic tar elimination during biomass gasification or pyrolysis—a review. *Renewable and Sustainable Energy Reviews*, **21**, 371–392.
29. Phuphuakrat, T., Namioka, T. and Yoshikawa, K. (2011) Absorptive removal of biomass tar using water and oily materials. *Bioresource Technology*, **102**, 543–549.
30. Ozturk, B. and Yilmaz, D. (2006) Absorptive removal of volatile organic compounds from flue gas streams. *Process Safety and Environmental Protection*, **84**, 391–398.
31. Paethanom, A., Nakahara, S., Kobayashi, M. *et al.* (2012) Performance of tar removal by absorption and adsorption for biomass gasification. *Fuel Processing Technology*, **104**, 144–154.
32. Chen, Y., Zhu, Y., Wang, Z. *et al.* (2011) Application studies of activated carbon derived from rice husks produced by chemical–thermal process — A review. *Advances in Colloid and Interface Science*, **163**, 39–52.
33. Zwart, R.W.R. (2009) Gas cleaning downstream biomass gasification, Status Report 2009, ECN-E--08-078.
34. Teislev, B. (2002) *Harboøre – Woodchips updraft gasifier and 1500 kW gas-engines operating at 32% power efficiency in CHP configuration*, Babcock & Wilcox, Volund R&D Centre.
35. Higman, C. and Van der Burgt, M. (2008) *Gasification*, 2nd edn, Elsevier Science & Technology, pp. 328–348.
36. Pröll, T. (2004) *Potenziale d. Wirbelschichtdampfvergasung fester Biomasse-Modelierung u. Simulation auf Basis der Betriebserfahrungen am Biomassekraftwerk Güssing*, TU Wien, Wien.
37. Könemann, H.W.J., van Paasen, S.V.B. (2007) OLGA tar removal technology; 4 MW commercial demonstration, 15th European Biomass Conference and Exhibition, Berlin, Germany.
38. Rabou, L.P.L.M., Zwart, R.W.R., Vreugdenhil, B.J. and Bos, L. (2009) Tar in biomass producer gas, the Energy Research Centre of The Netherlands (ECN) experience: an enduring challenge. *Energy and Fuels*, **23**, 6189–6198.
39. Di Carlo, A., Borello, D. and Bocci, E. (2013) Process simulation of a hybrid SOFC/mGT and enriched air/steam fluidized bed gasifier power plant. *International Journal of Hydrogen Energy*, **38** (14), 5857–5874.
40. Jand, N. and Foscolo, P.U. (2005) Decomposition of wood particles in fluidized beds. *Industrial and Engineering Chemistry Research*, **44**, 5079–5089.

41. Li, C. and Suzuki, K. (2009) Tar property, analysis, reforming mechanism and model for biomass gasification – an overview. *Renewable and Sustainable Energy Reviews*, **13**, 594–604.

42. Jand, N., Brandani, V. and Foscolo, P.U. (2006) Thermodynamic limits and actual product yields and compositions in biomass gasification processes. *Industrial and Engineering Chemistry Research*, **45**, 834–843.

43. Aravind, P.V. and de Jong, W. (2012) Evaluation of high temperature gas cleaning options for biomass gasification product gas for solid oxide fuel cells. *Progress in Energy and Combustion Science*, **38**, 737–764.

44. CEN/TS 15439 (2006) Biomass gasification - tar and particles in product gases - sampling and analysis, European Committee for Standardization.

45. Rapagnà, S., Gallucci, K., Di Marcello, M. *et al.* (2010a) Characterisation of tar produced in the gasification of biomass with in situ catalytic reforming. *International Journal of Chemical Reactor Engineering*, **8** article A30 http://www.degruyter.com /view/j/ijcre.2010.8.1/ijcre.2010.8.1.2188/ijcre.2010.8.1.2188.xml.

46. Dou, B.L., Shen, W.Q., Gao, J.S. and Sha, X.Z. (2003) Adsorption of alkali metal vapor from high-temperature coal-derived gas by solid sorbents. *Fuel Processing Technology*, **82**, 51–60.

47. Heidenreich, S. (2013) Hot gas filtration – a review. *Fuel*, **104**, 83–94.

48. Smid, J., Hsiau, S.S., Peng, C.Y. and Lee, H.T. (2006) Hot gas cleanup: pilot testing of moving bed filters. *Filtration and Separation*, **43**, 21–24.

49. Zhao, J.T., Huang, J.J., Wu, J.H. *et al.* (2008) Modeling and optimization of the moving granular bed for combined hot gas desulfurization and dust removal. *Powder Technology*, **180**, 2–8.

50. Xu, C., Donald, J., Byambajav, E. and Ohtsuka, Y. (2010) Recent advances in catalysts for hot-gas removal of tar and NH_3 from biomass gasification. *Fuel*, **89**, 1784–1795.

51. Di Carlo, A. and Foscolo, P.U. (2012) Hot gas filtration in the freeboard of a fluidized bed gasifier: development of a CFD model. *Powder Technology*, **222**, 117–130.

52. Smith, D.H. and Ahmadi, G. (1998) Problem and progress in hot-gas filtration for pressurized fluidized bed combustor (PFBC) and integrated gasification combined cycle (IGCC). *Aerosol Science and Technology*, **29**, 163–169.

53. Ahmadi, G. and Smith, D.H. (1998) Gas flow and particle deposition in the hot gas filter vessel at the Tidd 70 MWE PFBC demonstration plant. *Aerosol Science and Technology*, **29**, 206–223.

54. Mazaheri, A.R. and Ahmadi, G. (2006) Aerosol transport and deposition analysis in a demonstration scale hot-gas filter vessel with alternate designs. *Aerosol Science and Technology*, **17**, 623–639.

55. Papavergos, P.G. and Hedley, A.B. (1984) Particle deposition behaviour from turbulent flow. *Chemical Engineering Research and Design*, **62**, 275–295.

56. Cocco, R., Shaffer, F., Hays, R. *et al.* (2010) Particle clusters in and above fluidized beds. *Powder Technology*, **203**, 3–11.

57. Rapagnà, S., Foscolo, P.U. and Gibilaro, L.G. (1994) The influence of temperature on the quality of gas fluidisation. *International Journal of Multiphase Flow*, **20**, 305–313.

58. Zimmerlin, B., Leibold, H. and Seifert, H. (2008) Evaluation of the temperature-dependent adhesion characteristics of fly ashes with a HT-rheometer. *Powder Technology*, **180**, 17–20.

59. Tardos, G. and Pfeffer, R. (1995) Chemical reaction induced agglomeration and defluidization of fluidized beds. *Powder Technology*, **85**, 29–35.

60. Siedlecki, M., Nieuwstraten, R., Simeone, E. *et al.* (2009) Effect of magnesite as bed material in a 100 kWth steam-oxygen blown circulating fluidized-bed biomass gasifier on gas composition and tar formation. *Energy and Fuels*, **23**, 5643–5654.

61. Nagel, F.P., Ghosh, S., Pitta, C. *et al.* (2011) Biomass integrated gasification fuel cell systems–concept development and experimental results. *Biomass Bioenergy*, **35**, 354–362.

62. Heidenreich, S., Foscolo, P.U., Nacken, M., Rapagnà, S. (2008) Gasification apparatus and method for generating syngas from gasifiable feedstock material, PCT Patent Application no. PCT/EP2008/003523.

63. Simeone, E., Nacken, M., Haag, W. *et al.* (2011) Filtration performance at high temperatures and analysis of ceramic filter elements during biomass gasification. *Biomass and Bioenergy*, **35**, S87–S104.

64. Milne, T.A., Abatzoglou, N., Evans, R.J. (1998) Biomass gasification "tars": Their nature, formation and conversion. National Renewable Energy Laboratory (NREL) Technical Report, Golden, CO, Report NREL/ TP 570-25357.

65. Devi, L., Ptasinski, K.J. and Janssen, F.J. (2005a) Pretreated olivine as tar removal catalyst for biomass gasifiers: investigation using naphthalene as model biomass tar. *Fuel Process Technology*, **86**, 707–730.

66. Magrini-Bair, K.A., Czernik, S., French, R. *et al.* (2007) Fluidizable reforming catalyst development for conditioning biomass-derived syngas. *Applied Catalysis A: General*, **318**, 199–206.

67. Pfeifer, C., Hofbauer, H. and Rauch, R. (2004) In-bed catalytic tar reduction in a dual fluidized bed biomass steam gasifier. *Industrial and Engineering Chemistry Research*, **43**, 1634–1640.

68. Delgado, J., Aznar, M.P. and Corella, J. (1997) Biomass gasification with steam in a fluidized bed: effectiveness of Cao, MgO and CaO-MgO for hot raw gas cleaning. *Industrial and Engineering Chemistry Research*, **36**, 1535–1543.

69. Olivares, A., Aznar, M.P., Cabballero, M.A. *et al.* (1997) Biomass gasification: produced gas upgrading by in-bed use of dolomite. *Industrial and Engineering Chemistry Research*, **36**, 5220–5226.

70. Rapagná, S., Jand, N. and Foscolo, P.U. (1998) Catalytic gasification of biomass to produce hydrogen rich gas. *International Journal of Hydrogen Energy*, **23**, 551–557.

71. Orio, A., Corella, J. and Narvaez, I. (1997) Performance of different dolomites on hot raw gas cleaning from biomass gasification with air. *Industrial and Engineering Chemistry Research*, **36**, 3800–3808.

72. Simell, P.A., Hirvensalo, E.K., Smolander, V.T. and Krause, A.O.I. (1999) Steam reforming of gasification gas tar over dolomite with benzene as a model compound. *Industrial and Engineering Chemistry Research*, **38**, 1250–1257.

73. Rapagnà, S., Jand, N., Kiennemann, A. and Foscolo, P.U. (2000) Steam-gasification of biomass in a fluidised-bed of olivine particles. *Biomass and Bioenergy*, **19**, 187–197.

74. Corella, J., Toledo, J.M. and Padilla, R. (2004) Olivine or dolomite as in-bed additive in biomass gasification with air in a fluidized bed: which is better? *Energy and Fuels*, **18**, 713–720.

75. Devi, L., Ptasinski, K.J., Janssen, F.J.J.G. *et al.* (2005b) Catalytic decomposition of biomass tars: use of dolomite and untreated olivine. *Renewable Energy*, **30**, 565–587.

76. Rauch, R., Pfeifer, C., Bosch, K. *et al.* (2006) Comparison of different olivine's for biomass steam gasification, in *Science in Thermal and Chemical Biomass Conversion*, vol. **1** (eds A.V. Bridgwater and D.G.B. Boocock), CPL Press, pp. 799–809.

77. Fredriksson, H.O.A., Lancee, R.J., Thüne, P.C. *et al.* (2013) Olivine as tar removal catalyst in biomass gasification: catalyst dynamics under model conditions. *Applied Catalysis B: Environmental*, **130-131**, 168–177.

78. Matsuoka, K., Shimbori, T., Kuramoto, K. *et al.* (2006) Steam reforming of woody biomass in a fluidized bed of iron oxide-impregnated porous alumina. *Energy and Fuels*, **20**, 2727–2731.

79. Kirnbauer, F., Wilk, V., Kitzler, H. *et al.* (2012) The positive effects of bed material coating on tar reduction in a dual fluidized bed gasifier. *Fuel*, **95**, 553–562.

80. Kirnbauer, F. and Hofbauer, H. (2011) Investigations on bed material changes in a dual fluidized bed steam gasification plant in Güssing, Austria. *Energy Fuels*, **25**, 3793–3798.

81. Świerczyński, D., Libs, S., Courson, C. and Kiennemann, A. (2007) Steam reforming of tar from a biomass gasification process over Ni/olivine catalyst using toluene as a model compound. *Applied Catalysis B*, **27**, 211–222.

82. Świerczyński, D., Courson, C., Bedel, L. *et al.* (2006) Characterisation of Ni-Fe/MgO/Olivine catalyst for fluidised bed steam gasification of biomass. *Chemistry of Materials*, **18/17**, 4025–4032.

83. Di Felice, L., Courson, C., Jand, N. *et al.* (2009) Catalytic biomass gasification: simultaneous hydrocarbons steam reforming and CO_2 capture in a fluidised bed reactor. *Chemical Engineering Journal*, **154**, 375–383.

84. Di Felice, L., Courson, C., Niznansky, D. *et al.* (2010) Iron supported on calcined dolomite, CaO and MgO: a study of characterization and tar reforming activity for biomass gasification process. *Energy Fuels*, **24**, 4034–4045.

85. Di Felice, L., Courson, C., Foscolo, P.U. and Kiennemann, A. (2011) Iron and nickel doped alkaline-earth catalysts for biomass gasification with simultaneous tar reformation and CO_2 capture. *International Journal of Hydrogen Energy*, **36**, 5296–5310.

86. Virginie, M., Adanez, J., Courson, C. *et al.* (2012) Effect of Fe–olivine on the tar content during biomass gasification in a dual fluidized bed. *Applied Catalysis B: Environmental*, **121–122**, 214–222.

87. Rapagnà, S., Virginie, M., Gallucci, K. *et al.* (2011) Fe/olivine catalyst for biomass steam gasification: preparation, characterization and testing at real process conditions. *CatalysisToday*, **176**, 163–168.

88. Virginie, M., Courson, C., Niznansky, D. *et al.* (2010) Characterization and reactivity in toluene reforming of a Fe/olivine catalyst designed for gas cleanup in biomass gasification. *Applied Catalysis B: Environmental*, **101**, 90–100.

89. UNIQUE (2011) European collaborative research project, Integration of particulate abatement, removal of trace elements and tar reforming in one biomass steam gasification reactor yielding high purity syngas for efficient CHP and power plants, Project number 211517, Final Report.

90. Rapagnà, S., Provendier, H., Petit, C. *et al.* (2002) Development of catalysts suitable for hydrogen or syn-gas production from biomass gasification. *Biomass and Bioenergy*, **22**, 377–388.

91. Ising, M., Gil, J., Hunger, C. (2001) Gasification of biomass in a circulating fluidized bed with special respect to tar reduction, 1st World Conference on Biomass for Energy and Industry, James & James Science Publishers Ltd., London, pp. 1775–1778.

92. Toledo, J.M., Corella, J. and Molina, G. (2006) Catalytic hot gas cleaning with monoliths in biomass gasification in fluidized beds. 4. performance of an advanced, second-generation, two-layers-based monolithic reactor. *Industrial and Engineering Chemistry Research*, **45**, 1389–1396.

93. Ferella, F., Stoehr, J., De Michelis, I. and Hornung, A. (2013) Zirconia and alumina based catalysts for steam reforming of naphthalene. *Fuel*, **105**, 614–629.

94. Nacken, M., Ma, L., Heidenreich, S. and Baron, G.V. (2009) Performance of a catalytically activated ceramic hot gas filter for catalytic tar removal from biomass gasification gas. *Applied Catalysis B: Environmental*, **88**, 292–298.

95. Nacken, M., Ma, L., Engelen, K. *et al.* (2007) Development of a tar reforming catalyst for integration in a ceramic filter element and use in hot gas cleaning. *Industrial and Engineering Chemistry Research*, **46**, 1945–1951.

96. Provendier, H., Petit, C., Estrournès, C. and Kiennemann, A. (1998) Dry reforming of methane. Interest of La-Ni-Fe solid solutions compared to $LaNiO_3$ and $LaFeO_3$. *Studies in Surface Science and Catalysis*, **119**, 746–751.

97. Vargas, J.C., Libs, S., Roger, A.C. and Kiennemann, A. (2005) Study of Ce-Zr-Co fluorite-type oxide as catalysts for hydrogen production by steam reforming of bioethanol. *Catalysis Today*, **107–108**, 417–425.

98. Heidenreich, S., Foscolo, P.U., Nacken, M., Rapagnà, S. (2010) US patent application 223848.

99. Foscolo, P.U., Gallucci, K. (2008) Integration of particulate abatement, removal of trace elements and tar reforming in one biomass steam gasification reactor yielding high purity syngas for efficient CHP and power plants, 16th European Biomass Conference and Exhibition, Valencia, Spain, 2-6 June 2008, paper OA7.1.

100. Rapagnà, S., Gallucci, K., Di Marcello, M. *et al.* (2009) In situ catalytic ceramic candle filtration for tar reforming and particulate abatement in a fluidized- bed biomass gasifier. *Energy and Fuels*, **23**, 3804–3809.

101. Foscolo, P.U. (2012) The Unique project - integration of gasifier with gas cleaning and conditioning system, Int. Seminar on Gasification 2012, 18-19 October 2012, Stockholm, Sweden (http://www.sgc.se/gasification2012/).

102. Rapagnà, S., Gallucci, K., Di Marcello, M. *et al.* (2012) First Al_2O_3 based catalytic filter candles operating in the fluidized bed gasifier freeboard. *Fuel*, **97**, 718–724.

103. Rapagnà, S., Gallucci, K., Di Marcello, M. *et al.* (2010b) Gas cleaning, gas conditioning and tar abatement by means of a catalytic filter candle in a biomass fluidized-bed gasifier. *Bioresource Technology*, **101**, 7134–7141.

104. Vivanpatarakij, S. and Assabumrungrat, S. (2013) Thermodynamic analysis of combined unit of biomass gasifier and tar steam reformer for hydrogen production and tar removal. *International Journal of Hydrogen Energy*, **38**, 3930–3936.

105. Stemmler, M. and Müller, M. (2010) Theoretical evaluation of feedstock gasification using H2/C Ratio and ROC as main input variables. *Industrial and Engineering Chemistry Research*, **49**, 9230–9237.

106. Stemmler, M., Tamburro, A. and Müller, M. (2013a) Thermodynamic modelling of fate and removal of alkali species and sour gases from biomass gasification for production of biofuels. *Biomass Conversion and Biorefinery*, **3**, 1–12.

107. Stemmler, M., Tamburro, A. and Müller, M. (2013b) Laboratory investigations on chemical hot gas cleaning of inorganic trace elements for the "UNIQUE" process. *Fuel*, **108**, 31–36.

108. Ma, L., Verelst, H. and Baron, G.V. (2005) Integrated high temperature gas cleaning: tar removal in biomass gasification with a catalytic filter. *Catalysis Today*, **105**, 729–734.

109. DOE/NETL-2002/1179 By EG&G Technical Laboratory (2002) Fuel Cell Handbook (6th edn.), Morgantown, West Virginia.

110. De Diego, L.F., Abad, A., Garzia-Labiano, F. *et al.* (2004) Simultaneous calcination and sulphidation of calcium-based sorbents. *Industrial and Engineering and Chemistry Research*, **42**, 3261–3269.

111. Hu, Y., Watanabe, M., Aida, C. and Horio, M. (2006) Capture of H_2S by limestone under calcination conditions in a high-pressure fluidized-bed reactor. *Chemical Engineering Science*, **61**, 1854–1863.

112. Elseviers, W.F. and Verelst, H. (1999) Transition metal oxides for hot gas desulphurisation. *Fuel*, **78**, 601–612.

113. Zeng, Y., Kaytakoglu, S. and Harrison, D.P. (2000) Reduced cerium oxide as an efficient and durable high temperature desulfurization sorbent. *Chemical Engineering Science*, **55**, 4893–4900.

114. Patrick, V., Gavalas, G.R., Flytzani-Stephanapoulos, M. and Jothimurugesen, K. (1989) High-temperature sulfidation-regeneration of CuO-Al_2O_3 sorbents. *Industrial and Engineering Chemistry Research*, **28**, 931–940.

115. Wolf, K.J., Müller, M., Hilpert, K. and Singheiser, L. (2004) Alkali sorption in second-generation pressurized fluidized-bed combustion. *Energy & Fuels*, **18**, 1841–1850.

116. Diaz-Somoano, M. and Martinez-Tarazona, M.R. (2005) Retention of zinc compounds in solid sorbents during hot gas cleaning processes. *Energy Fuels*, **19**, 442–446.

117. Barisano, D., Freda, C., Nanna, F. *et al.* (2012) Biomass gasification and in-bed contaminants removal: performance of iron enriched Olivine and bauxite in a process of steam/O_2 gasification. *Bioresource Technology*, **118**, 187–194.

118. Kovalevsky, A.V., Yaremchenko, A.A., Kolotygin, V.A. *et al.* (2011) Oxygen permeability and stability of asymmetric multilayer Ba0.5Sr0.5Co0.8Fe0.2O3 ceramic membranes. *Solid State Ionics*, **192**, 677–681.

119. Hong, J., Kirchen, P. and Ghoniem, A.F. (2012) Numerical simulation of ion transport membrane reactors: oxygen permeation and transport and fuel conversion. *Journal of Membrane Science*, **407–408**, 71–85.

120. Antonini, T., Gallucci, K., Foscolo, P.U. (2014) Oxygen transport by ionic membrane conductors to a biomass steam gasifier: mass and heat transfer effects in the char burning process, iconBM: International conference on BioMass, 4–7 May 2014, Florence, Italy.

121. Xu, S.J. and Thomson, W.J. (1999) Oxygen permeation rates through ion-conducting perovskite membranes. *Chemical Engineering Science*, **54**, 3839–3850.

Index

absorption 45
acid gas removal 199
air separation unit 247
alkali metals 348
alkaline fuel cells 212
ammonia 221, 349
    absorber 258

BCFZ 91
bioenergy 331
bioethanol 219
biomass 243, 331
    gasification 332
Boudouard reaction 250
BSCF 90

candle filters 350
carbon capture and storage 8, 53
    oxy-fuel 9, 81
    post-combustion 9
    pre-combustion 9
    routes 54
catalytic conditioning 352
catalytic filters 363
cellulose 275
CFD models 352
chemical looping 117
    comparison (reactor concepts) 163
    interconnected fluidized bed
            reactors 124
    maximum temperature increase 137
    packed bed reactors 132
    process integration 144

rotating reactor 143
chemical potential 268
CHP 229
co-current 59
cold gas cleaning 335
collector efficiency 303
composite membranes 216
compound parabolic concentrator
            (CPC) 302, 304
concentration polarization 64, 222, 276
COS 348
counter current 60, 85
cryogenic separation 7, 10
    capture step 11, 16
    CO_2 avoidance costs 41, 43
    cooling duty 24
    cooling step 13, 20
    recovery step 12, 20
CTF 99

desalination 275
design optimisation 195
direct flow tubes 303
direct methanol fuel cells 212
direct steam injection 320
discrete random walk 352
dolomite 355
double spiral arrangement (heat
            exchanger) 287
dust filtration 349

ecological potential 294
electric potential 281

electromagnetic action 323
enhanced steam methane reformer 177
equivalence ratio 332
EREA 260
evacuated tube collectors 303

fast internally circulating fluidized
      bed 343
fick 222
film effectiveness factor 222
filtration 335
flat plate collectors 303
flux targets 216
Fresnel collectors 302, 304
fuel cells 211

galvanic bath 301
gasification 245
GIBBS 260, 268
glycerol 221
gradientless process energy supply 321

heat pipe tubes 304
heat transfer area 311
heat transfer coefficient 313
heavy metals 349
hot gas cleaning 343
humid air turbine 147
hydraulic pressure difference 271
hydrocratic generator 288
hydrogen chloride 349
hydrogen production 216
hydrogen separation membranes 216
hydrotalcite 180

IGCC 148
in-bed abatement 370
internal circulating bubbling fluidized
      bed gasifier 254

laminar boundary layer 222
Le Chatelier 176
LSCF 92

maximum theoretical power 285
mechanical filtration 275

membrane 47, 82, 321
membrane distillation 322
membrane reactor 53
    chemical looping membrane
        reactor 71
    fluidized bed 65
    hollow fibre 62
    micro-reactors 72
    packed bed 58
    types 57
methanation process 246
methanol 220
microelectromechanical systems 215
micromachining 224
microorganisms 323
microstructured reactors 214
micro turbines 346
MIEC (membranes) 83
    chemical stability 97
    fabrication 85
    integration 103
    scale-up 107
    sealing 87
    structure 83
MILENA 244
molten carbonate fuel cells 212

NGCC 144, 177

OLGA 249, 343
olivine 355
1-D model 14
optical losses 302
oscillatory flow reactors 317
osmotic power plants potential 290, 295
oxidative reforming 219
oxygen carrier 118, 120
    oxygen capacity 120, 125
    sulfur tolerance 123
oxygen transport membranes 81
oxygen uncoupling 119

parabolic trough collectors 304
pervaporation 323
phosphoric acid fuel cells 212
poly-generation strategies 331

power density 277
pressure retarded osmosis 270
pressure swing adsorption 192
pressurized fluidized bed
         combustion 350
primary catalytic treatment 355
process analysis 22
proof of concept 36
proof of principle 25
proton exchange membrane fuel cell 210
pyrolysis 351

renewable energy integration 308
reverse electrodialysis 279
reverse osmosis 272
reverse vapor compression 284
Reynolds 224

salinity gradient 267
salt rejection 275
salt-wedge estuaries 292
scrubber tar removal 257
SDIRK 14
secondary catalytic treatment 360
sedimentation 274
semipermeable membranes 275
SEWGS 177
sharp front approach 16
SHIP 305
Sieverts 222
sintering 351
site criteria 292
SOFC 199
softening temperature 351
solar process heat 299
   applications 300

collector development 302
industrial potential 305
system concepts 301
temperature gradients 319
thermal process heat 316
solid circulation flow 124
solid particulates 346
sorbent 180
sorption-enhanced processes 176
SPECCA 164
specific power output 287
spinning disc 315
spiral-wound modules 277
spray pyrolysis 85
stickiness 351
substitute natural gas 243
sulphur compounds 348
swirl cyclone-scrubbers 337

tar 332, 347
TARWATC 342
techno-economic evaluation 39
temperature swing adsorption 191
thermal cracking 354
thermal losses 302
turbine inlet temperature (TIT) 145

UNIQUE concept 363

van't Hoff 271
venturi scrubber 339

Wagner equation 84
well-mixed estuaries 292
WENO 14, 15
wet cleaning processes 336

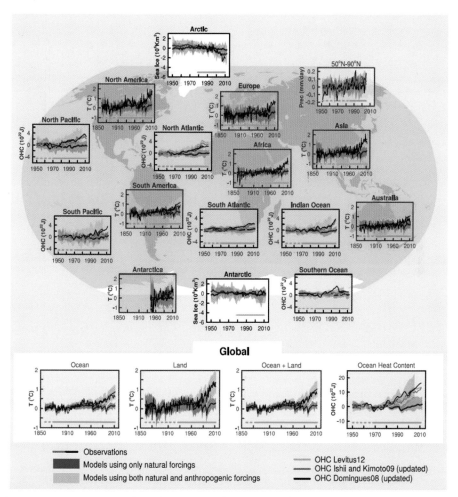

Figure 1.1 *Detection and attribution signals in some elements of the climate system, at regional scales (top panels) and global scales (bottom four panels). Brown panels are land surface–temperature–time series, green panels are precipitation–time series, blue panels are ocean heat content–time series and white panels are sea ice–time series. Observations are shown on each panel in black or black and shades of grey. Blue shading is the model time series for natural forcing simulations and pink shading is the combined natural and anthropogenic forcings. The dark blue and dark red lines are the ensemble means from the model simulations. All panels show the 5–95% intervals of the natural forcing simulations and the natural and anthropogenic forcing simulations. (Source: Extracted from the IPCC report 2013)*

Process Intensification for Sustainable Energy Conversion, First Edition.
Edited by Fausto Gallucci and Martin van Sint Annaland.
© 2015 John Wiley & Sons, Ltd. Published 2015 by John Wiley & Sons, Ltd.

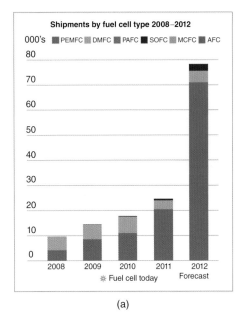

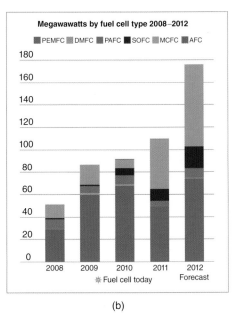

(a)   (b)

Figure 7.4 *(a) unit shipments by fuel cell type 2008–2012, (b) installed power by fuel cell type 2008–2012*